ACHIEVING STRETCH GOALS

Prentice Hall International Series
in Industrial and Systems Engineering

W. J. Fabrycky and J. H. Mize, Editors

Amos and Sarchet *Management for Engineers*
Amrine, Ritchey, Moodie and Kmec *Manufacturing Organization and Management, 6/E*
Asfahl *Industrial Safety and Health Management, 3/E*
Babcock *Managing Engineering and Technology, 2/E*
Badiru *Comprehensive Project Management*
Badiru *Expert Systems Applications in Engineering and Manufacturing*
Banks, Carson and Nelson *Discrete Event System Simulation, 2/E*
Blanchard *Logistics Engineering and Management, 4/E*
Blanchard and Fabrycky *Systems Engineering and Analysis, 2/E*
Brown *Technimanagement: The Human Side of the Technical Organization*
Burton and Moran *The Future Focused Organization*
Bussey and Eschenbach *The Economic Analysis of Industrial Projects, 2/E*
Buzacott and Shanthikumar *Stochastic Models of Manufacturing Systems*
Canada and Sullivan *Economic and Multi-Attribute Evaluation of Advanced Manufacturing Systems*
Canada, Sullivan and White *Capital Investment Analysis for Engineering and Management, 2/E*
Chang and Wysk *An Introduction to Automated Process Planning Systems*
Chang, Wysk and Wang *Computer Aided Manufacturing, 2/E*
Delavigne and Robertson *Deming's Profound Changes*
Eager *The Information Payoff: The Manager's Concise Guide to Making PC Communications Work*
Eberts *User Interface Design*
Eberts and Eberts *Myths of Japanese Quality*
Elsayed and Boucher *Analysis and Control of Production Systems, 2/E*
Fabrycky and Blanchard *Life-Cycle Cost and Economic Analysis*
Fabrycky, Thuesen and Verma *Economic Decision Analysis, 2/E*
Fishwick *Simulation Model Design and Execution: Building Digital Worlds*
Francis, McGinnis and White *Facility Layout and Location: An Analytical Approach, 2/E*
Gibson *Modern Management of the High-Technology Enterprise*
Gordon *Systematic Training Program Design*
Graedel and Allenby *Industrial Ecology*
Hall *Queuing Methods: For Services and Manufacturing*
Hansen *Automating Business Process Reengineering, 2/E*
Hammer *Occupational Safety Management and Engineering, 4/E*
Hazelrigg *Systems Engineering*
Hutchinson *An Integrated Approach to Logistics Management*
Ignizio *Linear Programming in Single- and Multiple-Objective Systems*
Ignizio and Cavalier *Linear Programming*
Kroemer, Kroemer and Kroemer-Elbert *Ergonomics: How to Design for Ease and Efficiency*
Kusiak *Intelligent Manufacturing Systems*
Lamb *Availability Engineering and Management for Manufacturing Plant Performance*
Landers, Brown, Fant, Malstrom and Schmitt *Electronic Manufacturing Processes*
Leemis *Reliability: Probabilistic Models and Statistical Methods*
Michaels *Technical Risk Management*
Moody, Chapman, Van Voorhees and Bahill *Metrics and Case Studies for Evaluating Engineering Designs*
Mundel and Danner *Motion and Time Study: Improving Productivity, 7/E*
Ostwald *Engineering Cost Estimating, 3/E*
Pinedo *Scheduling: Theory, Algorithms, and Systems*
Prasad *Concurrent Engineering Fundamentals, Vol. I: Integrated Product and Process Organization*
Prasad *Concurrent Engineering Fundamentals, Vol. II: Integrated Product Development*
Pulat *Fundamentals of Industrial Ergonomics*
Shtub, Bard and Globerson *Project Management: Engineering Technology and Implementation*
Taha *Simulation Modeling and SIMNET*
Thuesen and Fabrycky *Engineering Economy, 8/E*
Turner, Mize, Case and Nazemetz *Introduction to Industrial and Systems Engineering, 3/E*
Turtle *Implementing Concurrent Project Management*
von Braun *The Innovation War*
Walesh *Engineering Your Future*
Wolff *Stochastic Modeling and the Theory of Queues*

ACHIEVING STRETCH GOALS

Best Practices in Manufacturing for the New Millennium

Jonathan Golovin

PRENTICE HALL PTR
Upper Saddle River, New Jersey 07458
http://www.prenhall.com

Library of Congress Cataloging-in-Publication Data
Golovin, Jonathan.
Achieving stretch goals : best practices in manufacturing / Jonathan Golovin.
p. cm.
Includes index.
ISBN 0-13-376997-6
1. Production management. 2. Production planning. I. Title.
TS155.G58 1997
658.5—dc21 97-8431
CIP

Production Editor: *Rainbow Graphics, Inc.*
Acquisitions Editor: *Bernard M. Goodwin*
Cover Designer: *Scott Weiss*
Cover Design Director: *Jerry Votta*
Marketing Manager: *Miles Williams*
Manufacturing Manager: *Alexis R. Heydt*

Prentice-Hall, Inc.
A Simon & Schuster Company
Upper Saddle River, New Jersey 07458

Prentice Hall books are widely used by corporations and government agencies for training, marketing, and resale.

The publisher offers discounts on this book when ordered in bulk quantities.
For more information contact Corporate Sales Department, Phone: 800-382-3419, FAX: 201-236-7141, E-mail: corpsales@prenhall.com
or write: Prentice Hall PTR
Corporate Sales Dept.
One Lake Street
Upper Saddle River, NJ 07458

Printed in the United States of America

10 9 8 7 6 5 4 3 2 1

ISBN 0-13-376997-6

Prentice-Hall International (UK) Limited, *London*
Prentice-Hall of Australia Pty. Limited, *Sydney*
Prentice-Hall Canada Inc., *Toronto*
Prentice-Hall Hispanoamericana, S.A., *Mexico*
Prentice-Hall of India Private Limited, *New Delhi*
Prentice-Hall of Japan, Inc., *Tokyo*
Simon & Schuster Asia Pte. Ltd., *Singapore*
Editora Prentice-Hall do Brasil, Ltda., *Rio de Janeiro*

Contents

Introduction

1. Why Yet Another Book on Manufacturing?—20/20/20 Hindsight

The genesis of this book arises from my work over the last twenty years with leading manufacturing companies around the world. We have worked in over 20 industries, and in over 20 countries implementing manufacturing execution or plant floor management systems (MES). During that time many customers have asked me for a reference covering best practices on the plant floor and/or MES. While there were some excellent books aimed at specific topics—such as quality or costing—I never found a broader survey book with a comprehensive, practical, yet theoretically sound view of manufacturing. I believe that such a text is also needed for university courses in production management. As a result, I decided to write this book for use by our own customers.

The first draft completed in the early 90s covered the "theory." In fact, it was so dry, it was more soporific than enlightening. Upon awakening, I completely rewrote the book to capture the reality of trying to change a company to adopt best practices and continuous improvement. That led to the creation of the PCB Co, an amalgamation rather than conglomeration of many customers and situations. (I chose printed circuit board manufacturing because it allowed me to cover most types of non-continuous manufacturing operations from assembly to process to test.)

The book is divided by subject matter. Chapter one covers how to measure a plant and may be of most interest to plant management. Chapters two through six cover quality, scheduling, cost, technical or process capability, and speed. Chapter seven covers manufacturing execution or plant floor management systems and how they relate to and differ from ERP.

2. How to Use This Book

I suggest that the book be used as a change vehicle rather than purely as educational fodder. To this end, you can constitute an "improvement" team whether by manufacturing area, line, department or plant, function, or across multiple departments, etc. This team should be given responsibility for some step-wise improvement or "stretch goal."

The group can then read the chapter(s) most relevant to their task and measure the current performance (on absolute terms). Then it should discuss/brainstorm the five most relevant best practices/improvements to employ from the chapter and estimate the ROI or improvement potential. Then it can develop an implementation plan and begin the real work, fun and return.

The best practices and principles I have written about have all been tried and proven on many plant floors around the world. I hope yours is next!

Look at the principles, decide how they would apply to your situation, look at the simplest, most effective way to use them and learn about the king, ROI.

3. Comments for Continuous Improvement

I appreciate any and all feedback on: how to improve the book (for future drafts); your success or failure in applying these concepts; additional best practices; and other areas you would like to see covered that were omitted.

I can be reached at GOLOVIN@CONSILIUM.COM.

Acknowledgments

Any book that covers twenty-five years of education and practice has many "root cause" authors or creators. I would like to thank my PhD advisor, Dr. Arnoldo Hax, for instilling in me a passion for manufacturing and introducing me to the world of consulting and education. He is a remarkable role model of gentlemanly, courtly, scholarly, yet practical excellence.

Secondly, I would like to thank the two customers who introduced me into the world of semiconductor manufacturing—Terry Mudrock and Harry Hollack. They are two of the most creative people with whom I have ever had the privilege to work and were responsible for guiding most of our early pioneering work at Intel.

Thirdly, I have always had the privilege of working at Consilium with some of the finest and brightest people in this or any field. These ideas and systems were created with the talents of many outstanding minds—Steve Alter, Billy Chow, Doug Christensen, Dallan Clancy, Parris Hawkins, Phil Kurjan, Terry LeClair, Mike Leitner, Raj Mashruwala, Shantha Mohan, Sally Monk, Subhash Tantry, Carol Whitfield, and Huey-Shin Yuan to name a few of Consilium's finest.

And thanks to my partners in management who encouraged, supported and gave me time to write our story; in particular, Tom Tomasetti for his years of support and Frank Kaplan for his insights and help.

And finally, to those at home whose support is what makes it all worthwhile. To my mother, Inge Golovin, who ensured that her sons got the education she was denied in Nazi Germany. She personally financed Consilium when no one else would and her courage has always been only exceeded by her warmth and love of her family. To my boys who are the light of my life. And to Susan, my partner in life, who keeps us all whole.

1

Achieving Stretch Goals Through Best Practices in Measurement

The Goal: World-Class Competitive Manufacturing Capability

Everyone's goal today is world-class competitive manufacturing capability. There is no one left who wants to be the high-cost, low-quality, late, slow manufacturer. The goal of this book is to show you how to achieve world-class manufacturing capability through adoption of best practices. These best practices cover the best-in-breed practices we've seen at our customers representing over twenty countries and twenty-five industries.

To help illustrate the points and best practices in this book, we'll use a sustained case study of a printed circuit board (PCB) factory (the PCB Co., the PCB manufacturing arm of Diversified Electronics Company), a relatively easy to explain manufacturing process that has both process and discrete-type operations.

The examples in the book, however, represent an amalgamation of our experience with a variety of types of manufacturing including semiconductors, disk drives, fibers, plastics, films, electronic assemblies, food and beverage, pharmaceuticals, and medical devices, to name a few.

Each chapter will describe the best practices in a major area—plant performance measurement, product quality, customer service, cycle time reduction, cost reduction, and plant execution systems. Each chapter is divided into three sections. The first describes the "as is" situation at the PCB Co.—current practices, symptoms, and problems. The second section presents the best practices. The third follows the company trying to implement these practices—the difficult issue of changing to a culture that supports these new best practices.

We will focus on the use of these best practices to achieve major stepwise improvements in performance—the achievement of stretch goals.

Many companies today are trying to set stretch goals without knowing what is the "art of the possible." This process can send enormous shock waves of fear and confusion through the organization. In this book, we will show a methodology for setting stretch goals and then achieving them through best practices.

A book like this is a result of cumulative insight gained over years of work with innovative people in innovative companies. It is the result of continuous improvement in concepts and ideas as our customers have put them to the test in their manufacturing facilities.

The Two-to-One Performance Gap in Every Decade: Competitive Opportunity or Disadvantage

In the late 1970s and early 1980s, Texas Instruments' (T.I.) semiconductor wafer fabrication operations had line yield losses (scrap rates) of 20 to 30%. While this meant that typically one-fourth to one-fifth of all product was scrapped during production, such results were then considered excellent. As T.I. began benchmarking Japanese producers, such as Mitsubishi, they found that much to their surprise their competitors were targeting their scrap rates at zero! The Japanese operators viewed the loss of even one wafer as a serious avoidable mistake. What was standard operating procedure in one culture and factory was totally unacceptable in another. What was 100% of standard performance at one company was 70 to 80% performance at the other. What was 100% of standard performance at T.I. was a competitive advantage for Mitsubishi.

In the late 1970s, Ford Motor Company examined costs at their partially owned Japanese partner, Mazda. They were surprised to find that Mazda could manufacture and ship a car to the United States for $2,000 less than Ford could produce a similar car here.

Specifically looking at axles, Mazda could produce substantially higher quality parts at substantially lower costs. The difference was not in the direct labor cost but in every part of the cost equation. Again, what was 100% performance to *standard cost* at Ford was a competitive advantage for Mazda of $2,000 per car.

In the mid-1980s, Intel Corporation, in preparing themselves for increased competition, did a study of their factory "value-added" utilization—the percentage of time equipment was being used to produce product sold to customers out of a full 168 hour work week (7 days of 24 hour/day availability). They were surprised to find that in factories thought to be running "flat out" at 100% of standard *capacity* utilization, they were averaging around 35% "value-added" utilization. What this meant was that approximately only one-third of their capital was being used to make product they sold to customers! On their trips to Japan, they found semiconductor plants running at levels as high as 65%—getting nearly *twice* the output from the same factory and capital investment. Both plants were running at 100% of standard utilization; however, one of them produced twice as much as the other.

In each of these cases, there was at least a *two-to-one* competitive advantage to exploit on some measure of manufacturing capability.

More recent studies done by researchers at the University of California at Berkeley and Harvard add more examples of the wide range of performance among top companies. U.C. Berkeley's study of semiconductor manufacturing plant performance[1] encompassed 16 wafer fabrication plants in the United States, United Kingdom, Germany, Japan, and Taiwan. Companies participating in the study included IBM, Intel, I.T.T.–Intermetall, NEC, Toshiba, and Texas Instruments.

A summary of their findings are shown in Table 1.1 (the actual base data is given in Table 1.16 in the Appendix). They compared manufacturing performance on cycle time, quality (scrap), equipment productivity, total labor productivity, and customer service. What is striking about these results is that they represent a cross-section of major companies, many of whom are already seen as world class, and yet even within this group, there is typically at least a 2:1 ratio of best to worst performance. In fact, there is usu-

[1] Robert C. Leachman, Ed. The Competitive Semiconductor Manufacturing Survey: Second Report on Results of the Main Phase, Report CSM-08; September 16, 1994.

TABLE 1.1 Summary of Metric Scores: Ratios of Best to Average & Best to Worst Performance

Metric	Ratio of Best to Average Performance	Ratio of Best to Worst Performances
Cycle time	2.2	2.75
Scrap rate	6.5	10.7
Equipment productivity (steppers)	1.9	5.2
Total labor productivity (all headcount)	2.1	11.4
Percentage of missed orders	0% vs. 11%	0% vs. 24%

ally that great a ratio between average and best performance. The ratio between best and worst performance runs as high as 10:1 on measures such as scrap rates and total labor productivity.

In a similar vein, Professor Gary Pisano of Harvard University studied production process development time for 23 projects (13 in chemicals, 10 in biotechnology) at 11 major companies.[2] These projects involved the development of new chemical entities or biotechnological molecular entities. Again, his results showed approximately a 2:1 ratio between the time to develop a process in the chemical industry versus the biotechnology industry—80.15 months to 41.4 months for these 23 projects. The Best to Worst Performance ratio within these 23 would have been even higher.

Our experience over twenty years is that there is typically at least a 2:1 ratio between best and worst performance and that *there always will be!* The reason for this continual gap is that the best are rarely complacent. They continually raise the bar for what can be considered world class capability. Compaq recently discussed that they had lowered their ratio of indirect to direct labor at their personal computer plant to one-third to 1!

In the extremely price-sensitive P.C. market, it is not surprising then that they ended 1994 in the top marketshare position, with over 10% of the market, surpassing both IBM and Apple.

[2] Dr. Gary Pisano. Knowledge, Integration and the Locus of Learning: An Empirical Analysis of Process Development (forthcoming; Strategic Management Journal), HBS Working Paper, 95-006, Dec. 1994).

Blind—Without New Measures of Performance

Why were so many of these excellent companies blindsided by their competitors' clear manufacturing advantage? The reason lies in the way we measure (or don't measure) our factories. All of these companies relied on internal measurements of performance to standard. In reality, they were waiting to be measured against their competitors in the marketplace and that's far too late! In each case, a proper measurement system based on best practices in measurement would have first pointed out to management their real "value-added" levels of performance (usually hidden in the current financial systems) and second initiated action. It would have shown them the top candidate areas for stepwise improvement. It would have pointed out where they might achieve a temporary significant competitive advantage or avoid a temporary disadvantage.

Traditional Measures of Performance

When I first started consulting in manufacturing in the mid-1970s, there were a set of traditional measures of performance for measuring a factory, reflecting three major interested parties. Finance normally looked at financial performance—performance to cost standards—to see if the plant was and would continue to operate within preset budgets. Their selected measures were performance to standard cost and performance to standard leadtime (representing work in process inventory levels and investment). These measures really represented performance to annual budgets—set yearly—and were based on some theoretical product mix (that never occurred). Managers often spent more time explaining the variances due to actual product order mix or "lumpy" order size than working on improving their performance to them.

Marketing through production planning or production scheduling looked at delivery to schedule and leadtimes to see if "their" constituent, the customer, was getting product delivered on time. Delivery to schedule was normally measured in both volume (total units produced) and mix (customer orders delivered to schedule). Financially driven companies generally focused more on volume performance than mix. Performance was reviewed monthly after the fact.

Factory management measured: operator activity—measured either as earned hours (if there were official labor standards) or number of units produced/operator hour (a poor man's labor standard); quality—measured as

TABLE 1.2 Traditional Measures of Performance in PCB Factory in 1989

	This Month	Plan/Target
Standard cost	98%	100%
Average leadtime	5 weeks	5 weeks
Performance to mix	78%	90%
Performance to volume	102%	100%
Activity/operator hour	14 boards	14 boards
Yield rate (1-scrap%)	92%	90%
Rework rate	10%	10%
Performance to standard leadtime	98%	100%

yield (or scrap) and rework rates; work-in-process (WIP) inventory levels—measured directly as well as reflected in the leadtimes reported; and personnel measures such as safety incidents, turnover, reported absenteeism, etc.

Our first view of the PCB facility is the late 1980s. Table 1.2 shows typical monthly results for our PCB factory.

Leadtimes are measured in weeks. Problems exist in meeting individual customer orders, but the plant is successful in meeting total output volume. It employs a homegrown measure of productivity—activity measured in number of boards moved per hour divided by number of operators. Scrap and rework rates are measured in percentage, and the plant carries weeks' worth of WIP (and raw material) inventory. Overall monthly performance surpasses (even if just slightly) the planning or budgeted standards. These results are compiled monthly, available one week after the close of the month. Some of the results are manually tabulated weekly (performance to volume) and some daily (operator activity).

It is a factory that supplies corporate but is certainly not seen as a competitive weapon. It is a workhorse. Little investment has been made in it since it is still "productive." But it is a factory waiting for a problem—the outside world—to intrude. It is the picture of a factory that passes superficial top management muster—beating its standards—but about to fail dismally in world-class competition.

By 1995, not much had changed. They have continued to work on decreasing their cost per board but most of their efforts have gone into moving to surface mount, fine pitch, and other new technologies. They do no benchmarking against the outside world. They are still a central captive facility supporting all PCB manufacturing for all the parent company's product divisions. They do a large range of manufacturing supporting end product markets that range from military high-reliance to commercial to industrial to medical devices. They also supply several outside customers, a business that gives them additional economies of scale. They do extensive burn-in and testing of boards so end quality is perceived as good by their customers.

Mark Ritchards, the general manager of the PCB Co., has run the division for over ten years. He is about to present to the corporate steering committee a proposal for a major capital expansion at their plant.

Stretch Goals: The New Reality

Mark Ritchards felt slightly nervous before his appearance at the Capital Committee. He was proposing a substantial investment and expansion in their factory to add the latest capital equipment to their double-sided and multilayer throughhole and surface mount capability.

It was an important addition to their manufacturing capacity and technical capability, allowing them to manufacture the latest designs for the corporation. He didn't relish the trip to corporate. He preferred to keep a low profile and just "get the work done." He saw himself as an operations type—someone far removed from all the politics he perceived existed at corporate headquarters. In his mind, he was not skilled in maneuvering through the thicket of organizational briar patches without getting stuck.

Once a year he spent months preparing an acceptable budget. He always started out lower than last year in real cost/unit, but above what he knew he could achieve. Every year he let himself be negotiated down to what he had originally thought he could do, and therefore was viewed as a good soldier. Quarterly he went to corporate to present his last quarter's results, explain his variances, and give his next quarter's projections. He presented slowly, carefully, and usually with few questions. This was due to his careful preparation—and the fact that almost no one was that interested in manufacturing. Most of the quarterly corporate discussions he attended were centered around the marketing projections, competitive updates, new products in the pipeline, and major customer issues. He was rarely challenged, and likewise, rarely challenged others. He saw himself as a service organization that met the marketing and sales needs of the other end product divisions.

Every three or four years when one of the other divisions fell flat on its face, he agreed to increase his margin contribution by foregoing some capital he had been promised or delaying hiring some needed replacements. He saw himself as staying out of the gunsights of whomever was the latest hotshot out to change the world. They came and went while he carefully did his job meeting or exceeding the pre-negotiated expectations he'd set. His philosophy was "keep your head and overhead low." He was widely known as a manager who kept his commitments and, if not exciting, was dependable and thorough.

Therefore, while not totally comfortable, he was also not prepared for what unfolded at the meeting.

He could see from the silence when he entered the room that either the previous discussion had not gone well or that the next was not going to (or both). The corporate vice-president of finance wouldn't even look at him. Since Mark had always done what he'd asked, and like any good soldier, done a preclose on him a month ago, something had gone wrong. He looked around the room and saw a few new faces. He had a sinking feeling. Maybe he had not done enough analysis, but the facts seemed convincing. This investment would allow production of printed circuit boards for their newest products at his plant. They'd always had acceptable margins on the products they'd produced—at least for the whole product line.

He stood there for a few minutes waiting for the president of Diversified Electronics to finish a whispered conversation with their vice-president of marketing. Then he was asked to give his presentation.

He went through their historical performance on the current PCB product line—showing performance meeting or beating their budgets and standard costs for eight years running. Then he presented a review of their current capabilities for throughhole and surface mount production, one-sided and double-sided, and multilayer boards. Then he presented the projected three-year product mix and how the percentages of double-sided and multilayer were going to represent over half of their projected volume. He then showed their existing capacity at standard utilization as being inadequate beyond the current year. Given the leadtime on getting equipment, qualifying it, and getting the operators and maintenance technicians trained, they'd have to make a decision within the next six months at the latest, and preferably now, on additional capital equipment and facilities.

They all had a large package in front of them with the detailed breakdowns of the new products, their projected product volumes and mix, existing equipment and capacities, and the proposed equipment specifications. The analysis reported proposed capital costs, operating costs, and projected margins.

He was puzzled. He had been through similar presentations and proposals. He knew the company was reasonably flush now; they had just purchased another printed circuit board operation recently and still had considerable cash reserves. Their stock was doing well. The vice-president of finance had already told him that the proposal looked reasonable assuming projections for sales actually occurred, and that there were no unexpected events or price fluctuations (the closest he ever came to a positive position).

So why the silence and why no questions? He would have preferred a barrage of questions to this polite wall of indifference.

Finally the president spoke. "Mark, that was an excellent presentation."

Now Mark knew he was in trouble.

"But we as a group have been having some discussions before you came in that have significant bearing on your proposal. To be quite frank, we aren't sure at this point what to do with our manufacturing capability. As you know, we recently purchased another PCB company specializing in small boards for hand-held or portable devices.

"During that exercise, several significant questions arose. The first question was whether we should do manufacturing of printed circuit boards at all or subcontract it to these contract manufacturers. Even IBM is using outside services for manufacturing. The services have whatever technology we want, we can shop around on price, and there is no capital investment required."

Mark felt his heart sinking and his stomach's contents rising. They met in a heartburn meltdown somewhere midrange.

never expected it in their market. They were a medium-sized diversified electronics company; his plant produced boards for their many different divisions—aerospace, medical products, and consumer. One division was always up, one was always down and they averaged out to slow, steady growth.

The president was quiet for a moment. Silence filled the room. Strangely, in that moment, Mark felt the resolve in his boss's words. Something had changed. Mark didn't want to be defensive. He simply didn't know exactly what to say. He wasn't sure if they could even come close to these new stretch goals. Personally he doubted it.

If they could have, they would have. And yet, he didn't want to give up without trying. Or did he? Did he really want to go through the wrenching change that had to come with such dramatic improvement goals. He knew that his present team would go nuts when he told them of management's response. "Say what? Are they crazy? What do they think we're doing now?" Did he have the leadership to get them through this crisis? Or was he part of the problem? Did he have what it took to find out the answer to his boss's question? Part of him doubted it. He was challenged every day at work but with the same type of challenges he'd had for years—a labor problem, a supervisor who couldn't manage, and equipment that seemed unreliable. He knew how to manage these problems and incrementally improve things. But this wasn't incremental improvement—this was a massacre. This was wholesale transformation. This was a metamorphosis. He wasn't sure if he could do it.

The silence lingered for what seemed minutes but was seconds.

Mark finally nodded "I understand. I'm not sure how we'll do it, frankly. But I understand what you're asking of me."

But he really didn't understand. Why now and why all at once stepwise improvement? Why the sudden change in the rules? And more importantly, most importantly, he had no idea of how he could do it. But that didn't seem like a wise thing to say at that moment. Though realistically, they probably knew that and weren't sure he could do it either.

The president spoke again. "Remember how you implemented the Total Quality Management (TQM) program in 1987? At that time, it didn't seem possible to improve quality as much as you did either. But it was—with new thinking and new approaches and new tools. This is no different.

We were a little late to adopt TQM and we're probably a little late on this too. But it's time. And if you want some help, I've gotten a recommendation for a consultant, a guide."

And that was the end of the meeting for Mark. What was he going to tell his team? As the president had explained, in six months they'd make a decision on whether to continue manufacturing internally or subcontract their printed circuit board manufacturing. The decision would be based on what he called "the art of the possible" at their current plants. He had six months to present what they *could* attain in performance and the plan to do so. If they showed that they could manufacture competitively, then they would be targeted for the capital expansion for new products. If they couldn't, they'd be phased out over time.

So instead of a capital expansion, he had returned with potential capital punishment. Not a good start to the day. He wondered if it could get any worse.

The plane ride back from corporate was a private hell. How had they ended up at this point? No, how had *he* ended up at this point? He felt sorry for himself. He'd worked hard, very hard, to provide a good return for his company, to build factories that ran smoothly and efficiently, and this was his reward for his loyalty and dedication.

What was he going to tell his team? They'd never be able to achieve those goals. Stretch goals? They were called stretch goals because they were like being put on a rack and stretched. They should be called stretch of the imagination goals.

He thought through the alternatives. Maybe they should just lay low and let the other guy fail. No one could meet those goals. Anyone who accepted them was being set up for failure. Lay low, keep doing their job every day, and be there to pick up the pieces when it all fell apart. Ready to be a hero then—ready to emerge and save the day. They wouldn't close their plant, just keep it the way it was. That wouldn't be so bad. These fads came and went; management came and went. Let someone else go first and fail. He'd learn from what they did wrong and do it right. Not first maybe, but without the wrenching changes that he saw in his new challenge. Realistically, he knew that his team couldn't do it. There was too much resistance to change, too much N.I.H. (not invented here) syndrome. His head of planning was already at war with production. Cut leadtimes?! All that inventory was the only way they got anything shipped now as it was. He couldn't imagine explaining to his team how he had left the meeting without emphatically stating they were already at 100% of capacity utilization—their numbers showed it. He had requests on his desk for another 16 in headcount—and now they wanted him to cut the existing number in half!

As he landed at the airport, his mood was beginning to swing towards toughing it out, keeping their heads down and waiting for someone else to fail. The company wasn't shutting them down. It was just putting them into a wait and see hold. They could forge forward and just keep doing their jobs. It was too much change, too much to ask of his team, too uncertain. If he really knew how they could do it, he'd know how to lead them there. But he really didn't know this time. He'd be stepping out there with no security net and a long ways down, all the while fighting half of his staff. He just didn't see how he could pull it off.

That swung it—he'd just tough it out the way they were. They would make it. The company had too much invested in them. They were too small to really get that much scrutiny. They would just be prepared to tighten their belts a little if they had to—pare the capital budget and keep their heads low and let someone else fail.

He had scheduled a meeting with his staff for the day after tomorrow to discuss his direction and see if they disagreed.

Reality

He had trouble sleeping that night. In the early morning he got up, brought in the newspaper, and started to focus on what surrounded him: his car, television, camera, stereo, and golf clubs, all made in Japan; his daughter's car made in Korea; another restructuring announced this morning in the newspaper; a $1.4 billion writeoff by Philips, 10,000 in layoffs, and closings of unprofitable operations; and a General Instruments leveraged buyout. Reality was all around him. His head was spinning, opening up to what others must have already had to face up to—massive relentless competition.

Who was he kidding? You couldn't hide in what had become worldwide capital and manufacturing markets. If his company could get their boards more cheaply somewhere else, they would. He couldn't hide from his competition—it would come and seek his customers out. He had no patent protection. There were no captive customers anymore. The race was going to go to the most competitive, not the most complacent. He could pretend to be an ostrich, but it would not change his reality. The automotive, consumer electronics, semiconductor, and shipbuilding industries had proven it. There was no place to go except to work.

He owed it to his workers to give it 100%. Otherwise, all their jobs were in jeopardy. If he didn't have the expertise, he would have to get it—just like he would if he needed a new head of engineering, maintenance, or facilities. He would not shut down the factory for lack of that expertise—he would find it. And he'd do the same in this situation. That's what they paid him for—to lead, not to wait and watch. He was nervous about it, but he was nervous about many things. Better to grip the bull by the horns instead of hanging onto its tail.

Getting Started—Again

He called the president back for the name of the consultant he had mentioned. If they didn't have all the expertise in-house, that was no excuse for not moving forward—it was, in fact, more impetus because it meant that they were further behind. He thought about what he needed to do to be prepared for the next day's meeting—he had to pass along his new urgency and commitment in terms they could relate to.

Factories With a Future

At his staff meeting the next day, he announced that there were two types of plants—those that had a future and those that had only a past. They were going to be a factory with a future and that meant change. Since they did not have all the expertise they needed in-house, he had already called in a consultant who had helped many other factories attain world-class performance, including several other PCB facilities. He showed them reprints of an article comparing characteristics of each type of plant and asked each of them to read it right now. He noticed that one or two of them seemed to just scan it and he made a mental note of that. Then he asked each of them which plant they wanted to be—and if they would fight to be that one. It was going to be a struggle, but one they could win if they all fought to win as a team. They all had to keep the big goal in mind—being a plant that controlled its own competitive destiny instead of leaving it in their *competitors'* hands.

He could see that several of them were either unconvinced, quiet, afraid, or more likely, waiting to see which way the wind was blowing.

They said the right things but without any conviction, simply saying, "I agree with Bob" (the last guy to have spoken), or "We'll give it a try" or "Sure, we can do that if you want us to." He would need help to get them on board. This was going to be a rocky enough ride without passive onlookers or naysayers at every turn. His gut told him that a few of his people wouldn't make it—the changes were too threatening or disturbing or hard after what they were used to—but they had to be given a fair chance. Everyone had to see that management was committed to getting everyone through the change process but, alternatively, would not let anyone stand in the way.

"We are all going to help each other," he said. "For my part, I'm going to provide the best consultant to help train and organize this improvement process. But since we have to own it, I'm personally going to chair the improvement committee and assign each of you to lead up one improvement team. It's our responsibility to make this work—to give our customers the best product we can with the most competitive facility possible so our jobs are secure. No one else can get the job done for us—unless we want to let another plant handle this.

"But from this point onward, we have to change the way that we think. From this day forward, we are going to have to continually improve every week. So we are going to have to view this not as a project or one-time exercise, but as a change in our way of doing business. We are facing the equivalent of a business heart attack—we have gotten a bit fat and lazy. And it's my fault. Look at our costs." He put up an overhead with their cost/board—labor, materials, and overhead—faxed to him by the president, along with some other data he had compiled. "We can lower the direct labor cost by 80% by moving production to Mexico; we can lower the cost of materials by redesigning a board with ASICS, but our *overhead* is still 60% of the total. That's where our biggest opportunities lie and the place where we've done the least. We need to look at the way we are doing business. Look at our competitors." He put up a second overhead. "I also called up several contract houses yesterday and a few of our vendors. They each sell a similar board in number of layers and density for anywhere from 1.08 times our cost to 60% of it!" Now there was silence. "Would you buy a board from someone who was over one and a half times more expensive?" I'm not going to set our goal yet—but over the next week, we're going to look at our operations and then set a joint target. And then we're going to hit it."

He didn't want to set their stretch goals. He felt he needed the group to collectively come together and agree on them. But he needed them to start the exercise with their eyes open. It wasn't time yet to tell them the goal; it was time to lead them to it. But it was time to explain the stakes. Change had to be motivated; urgency had to be created through clarity and awareness.

"So why are we having his discussion?" All of a sudden he could see several nods and a few grim expressions. The intensity in the room rose immediately.

"As you know, I went to corporate with a capital appropriations request for our expansion. It was carefully reviewed. Here's the bottom line. If we can prove to them that we should be the preferred supplier to the corporation, then it's ours."

He waited for the team to pick up the point. This was the time to start to share the responsibility, the challenge.

Bill, his head of engineering and a quick study, was first as he would have bet.

"And if we can't?"

"Well, let's ask Bob," Mark replied. Bob was the head of finance. "What would you do, Bob, if you had a choice of vendors?"

"I'd pick the cheapest one who could meet my specs reliably," he replied.

"And how are we supposed to prove that we should be the preferred supplier? Are they looking at it as of today? Or do they want a plan? Is it cost or technical capability or reliable supply or turn around time?" Bill continued his questioning in a logical path.

"Yes, yes, yes and all yeses. It's all of those and we have less than six months to prove it. And first we can start by proving it to ourselves."

The group was very silent and there was not a smile for sale at any price. Grim had replaced grin.

"And if we can't prove it?" asked Rod, their head of manufacturing.

"Then we don't get the new products. They'll manufacture them elsewhere and leave us as a grandfathered product facility until those boards are replaced by newer models."

It was very silent. Funny how eloquent silence can be.

As the meeting broke up, he felt that his goal had been partially accomplished. He had passed along some of his urgency without mandating a result. He had introduced the approach of using an outside consultant. But he could feel the various reactions that bubbled below the surface—surprise at the sudden change in the rules, anger at being questioned on their performance, bewilderment as to how real and permanent this situation was, and a trace of fear. He could see it in almost everyone—a sudden jolt of concern about how secure their jobs really were, what a loss of income would mean, and how marketable they really were in today's economy.

He had no doubt that whether ultimately this was a positive event or a disaster, it was for real! He had read the business section for too long to believe that this was an aberration. Now that he was through the initial shock, he saw this as maybe the inevitable that had finally shown up here as well. Now there were the three inevitables—death, taxes, and worldwide competition.

Taking the First Steps: The Re-education of the Plant Management

The consultant turned out to be fairly experienced with PCB manufacturing. This was the fifth plant he had worked with, and he gained interest and some respect by discussing some of the results he had had. "We decreased work in process by 85% and increased first-pass quality to 98% in one plant. Another increased customer service to 93% on mix and first-pass quality to 97%. A third decreased their leadtime by 60% and improved process capability from 4 mil to 2 mil. Each one was different in their number of layers, density, amount of surface mount, and so on. So, until I see your product line, factory, and manufacturing process, I can't say what reasonable goals or targets are, but there's probably something we can do."

A plant tour was the first order of business. The workers were used to quick tours for customers and top management visiting the plant. They were unprepared for the detail the consultant took them through first on a blackboard and then on the tour itself. For each step in the manufacturing process he covered:

- The manufacturing process itself at that operation
- Typical WIP levels and leadtimes through the operation versus actual processing time
- Typical yield and rework levels
- Tests for quality carried out before and after the operation
- Setup times, calibrations done
- Amount of equipment downtime and causes
- Frequency of engineering change notices (ECNs) for work instructions
- Whether this operation was a bottleneck
- Labor absenteeism, lateness
- Number of processes, products run there—frequency of new or custom product additions
- All data recorded at the operation, all paperwork required at the operation
- Level of automation—current and possible
- Sensitivity to facility conditions (if any)
- Process capability versus specifications (typical process variation versus allowed in the process) versus best performance—which equipment had the best performance and which the worst and why
- What made this operation difficult from a quality, scheduling, or speed perspective—for example: lengthy setups for product changeovers, frequent equipment breakdowns, operator errors, frequent changes to work instructions, missing or limited tools/dies/parts, changing customer orders or rush jobs, and so on

This process took a long time as the consultant flowcharted their manufacturing operation in detail. He frequently asked for estimates of the utilization of the operation—what percentage of the day they thought they made boards for end customers, the frequency of breakdowns, how long a breakdown lasted, how long it took to set up, and how the operation was scheduled. His questions covered so many areas of manufacturing that finally a group went along on the tour—a process engineer, a maintenance tech, the operator at each operation, and Mark, curious about this process.

Mark Ritchards became more and more discouraged as the tour went on—he had never realized how poor their productivity sounded to an outsider: all the downtime for setups, the actual first-pass yield before rework, the number of ECNs, and the level of work in process that accumulated. He was feeling defensive but swallowed his pride. This was an indication of how competitive his factory was right now. It was reality. Better to get it on the table if it might help them find the levers for meeting their stretch goals, but it certainly was unpleasant.

At the end, the consultant thanked everyone and complimented them on their cooperation. It had taken most of two full days—starting at the inventory areas, moving through each operation, including the rework lines, storage areas, tool cribs, and finished goods areas. For the rest of the week, he had asked to meet with representatives of each department—maintenance, production, scheduling, materials, facilities, process engineering, product engineering, field service, top management, and finance and marketing.

He had given each of them an agenda of what he was going to cover during the meeting:

- Their job responsibilities/tasks
- How they were measured—quality, cost, speed, etc. and current performance on those measures
- Current reports, data they got—formally or informally
- How to improve their quality
- What were their biggest operational concerns, problems
- How they thought their performance compared to competitors—where they were stronger and where they were weaker
- How they could make the most dramatic improvement in facility performance/operations—not just limiting themselves to their own department

Mark did not go along on these interviews to avoid biasing the responses. The results were rather striking, however, when the consultant summarized the results to him privately. What was clear was that each department viewed most of the problems affecting the facility as residing outside of their own department. Almost none of the departments outside of production measured their quality or customer service or speed. Manufacturing focused on quality and total output but not customer service in terms of individual orders. Engineering continually sent out ECNs to tune new processes. Equipment downtimes for setup changeovers and equipment jams were significant.

Mark had a sinking feeling. How could things be so out of whack? They were not doing poorly on their standard costs. They had run the plant this way for years without a problem. Doubt welled up—maybe the consultant was off base. But he realized that the consultant had not really commented on his findings—he was just acting as a tour guide— reflecting, as in a mirror, the plant's *reality*. He was simply showing him what his team was saying. Scheduling said that their job would be easy if production simply worked on the right jobs at the right time. Production said it would be easy if scheduling did not continually jerk them around, if maintenance would keep the machines running properly and if the product engineers did not design boards so that no two were alike and each was harder to build than the last. Maintenance said that it would be easy to keep the equipment up if production did not abuse the equipment and if process engineering did not keep tweaking the process. Process engineering said that the operators did not follow the work instructions and the raw materials varied too much. No one even came close to saying that they themselves owned part of the problem. Almost none of them even thought about measuring their *own* quality or performance. But they certainly had a lot of suggestions for everyone else.

For the second week, the consultant went off and did one-day studies—one at the insertion equipment, one at wave solder, one with process engineers, one with schedulers, and one with a supervisor and operator team. Mark tried to be patient and not constantly ask the consultant how it was going. But he realized, watching the consultant, how little of their collective time was spent trying to improve things or see them in a different light. Most of their time was spent managing the status quo daily to meet their budgets and schedules. It seemed like a full-time-and-a-half job just meeting the current expectations, much less beating them with stepwise change. It was edifying to see their operations through a stranger's eyes—and an expert stranger who had seen many plants. Part of him, his pride, wanted the consultant to come back and explain that their performance was already excellent in most areas and a credit to their company. Part of him, his practical side, knew that there had to be many opportunities for improvement. And part of him, his optimism, hoped that the consultant could help them reach the mandated corporate stretch goals or at least teach them how to evaluate and prove what was the art of the possible.

What was already clear to him was that they, as a team, spent too much time on operational issues and not enough time on improvement and new vision. They didn't bring in enough fresh ideas to continually rejuvenate and innovate. If he got through this, he promised himself that he would spend at least ten percent of his time looking for new and better ways to do things instead of 100% ensuring the past remained the present. Now he needed to ensure that the present had a future.

Factory Facts

After the two weeks of study, he asked the consultant to present his initial findings and any other material he wanted to at an offsite meeting. While his team filed into the meeting, Mark read their faces and tried to imagine what their reactions were going to be. He assumed that some were now feeling many of the same things he had—defensiveness, fear, confusion, hope, and despair. Others were probably into denial or anger. He had previewed the presentation and felt it was right on the mark.

The consultant began. "Many companies I visit with today are perplexed. They have run for years with a set of measurements and goals that come out of their yearly budgeting exercises and against which they measure themselves and award bonuses. They use them religiously to judge where they are against their budget and demand forecasts as the year progresses."

The consultant put up the first foil.

"As I collected your current measures of performance, I found that they are just like everyone else's! You measure yourselves, first and foremost, against standard cost to track your performance to budgets. You measure performance to customer orders—in volume, both dollars and units—and mix, as measured by the percentage of orders shipped within a week of the *accepted* order date and within 95% of the *accepted* order quantity."

Mark wondered about the emphasis on the word "accepted" but didn't want to interrupt.

"You have a measure of direct labor or operator productivity—the number of boards produced or moved per operator hour. You have a quality measure of first-pass and second-pass yield—the percentage of boards that make it through first inspection, and the percentage of boards that make it through inspection after being reworked—as well as monitoring the rework and scrap rates. You have a measure of leadtime—based on the average work in process in boards divided by the order rate—to estimate the number of weeks of orders in the shop. You also have many health, safety, and personnel measures. They are more H.R. related."

There was silence, but so far no real disagreement. Everything he had said had been accurate. People were waiting for the "but."

"And every month you generally meet or exceed your measures. In fact, as far as I can tell, you've done an excellent job of this for years now. Each month you've consistently achieved your goals." He paused.

"So why aren't you happy? Why is there any reason to continue this meeting? You're meeting *your* goals.

"I see this in every company I visit—they're meeting their goals, achieving their standards, but they are uneasy. They don't feel that they are fast enough, that they are the low-cost, high-quality producer. They worry about competitors. Their customers are not completely satisfied. They hear things about other companies using new approaches and aren't sure which are fads and which are significant important new techniques. They are barraged with articles on how companies have cut leadtimes by 95%, achieved defect rates in parts per million, and doubled their output with less headcount. It doesn't seem possible or plausible. In a word, they are confused. How can they be so good on the one hand and so bad on the other? Are any of you feeling that right now?"

A few hands went up, then a few more until every hand was raised. Everyone was listening now with more open body language.

"Well, the times are not 'a changing.' They *have* changed! We live in a different world where you can no longer measure yourself against yourself—you have to measure yourself against the possible, the best, the achievable. And we're going to look at a whole new way to measure based on absolute or value-added measurements, an approach that will really give you some insight into what you can hope to attain.

"What you've been doing is using performance to standard—*your* standards—as a measurement. Typically these measures are inadequate to lead improvement or change. Generally they hide the real opportunity or improvement and actually stifle the incentive to change."

A Value-Added Operational Review Using Best Practices Measurement

"Let me share with you my snapshot view of your factory using some new principles of measurement and show you some initial findings. Remember, these are based on limited data, so let's not draw any conclusions.

"But they do illustrate a new way to look at your factory. Right now your measurements are focused on the work order or lot or batch you're making. You focus on how much it cost to make, how long it took to make, and its quality and performance to schedule versus your plan or budget or standards.

"Let me go back through your measures for speed, cost, quality, customer service (Table 1.2), and show you how I'd measure them in a typical best practices weekly or monthly operational review.

"Right now, as you can see, you're basically doing great. You're under budget, you're shipping more than was ordered, your operator productivity and leadtimes are on track, and your quality is above plan. The only area you're under plan is on customer service to mix, and as far as I can tell you always are. So what could possibly be wrong?

"To explain how I'd measure you, first let me put up the basic principles for best practices measurement and then we'll apply them."

The Principles of Best Practices in Measurement

1. Measure value-added time components of the activity or asset.
2. Measure to the level of detail, if possible, that exposes the root cause and owner responsible for non-value-added components (what can be changed or corrected?)

3. Measure at the time of the activity with objective measurement tools that are replicable and accurate.

4. Measure all asset utilization and activities and their value-added component until no improvement is possible/necessary/or justified competitively.

5. Measure for predictive control and not for reactive postmortem (to avoid a problem or to detect and correct versus to search for the guilty).

6. Measure frequently enough to support Principle 5.

7. Measure what you need to deliver—customer satisfaction or a competitive advantage.

8. Measure your rate of improvement on these measures over time—it is a measure of your rate of learning or speed of improvement.

9. Measure the variability of the component over time—not just the mean—to see if it is under control and to find best and worst performance.

Measuring Speed Using Best Practices

"Let's start with production speed or cycle time or leadtime measurement. Currently we know that our standard leadtime from order release (by production planning to the floor of an authorized work order) through delivery into finished goods inventory of the completed work order is 5 weeks for your two-layer boards. If your leadtime were measured in minutes or a few hours, we would not even bother with this measure; we'd simply monitor the output rate/hour. But when leadtimes are measured in days, we need more visibility.

"Your performance has been about 98% over the last two weeks as you measure it. This means that the average leadtime for the orders completed over the last 2 weeks (from the work order tickets I read) was about 5 weeks. So you are right about on plan, operating to your standards.

"But does that make you feel competitive? Fast, slow? World-class? Do you know? Let me show you, based on a sample of work orders, the way I'd measure leadtime. Let's start with Principle 1 and now look at our *value-added* speed versus non-value-added components."

TABLE 1.3 Average Leadtime by Component

	Hours	Percentage of Total Hours
Processing time	16 hours	2%
Burn-in time	48 hours	7
Setup time	8 hours	1
In-transit time	1 hour	0
Queue time	4.33 weeks	87
On-hold—no parts	16 hours	2
On-hold—MRB (material review board)	7 hours	1

He put up a slide (Table 1.3).

"Currently, you run your factory three shifts per day, six days per week. If I look at what percentage of the time you're doing value-added activities in your current leadtime or cycle time—it's 64 hours of processing time and burn-in or eight shifts out of 5 weeks actual leadtime or 90 shifts (18 actual shifts/week × 5 weeks). This is about 9% value-added time—assuming we can't find an alternative to a full 48 hours burn-in. If I throw in your current setup and in-transit times (for what is attainable without changing the factory floor layout or redesigning the setups), that adds another shift of cycle time and brings us up to 10% of your current *standard* leadtime. We can see, at least at first glance, that your current five-week leadtime is grossly over a value-added leadtime of 64 hours or theoretically currently attainable leadtime of 73 hours. So each year when you cut 5 or 10% off your standard leadtime, you're not really confronting the art of the possible. We're confusing improved performance with a potential competitive advantage. The first competitor who doesn't make that mistake, in fact, gains a temporary competitive advantage. They can accept and process orders in less than a week to your five weeks; they have four weeks less work-in-process inventory-reducing working capital tied up—along with greatly reduced storage, planning and scheduling, and expediting costs. They benefit from major reductions in any *indirect* costs associated with a longer manufacturing scheduling horizon, more orders, and more work-in-process in the factory.

"By the way, this is a common result we'll find over and over again. As we eliminate a non-value-added direct activity, we also reduce or eliminate supporting indirect activities, giving us a multiplier effect beyond our first expectations.

"So at first glance, we should be able to cut our cycle time by as much as 80%. What seemed like acceptable performance now looks slow and inefficient.

"Clearly, by refocusing measures onto your real value-added performance, we are continually confronted with the reality of our performance. Management always has a clear picture of where the greatest opportunities are to reduce costs, improve quality, and increase speed—where investment can have the greatest payback. We see where our facility performance really stands—what is the art of the possible."

The Devil Is in the Detail

"Now what is the right level of detail at which to collect this data? Let's apply Principle 2—we want to collect it at a level of detail that corresponds to the root causes of the non-value-added components and to the owner responsible for improvement. What we want to be able to do is have visibility into all the *key* non-value-added categories of cycle time that can be matched up to an owner.

"In this case, the largest non-value-added components are queue time, setup time, and on-hold time. In-transit time is relatively small. This level of detail also allows us to match these up to an owner—scheduling (for queue time), equipment engineering (for setup time), and process engineering (for on-hold time). Now we can assign responsibility for our stepwise improvement to the appropriate people."

Know It Now or Know It Later?

"Right now you measure actual job leadtimes after they are completed. You calculate an approximate 'instantaneous' leadtime by dividing the total work in process in the line by the daily output rate. Principle 3 says measure speed at the time that each job is processed and at each operation. Why? By monitoring a 'real time' view of speed we can instantly see trends (queues building up) or exceptions (jobs going on hold) as they occur. This allows us to be alerted to incipient problems and correct them before they disrupt our deliveries. If we wait for the jobs to be completed before we measure our speed, we are seeing the factory as of five weeks ago. We don't know of problems until the customer asks where his order is. If we use a factory average, we don't see the work in process building up at operations. The total work in process may be correct; it just may not be balanced properly."

But We Don't Have That Data

"While this may seem like an obvious way to display the data on leadtimes, most companies don't collect their leadtime data this way. We've discussed that they look *retrospectively* at the leadtimes of jobs being completed or calculate an instantaneous measure. Since they don't monitor the progress of the jobs within the factory in detail, they can't measure the non-value-added leadtime components in detail. As a result, they must guess at the likely causes and amounts. As we've seen from many customers, typical mythology is often tainted by significant single occurrences of a major problem—"our vendor missed a shipment and all the jobs sat'; 'the equipment was down for three days and production came to a standstill'; or older experiences—'the problem is always at insertion, we continually have a problem with bent leads'—even after they've been corrected.

"There are new paperless manufacturing systems today that some companies are moving to continually collect, display, and analyze this data that right now goes onto your paper travelers. These manufacturing execution systems (MES) ensure accurate and timely collection of this data (more in Chapter 7).

"So now we have a best practices view of your speed. Let's move on to your cost performance."

Standard Cost Versus Standard Waste

"Right now you measure your performance to standard cost religiously. It is the basic tool for monitoring your performance against budgets. Like all companies, you have standards for direct labor, direct materials, and overhead at each operation, and you accumulate them as the batch moves with variances taken for higher or lower scrap, rework, material, or labor usage. What I want to focus on, however, is your overhead, the category with the least visibility that now accounts for over 60% of your costs! Each job certainly has material and direct labor content. Most of our variance there should come from quality problems. But what about your overhead costs? Are they under control or understood?

"Your job cost is really the sum of the costs of each of the resources applied or used. These resources are the equipment, people (direct *and* indirect), facilities and materials, or ingredients (direct and indirect). What is clear to us is that we have very little visibility into the indirect cost components and that these indirect costs are typically greater than direct labor and materials costs in most high-tech companies today except for simple high-volume assembly operations.

"So let's relook at our approach to cost measurement now, based on Principle 4—measurement of our asset utilization and indirect activities value-added components. It is clear how many components you need to build a board. It is relatively clear how many operators may be required. But is it equally clear how many schedulers, process engineers, supervisors, maintenance technicians, and so on are required? Is it clear how much equipment is required? You may think so based on historical standards. But now let's look at a best practices way to look at your costs—based on value-added utilization of all your resources. I've looked at three types in my survey equipment, facilities, and indirect labor.

"Let's start with equipment as I know that you're facing a major plant expansion."

Equipment Utilization: Myth—Interpretation of Standards

"Today most companies measure their fixed assets (such as equipment) capacity utilization against some standard measure of capacity utilization required per unit produced. For example, you have basic standards for how many boards/hour you can do at insertion, wave solder, and burn-in. Your plan is based on those standards, and you schedule up to 100% of *standard* utilization. If you achieve the plan, you've achieved 100% of standard utilization. That's why you feel that you're basically out of capacity at insertion and wave solder and close at burn-in (Table 1.4).

"Most companies don't have standards for their indirect labor. They staff by some rule of thumb and inevitably seem to need additional headcount.

"So, based on the added volume of new products, you obviously see the need to add equipment. That's logical since you're already nearly at full capacity utilization and right on budget to standard cost.

"But, let's now apply Principle 4 for measurement to your equipment. Here is some rough data (Table 1.5) that I collected at insertion for actual time in status by activity over several shifts (projected out for a potential seven days per week, three shifts per day or 168 hours).

TABLE 1.4 Standard Capacity Per Month Calculations

Area	Projected Volume (Boards) = 80,000/Month	
Insertion	Standard hours/board/insertion total standard hours	.5 hrs/board
	Total standard hours required	40,000 hours
	Total hours available	42,000 hours
	% Standard available	95%
Wave solder	Standard hours/board/wave solder	.033 hrs/board
	Total standard hours required	2,640 hours
	Total hours available	2,800 hours
	% Standard utilization	94%
Burn-in	Standard hours/board/burn-in*	.17 hrs/board
	Total standard hours required	13,000 hours
	Total hours available	16,000 hours
	% Standard utilization	85%

*(Based on 48-hour burn-in and 288 boards/oven)

TABLE 1.5 Actual Utilization of Insertion Equipment

Actual Time in State/Activity	Actual Clock Hours	Percentage of Utilization
Running production	79	47
Running rework	7	4
Running scrap at insertion	2	1
Lost production due to scrap* at downstream operations	6	4
Running engineering test runs	4	2.5
Setup/changeover	14	8
Idle—equipment not scheduled to be run**	24	14
Idle—no jobs available	12	7
Broken—waiting repair	3	2
Broken—in repair	4	2.5
Idle—while boards are visually inspected	6	4
Idle—while boards are counted and paperwork prepared	2	1
Miscellaneous/other	5 Hrs/Wk	3
Total	168 Hrs/Wk	100%

*Hours of insertion scrapped at any future operation after insertion.

**Sunday. No shifts scheduled.

"You see that once we have *real* visibility on what actually goes on and do not have to depend on standards, a vastly different picture emerges. The *real* percentage of time (based on the *real* available hours in a week) that we utilize the equipment to make product for customers is under 50%! Quality problems here and at downstream operations cost us 15 hours of capacity. Not running on Sunday costs us 24 hours. Equipment quality problems cost us 7 hours. Setups cost us 14 hours. The way we manage the operation—having the operator inspect some of the completed boards and fill out the paperwork before starting another job costs us 8 hours. Scheduling issues that don't keep the equipment fully utilized cost us 12 hours.

"This value-added actual utilization view has several major differences from our 'standard utilization' view. First of all, it shows us the key potential areas to explore for improvement. Second, it shows us the potential magnitude of improvement possible. Third, it gives us more specific areas to benchmark to see how other companies may be running their operations. Finally, it truly galvanizes us to improve. Seeing that *less* than half our asset is being productively utilized motivates us to improve it.

"So for this operation, while we are at 100% of standard utilization and standard cost, we are under 50% of value-added utilization. The former tells us where we are against our budget. The latter tells us where we are against the art of the possible. The former doesn't give us any indication of the magnitude of improvement possible. The latter suggests what capital and cost savings may be possible. We certainly need budgets, but we also need a basis for setting them that goes to the *root cause* of our costs—value-added vs. non-value-added activities.

"There is a basic lesson in both these examples on speed and utilization measurement.

"Without accurate and timely visibility into our performance at a level of detail that relates to our *individual* activities or responsibilities, we can't link our actions to possible results. Good measurement schemes lead to performance gains. Poor measurement schemes lead to status quo. If there is an overall standard for equipment utilization that lumps *all* of our activities together without any more detail, then no one is really responsible for improved performance. This quote in Figure 1.1 really sums it up.

"When performance is measured,
performance improves.
When performance is measured and reported back,
the rate of improvement accelerates."

Therefore,

"If you want to improve the quality of performance in any area,
you simply improve or increase
the frequency of feedback."

"Feedback is the breakfast of champions."

—Thomas S. Monson

Figure 1.1 The Breakfast of Champions

"This approach to measurement now lets us set a stretch goal for improvement. In this case, we could certainly look at anywhere from a 50% or higher gain in output potential."

Facility Utilization: The "Invisible" Elephant

"I've done the same analysis for your facilities—looking at the *actual* value-added utilization of your floor space (Table 1.6).

"We all take floor space for granted—it just exists. But the key for any competitive company is to transform your cost of production into profits with the minimum lag and the minimum assets. Facilities are an asset. If you have non-value-added space, you can use it for expansion, potentially rent it out, or move into a smaller facility. In the meanwhile, while you accept it as a given or fixed asset, you either pay rent on it or depreciate it, heat it, light it, clean it, secure it, walk through it, paint and maintain it, have insurance on it, and so on. A facility with excess space is like a runner who is overweight. The excess weight makes you less competitive. You may think of it as not being controllable, but in a competitive world, *everything* needs to be considered controllable. And you currently waste somewhere between 25 and 40% of your space and with it all of its associated costs.

"If we look at how you're using your facility asset: 13% is used for your equipment; 37.5% is holding inventory—inventory that adds no value

TABLE 1.6 Actual Facility Utilization

Floor Space Use		Actual Square Footage	Percentage of Total
Production equipment		62,000	13%
In-transit walkways		62,000	13
Rework area		31,000	6.5
Inventory storage—raw materials		83,000	17
Inventory storage—work in process		41,000	8.5
Inventory storage—finished goods		60,000	12
Tool crib		25,000	5
Idle		120,000	25
	Total	484,000	100%

and takes up space and needs handling; 6.5% holds your quality problems; and 25% is simply idle due to poor layout or huge hallways or eventual planned expansion. I could put 4 to 5 factories within this space. If I were a competitor, I could manufacture at 20 to 40% of your facility costs. And I can assure you, someone already is.

The Indirect Mystery of Life

"How about your indirect labor value-added utilization? Here again are some very rough figures, ballpark estimates for a few engineers, schedulers, and maintenance technicians.

"Most companies don't have *any* view into their indirect labor productivity. They don't have any standards for most of their indirect positions and only a vague ratio of indirect labor required to the number of products, equipment, or work centers.

"Table 1.7 represents three shifts of data for an engineer, scheduler, and a maintenance technician based on logs I asked them to keep. What do these logs tell us? That these are *knowledge* workers. Their raw material and output are data, knowledge, and ideas. In this case, they clearly are spending less than half of their time doing the work you pay them to do. They are spending nearly a fifth of their day and your budget just getting the data, the raw material that they need to do their jobs.

"For example, your schedulers walk through the factory to find out the current equipment status, the actual job locations, who's in, and who's

TABLE 1.7 Indirect Labor Productivity

Activity	Time Spent (Hrs)	Percentage
Work activity	14.5	36%
Rework of work activity	3	8
Getting data needed to support work activity	7.5	19
Getting approval of work activity	4	10
Explaining results of work activity	4	10
Transfer of activity to another person	2	5
Follow-up on success of work activity	2	5
Meetings—general	2	5
Training	1	2
Total	40	100%

not that day. They go to the stockroom to find out the exact raw material availability at that moment. They call purchasing to find out when additional shipments may be arriving.

"Your engineers hunt down where the jobs they need are physically located along with the paperwork they need. They go to the maintenance departments for equipment records. They interview operators about equipment performance. They look up raw material specification sheets. And then, because frequently the data they get is old or inaccurate, the work has to be redone or updated or even scrapped. Put that down for 8% of their week.

"Most of them need to get their work approved. They need to bring it over, get it signed, explain their results or ideas, and wait for the approval. That takes another 10% of your budget.

"Then decisions or plans often need to be explained to someone else. Some tasks go over a shift or must be executed by another department. There goes another 15% of their time.

"The bottom line is that indirect labor is often the fastest growing and least measured or understood expense. Whenever I see a manufacturing facility tracking production using paper I know 10 to 20% of its indirect labor productivity is *instantly* lost. When you couple that with indirect labor departments spread out over several buildings, separate from the manufacturing floor, add another 5 to 10% for coordination and distance. So right now your actual value-added indirect labor productivity is about the same as for your equipment—a little over one-third.

"This is just a snapshot, but a reasonable stretch goal here would be a 50% productivity improvement. Once again, with visibility, we have a rational way to set a stretch goal.

"Right now, you try to control all of these indirect costs—facilities, equipment, and indirect labor—using standard costs. You have *allocated* whatever it *currently* takes to run the plant today. Unfortunately, by doing it that way you really are blind to any areas for improvement. The standard cost system is really hiding the factory. Why? Because it really only focuses on direct materials and direct labor."

Seeing the Full Factory—Taking Off the Blinders

"If we were to draw a picture of the factory based on what you're currently measuring through your Material Requirements Planning (MRP) or Enterprise Resource Planning (ERP) system we would see a factory that only consisted of jobs and direct operators. There would be no tools, no equipment, no facilities, and no indirect personnel (maintenance, engineering, planning and scheduling). Our view into the factory and its costs, speed, and quality would clearly be incomplete. And if we had to physically walk around the factory based only on that information, we would continually trip over what we don't and can't see. This is metaphorically what happens, in fact, in our factories. We are running *blind* to our equipment utilization, our tool life, our time to repair equipment, our indirect labor productivity, facility utilization, and so on. We are running blind to what in most factories, including yours, is typically two-thirds of your total cost structure, all lumped under indirect standard cost.

"When most manufacturing was still a labor-intensive assembly of raw materials, the standard cost view used to reflect or account for two-thirds or more of total manufacturing costs. However, in today's more capital-intensive, multiple-product line, and customized product environment, these measures are not adequate to take the full value-added measure of your factory. They simply don't allow us to see the room for improvement, and set and assign stretch goals to the responsible parties.

"The current financial system is *not* going to be of any use in moving off of the status quo. Standard costing is a remnant of the 1940s and one of the great *inhibitors* to change. We really should call it substandard costing. And don't expect ABC to necessarily be better. I call it a more equitable way to allocate the waste.

"We need to move to absolute value-added measures of our factory based on the principles I've listed." (More in Chapter 4 on best practices in costing.)

Customer Service—Not

"Right now, you measure your customer service on performance to volume, in dollars and units, and secondarily on performance to mix. Actually, that's in reverse order from a customer's viewpoint. The customer only wants to know if they got their order when it was promised with the quantity promised to their specifications. Again, that focus on volume performance is a focus on your goals for revenue or costs, not on your customers' goals. If you're not meeting their requirements and someone else can, you'll soon be losing customers. You need to focus on mix performance when you measure customer service. Volume performance is a remnant of the days when hitting customer orders was a theoretical concept and you could ship orders when *you* completed them.

"Leading-edge companies are now even measuring the percentage of orders they can accept based on the *customer's* first request date. All previous measurements of customer service were based on performance to your actual committed date (and usually based on shipping 95% of the order quantity requested within so many days of the committed date). But companies in highly competitive markets with substitution possible know that to thrive they need to meet the customer's request, not their internal response to it.

"But looking at your current performance, you ship a lot but not to the customers' committed dates. That's running under 80%. So clearly, there's room for improvement. How should we measure customer service?"

Prediction Versus Postmortem

"Most schedule performance measurement is done postdelivery. We either make the schedule or we miss it. Performance is often determined during a crazy last day or week in your financial or shipping period. Stand too close to the end of the line during month end and they'll ship you too.

"When you do miss a schedule (and you do), you have little visibility into why—we just know the order was late or short. A first step to improving measurement would be to use Principle 2 and explain the root cause of

TABLE 1.8 Primary Causes of Missed Customer Shipments

Percentage of Missed Orders	Due to:	%	Root Cause
7%	Insufficient quality	5	Excessive scrap
		1	Rework left behind
		.2	Inadequate piece parts
		.2	Additional units on engineering hold
		.6	Miscellaneous
21%	Missed schedule ship date	13	Excess queue time
		3	Waiting for missing piece parts
		2	Started late—customer specifications rec'd late
		2	Started late—order promised in under standard leadtime
		.5	Engineering hold for quality
		.5	Miscellaneous

the missed schedule. Here's an example of what that information might look like (Table 1.8).

"While this analysis is helpful in assigning responsibility for improvement, it violates another principle of measurement—measure to avoid a problem or detect it at the earliest moment. In this case, we need a different approach—we need a *predictive* measure of schedule performance.

"We need some form of simulation or scheduling system that will, proactively, tell us *whether* we will meet our schedule. We can think of this as 'customer service process control,' the equivalent of statistical process control in the quality world. What we'd like is a scheduling simulation with the ability to consider various corrective actions such as overtime, changing priorities, or reassigning resources to correct any predicted missed shipments. Again, what such a scheduling capability will need accurate real-time status data reflecting current job locations and quantities, equipment status and availability, and operator status and availability. Without up-to-the-minute accurate data, we're steering our course based on yesterday's position. This is an example of the application of Principle 6. Measurements must be timely and accurate enough to support our management of the factory. Today's shortened cycle times demand more frequent measurement. When you're walking, you don't need to look at the road every second.

When you run, you need to look ahead more frequently. When you fly a jet, you need to continually see miles ahead. As your cycle times decrease, the picture of your factory changes more rapidly and you need updates more frequently to stay on course. Measurement systems must provide continual vigilance, early warning, and aid in course correction."

Quality

"We can apply these same measurement principles to quality. Right now you're running at a 92% yield rate. One way to record quality problems is by observed problem—such as a cracked or broken component, a bent lead, an open circuit or a short, a missing component, a trace too wide or too narrow, the wrong component, and so on. This is certainly useful information.

"But a more useful approach would be to also record the *root cause* of the observed problem and categorize it into one of three categories:

- Preventable if the current specifications had been followed correctly (human error)
- Preventable by adding to the current specifications (understood cause but not now covered by existing instructions)
- 'Mystery variation' periodically observed (cause not yet understood)

"The reason for this breakdown is to look at quality problems in a way that suggests how to prevent them. For quality problems caused by preventable errors, we need to make sure the existing specifications are clear and followed precisely. This may involve more operator training, and/or rewriting the instructions if not unambiguous, and/or display of work instructions, and/or active checking of compliance by a computer system (see Chapter 7 on MES). For common quality losses caused by known but not currently specified problems, we need to extend the specifications to cover these potential problems and train our people accordingly. For 'mystery' variation we need to try to analyze the underlying cause and eliminate it (see Chapter 2 on Best Practices in Quality).

"Knowing the percentage in each of these categories also leads to our understanding the improvement possible. Certainly, most errors caused by not precisely following the current specifications can be eliminated as can

TABLE 1.9 Current Quality Report: Yield Loss at Test by Cause

Observed Cause		Percentage Lost
Bent leads		1
Opens		.8
Shorts		.8
Broken component		1.5
Wrong component		.5
Trace too wide		.3
	Total % losses	7%

errors preventable by extending the specifications. If we look at your current breakdown of quality (Table 1.9), we're not sure of whether the problem is 'preventable' or 'mysterious' (within the currently expected process variation).

"If we now report on quality by root cause (Table 1.10), we see that 3% (of the 7% you're currently losing) is potentially preventable if we could ensure that the specifications are followed.

TABLE 1.10 Proposed Quality Report Yield Loss at Test by Root Cause

Root Cause/Category	Observed Cause		% Lost
Operator error	Wrong component		.5
	Broken component		.8
	Shorts		.2
	Opens		.3
		Subtotal	3%
Specifications not complete	Broken component		.6
	Bent leads		.8
		Subtotal	2%
Mystery	Trace too wide		.3
	Broken component		.1
	Shorts		.6
	Opens		.5
	Bent leads		.2
		Subtotal	2%
		Total % losses	7%

"An additional 2% could be prevented by improving our specifications to cover additional causes of losses. And finally, there are 2% of losses where we're not yet sure of the root cause and therefore how to prevent them.

"A question that I'm often asked is how we know *what* we should be measuring. We have talked about characteristics of *how* to measure; now let's now explore *what* to measure."

Measure Your Competitive Position

"In the early 1970s, most companies measured cost and customer service (as delivery to schedule). The 1970s showed the manufacturing community that measuring cost without measuring quality was a strategic error—that a company could gain a strategic competitive advantage by supplying a measurably higher quality product.

"This change pinpoints Principle 7 of best practices in measurement. Companies need to measure all attributes of their total customer satisfaction—the attributes by which their customers grade them. It is within those measures that a strategic competitive advantage will exist, a way to differentiate from competitors. In a world-class company, manufacturing is used to gain a competitive advantage through some key differentiation.

"Today, we see six competitive factors potentially provided by manufacturing—differentiation on cost, quality, speed/fast turn-around, customization/flexibility, technical prowess, and/or customer service/on-time delivery.

"While cost is only one competitive factor, and as the Japanese have proven, customers will pay higher prices for higher "value" (in quality or other measures). Manufacturers will still face constant pressure to reduce costs. This pressure will come from an increasing global market where competitors with lower cost structures will use that advantage to enter worldwide markets. We can no longer simply worry about competitors in our own country (as we did in the 1960s), in Japan and Europe (as we did in the 1970s), or in the Newly Industrializing Countries (NICs) of Taiwan, Singapore, Hong Kong, and Korea (as we did in the 1980s). We will see competitors in Eastern Europe, India, China, Mexico, and South American countries as they move to industrialize and capitalize (in both senses of the word) on their lower labor, raw material, or capital costs during the rest of 1990s.

"Quality is a key competitive measure that we have all come to expect and accept as a given. Every manufacturer has seen the results in the automotive and electronics industries of the low-quality producers losing marketshare and profits to the high-quality manufacturer. This picture is not going to change. Quality is a fundamental measure of all competitive products, and the hurdle for attaining a competitive advantage is ever increasing. Today, in Japan, the goal is zero defects—not one in a thousand or million or billion, but zero. Certainly that is a goal that few manufacturers today have accepted—yet it is now their competitors target.

"Speed is a competitive advantage that became a key differentiating factor in the mid- to late 1980s. As companies reduced their work-in-process levels, they also dramatically decreased their manufacturing cycle times. This meant that customers could get orders filled in days (or hours) instead of weeks, especially for made-to-order products. This, in turn, let them be more responsive to their customers. This move toward flexibility and responsiveness (a high-quality product at the right cost *and* at the right time) again created a true competitive differentiation. Speed will be one of *the* key competitive advantages in the remainder of the 1990s and probably lead to more local manufacturing—to be responsive to customers—as we have seen in the automotive industry. Parts suppliers move their operations to be near their customers' plants to allow just-in-time deliveries (speed is discussed in Chapter 6).

"Technological prowess is the capability to manufacture products with processes that have unique characteristics—finer tolerances, stronger bonds, stronger materials more resistant to failure, and use of less expensive materials or equipment. This competitive advantage is harder to attain in many industries because the equipment and raw materials used are purchased from outside vendors who supply competitors as well. In any major industry, there are few companies who can continue to surpass the technology and/or cost of vendors supplying a larger segment of the market. We have seen this dramatic shift in the semiconductor industry. Over a short period of time, the source of the equipment and raw materials used changed from a majority being supplied in-house to nearly a complete reliance on external suppliers.

"This means that manufacturers must focus very carefully on what aspects of the manufacturing process could give them a competitive advantage and focus on those—either for in-house development or modification of equipment, materials or specifications, or joint development with vendors. Texas Instruments still produces or modifies some key semiconductor equipment. IBM has many advanced co-development projects with key vendors. Japanese companies often modify key equipment to improve performance.

"Companies also can gain competitive advantages by focusing on improving or developing new manufacturing processes. This involves both research into new techniques and continuing experimentation with existing processes to completely understand and characterize their performance over a range of operating conditions. For example, by studying painting processes, we can continue to refine specifications on allowable moisture, electrical charge, cleanliness, drying rates, and other operating conditions that affect cost and quality. It is these activities that will allow competitive advantages when equipment and materials become widely available and not a differentiating advantage (technical capability is discussed in Chapter 5).

"Customization/flexibility is clearly becoming a second major competitive thrust in the 1990s. Japan has already moved away from a commodity product strategy into one of a diverse mix of products. As more countries target the commodity market, a real competitive advantage is gained by offering more product differentiation—either through more product options or customized product to customer specifications. The capability to offer more options and/or customization depends on both improved design processes (of product and process), communication of those product and process specifications to vendors (for parts, tools, or testing) and the factory floor, and scheduling of the potentially more complex product mix.

"Finally, customer service is the ability to deliver to customer commitment dates as promised. We believe that this measure will be dominated by the measure of speed—the ability to rapidly meet orders. A vendor who can respond in hours or days will still be preferred to a vendor who is consistently accurate on meeting committed shipment dates but has order lead-times measured in weeks. However, it is a critical measure for our factories in many other ways, as we will explore—and longer term, when vendors are all able to meet short turn-around delivery, it will obviously regain key importance. However, if you could call two cab companies, one that will be there within five minutes (plus or minus ten minutes) or another who requires advance notification of four hours but arrives exactly on time, it seems obvious which company would thrive (and perhaps learn to improve on its on-time arrival)."

What's Good for the Goose

"And, by the way, you should measure speed not only for production but for *all* your manufacturing activities—maintenance, repair, training, cleaning, and setup. And the same is true for quality and cost, and customer service. These metrics apply to all direct *and* indirect activities.

"If you miss a maintenance or do it incorrectly, or it takes longer than necessary or is more expensive than needed, it makes you less competitive. It means your value-added utilization is lower, your costs are higher, and product orders are delayed. We'll need to use these measurement principles on *all* activities to be world-class."

Keeping a Competitive Advantage Means Constant Change

"All of these measurements are valid ways to benchmark our manufacturing performance—and always have been. As we've seen, competitive measures have changed significantly over the last 30 years. This should not be surprising; in fact, we should expect it and initiate it. The business strategy of any successful corporation is the continual search for a temporarily sustainable competitive advantage. Any new entrant into a market must look to differentiate themselves along some new 'marketable' or 'value to customer' axis. Because we have seen a dramatic change in competitive axes over the last 30 years, we should not expect a less dynamic world over the next 30. This means new competitive measures as well as new standards for acceptable or competitive performance on existing measures. One potential new measure we may see in the second half of the 1990s is 'relationship to customer.' We have seen ever closer relationships between vendor and customer—from joint research and development (IBM/Perkin/Elmer) to equity investments (Compaq and Conner Peripheral) to linked information systems and direct electronic communication (EDI) to technical assistance (GM/IBM) to actual creation/spin-off (Claris/Apple).

"Another likely new measure in today's environmentally sensitive world is degree of 'green' as measured by the percent of recyclable material in your product and the variation from zero emissions.

"It is possible that by 1998, we'll add a measure for this 'green' capability as well as one for going beyond providing product to customers to providing services to customers (design, consulting repair and maintenance, etc.) on how to best use your products (as IBM and DEC are increasingly doing). Another measure may be required for global manufacturing. A measure needs to be broken out separately when it *by itself* becomes a significant competitive advantage aside from its advantages already represented in other measures (such as in cost, speed, etc.).

"The point is that the measures we use are not static because our manufacturing strategy isn't static. It must adapt to new or changing technology, regulations, competition, costs and resource availability, and customer needs and markets."

Keeping a Competitive Advantage Through Our Rate of Change

"We've discussed the attributes and scope of 'full' measures of your manufacturing capability. However, as we have just seen, manufacturers compete in a *dynamic* environment where customers, competitors, and environmental factors are not constant. Therefore, we need to discuss how to use measures as active targets in the dynamic world we should expect and anticipate.

"When we compete with our manufacturing capability, we have to expect that our competitors are actively moving to improve their manufacturing capabilities. Similarly, we have to anticipate our customers' needs for improved performance in quality, customer service, cost, speed, etc. to meet *their* improvement plans. Therefore, we need to add another concept to measuring our factory: our *rate of change* in performance, representing how fast we are learning or improving our manufacturing capability (the eighth principle of measurement).

"In Table 1.11, we see an example of current performance from a supplier of our integrated circuits versus their competitor. Their manufacturing performance relative to their competitor still gives them a competitive advantage, even if slight.

"However, a key question which this static picture doesn't show is whether they are catching up, falling behind, or staying even. What is missing from this view is the *rate of change* in improvement—a projection of our competitive position over the *next* 6 to 18 months based on our improvement programs or technical roadmaps.

"If we look at Figures 1.2 and 1.3, we see two very different scenarios that Table 1.11 could represent—a snapshot of a company gradually losing

TABLE 1.11 Current Performance of Preferred Vendor Versus Second Choice

	Preferred Vendor	Second Place Vendor
Quality	.98	.96
On-time delivery	87%	83%
Cost/unit	$.63	$.63
Leadtime/speed	8 weeks	8 weeks
Customization possible?	No	No
Technical capability	.25 micron	.35 micron

its competitive position or actually adding to it. As we see in Figure 1.2, the second-place vendor is learning or improving at a *faster* rate than the preferred vendor in quality, cost, and customization. Within 12 months we expect him to surpass our preferred vendor and replace him. In Figure 1.3, the reverse is true. Our preferred vendor is actually gaining an even greater competitive advantage in cost, quality, and on-time delivery.

"The point of these examples is that competitive advantages are not static, and therefore measures of performance cannot be either. We must look at our rate of change and try to gauge it versus our competitors based on their past performance, market intelligence on performance or capability being promised to customers, new equipment being purchased, research and development efforts, or any other data available.

"We need a time-based competitive view of your manufacturing capability—today and at six-month intervals both forward and backward versus your two closest competitors. We know from your market research with suppliers of processing equipment and customers that one of your competitors has just bought equipment capable of 2-mil line widths and surface mounting. This investment should allow their technical capability to leapfrog yours without a similar investment in both equipment and manufacturing R & D. The other competitor has invested in a facility in Malaysia for their high-volume boards which we estimate will decrease their cost/board by at least 24%.

"Without this forward look, done at least yearly, if not biannually, your measures will not alert you to what the marketplace will eventually tell you—that you no longer may be competitive. By then it is already too late to anticipate and correct your weaknesses.

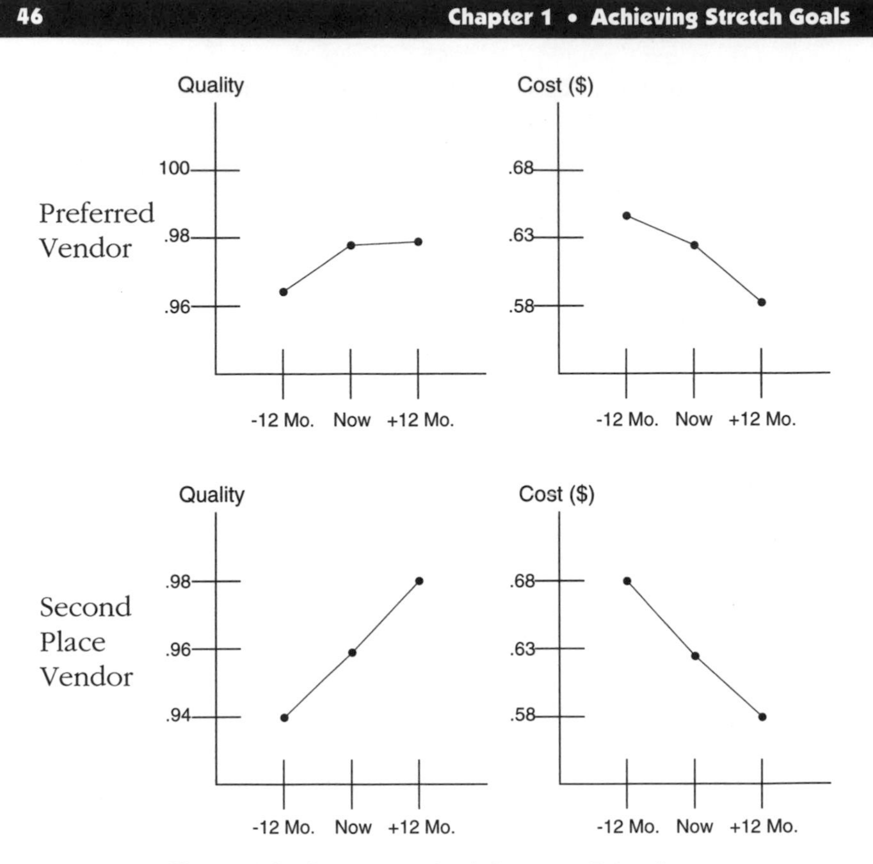

Figure 1.2 Scenario 1: An Advantage Being Lost

"If we also had a comparison to your key customers' (or target customers') desired requirements over the next 12 to 18 months, your view would be nearly complete."

Average Performance Versus Variability or Uncertainty

"The average family in America has 2.2 children. Clearly there is no average family in America. The average job in your factory takes 5.0 weeks to go through the factory and has a yield loss of 8%. Probably there is no average job either. Some jobs take longer than 5 weeks, some less time; many jobs lose no boards, some one; a few are completely ruined and scrapped. Strangely enough, while almost everyone only reports on average leadtimes, quality, rework, or other operating measures—no one except

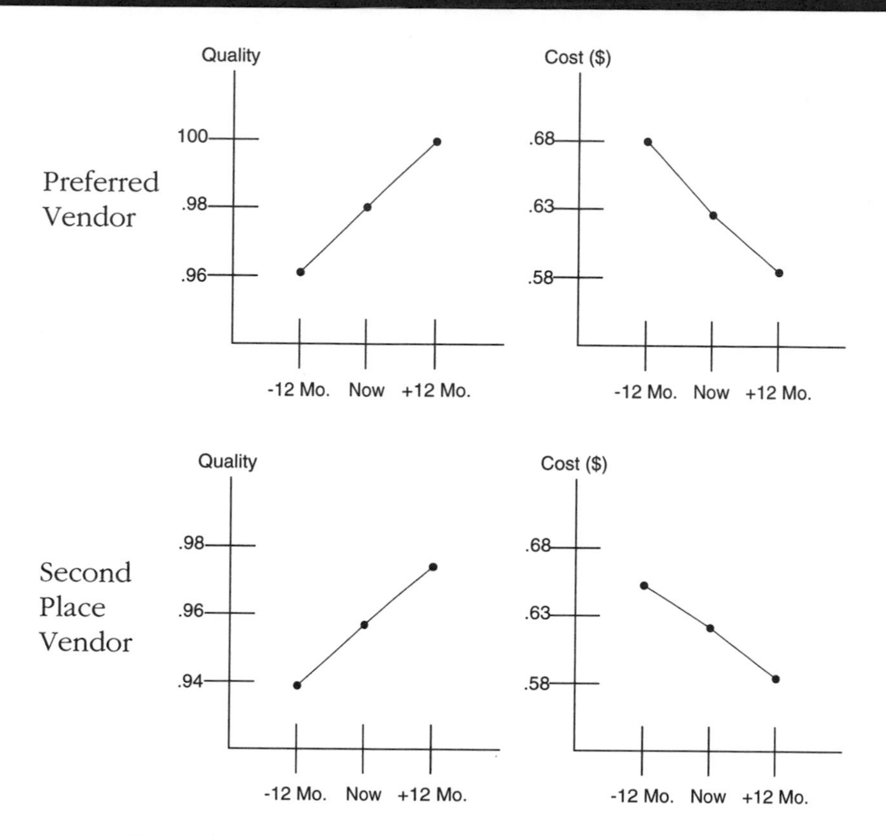

Figure 1.3 Scenario 2: An Advantage Being Strengthened

management is affected by the average. What each customer cares about is what happens to one's specific order. What most operating personnel have to worry about are the worst operating conditions or the best operating conditions—the degree of variation.

"For example, our facility has a goal to deliver 95% customer service to quantity and due date. Currently, we are running at 78%, where the facility has been operating for years.

"What is the problem? We are scheduling using the average leadtime and the average yield! If our distributions of leadtime and yield are reasonably symmetric (Figure 1.4) that means, however, that we exceed the average leadtime 50% of the time and have higher yield losses 50% of the time. In fact, we should theoretically be at or below 50% customer service (if leadtime and yield variance were independent, we would be at 25% customer

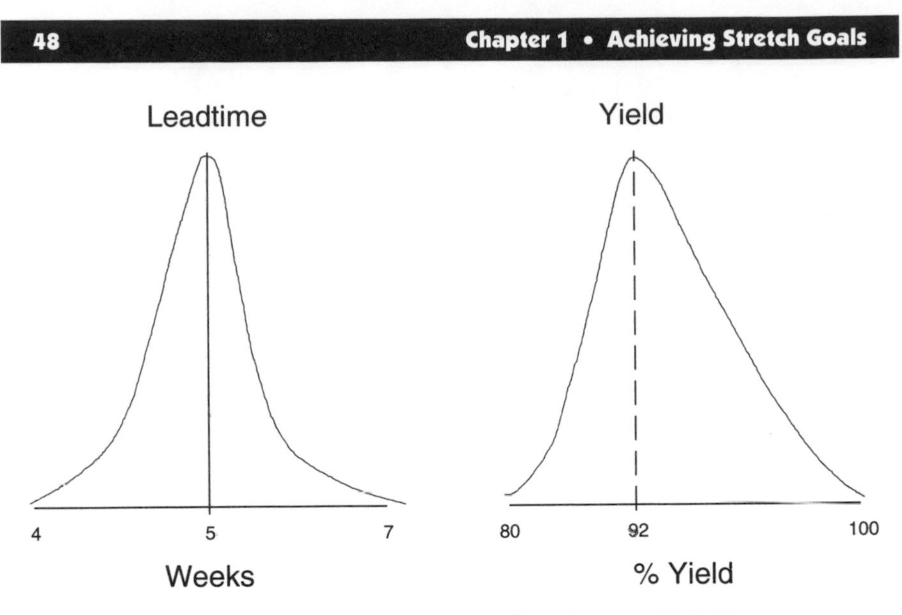

Figure 1.4 The Distribution of Leadtime and Yield

service, but they are often partially correlated—problems with quality slow the job). We get above 50% customer service by expediting late jobs, working overtime periodically and allowing ourselves a grace period of several days in our customer delivery measure of on-time performance.

"So what determines our customer service performance is not the mean or average leadtime or quality; it is the *variability* of the leadtime and quality. If you really wanted to provide 95% customer service, you would need to plan using a yield you exceeded 95% of the time and a leadtime you could beat 95% of the time.

"Similarly, if you wanted to know what was attainable in quality, you would want to know your best performance, not the average performer. By focusing on averages, we are losing sight of how good and how bad our performance really gets. Variability in our performance inevitability leads to waste. It becomes uncertainty to someone else who is our 'customer' and must buffer himself against our lack of consistency. Production planning uses safety stock to buffer against variation in quality. Production keeps high work-in-process levels to buffer against machine breakdowns (so other downstream operations are decoupled from a machine failure, and direct operator productivity is not affected). Maintenance uses overtime to buffer against varying breakdown/repair rates, as does production to handle varying workloads at a work center.

TABLE 1.12 Leadtime Variability—Fractiles

Fractile	Value
.05	4.2 weeks
.25	4.7 weeks
.50	5.0 weeks
.75	5.3 weeks
.95	7.0 weeks

"In general, variability is the enemy of efficiency. The amount of variability is an indicator of whether a process is under control and predictable or subject to wide variation. It is a basic cause of waste that brings us to our last principle of measurement (Principle 9) that we need to measure not only average performance but also variability. So when we look at leadtime, quality, and other operating measures, we need to look at the distribution or range or some measure of their variability. In Table 1.14, we report on the 5, 25, 50, 75, and 95 percent fractiles of leadtime.

"Then we can examine the specific reasons for the instances of extremely good and extremely poor performance (as Principle 2 requires) as a Pareto chart. In Figure 1.5 we can see that for jobs in the 75th and above fractile, the leading causes for the delay were longer waits for missing components, broken equipment, and rework times. This tells us that scheduling

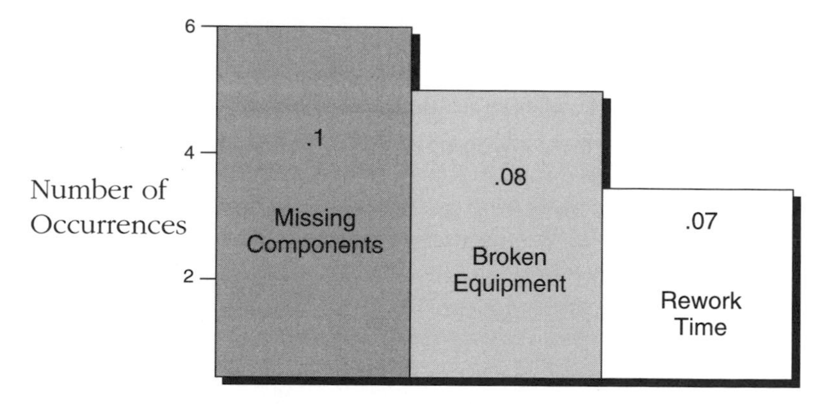

Figure 1.5 Pareto Analysis of Leading Causes of Delay Above the Average Leadtime

rules alone will not solve leadtime variability—we must improve our release rules (so that jobs without all raw materials are not released) or JIT vendor relationships as well as equipment uptime and quality."

The consultant wrapped up. The meeting had taken over three hours. "Today I really wanted to address the issue of what *might* be possible. What order of magnitude improvement or stretch goals are reasonable to consider. That was the easy part. If you want to move forward and improve these, then we need to start looking at best practices in manufacturing management—in asset utilization, quality improvement, scheduling for customer service, cycle time reduction, and so on. We've only covered best practices in measurement—which *is* the right place to begin. If we don't use the right measurements, we have a much lower probability of success.

"Now you, as a group, have some real decisions to make." And he was gone after shaking hands with each person.

Mark looked around the room at his team. As best he could tell there were a variety of reactions to the consultant's analysis. Bill, his engineering manager, had been nodding attentively, taking notes and seemingly excited by the new insights. Bob, his finance guy, had been turned off since the attack on standard costing. Bob had just spent a small fortune on a new state of the art MRP system and was probably feeling like a bit of a sap. Mark wasn't sure he had even listened to the rest of the presentation. Rod, his vice-president of operations, had been squirming a lot, trying not to show his discomfort. But it had been obvious during the discussion of equipment and indirect labor utilization.

Tom, his Information Technology (I.T.) Director, had been attentive, alert, and watching for Mark's reaction. All in all, it was a group divided, he guessed, all worried about their jobs, some defensive, some open to new approaches. It was up to him to lead them to a unified position. Otherwise, it could deteriorate and retreat into sporadic sniping between departmental bunkers. What made it difficult to achieve was that he too was feeling many mixed emotions.

The presentation, which he had previewed (the consultant was too savvy to surprise his client, though Mark was fairly sure that their president would also get a copy and be briefed) was a classic case of good news/bad news. The good news was that there *was* a lot of room for improvement. The bad news was that there was *a lot* of room for improvement. He also felt responsible for their situation. He and they had lost track of the art of the possible. He had not spent enough time as an individual, team, or division keeping up with the current thinking and trends in manufacturing practices as a science. In fact, he had not seen it as strategic. He had focused on manufacturing equipment and processes innovation—new surface mount techniques, new wave solder equipment, new testing approaches—and not on manufacturing management.

"So," he ventured, "What do you think? What are your reactions?"

Bob didn't waste any time. Mark had correctly guessed his reaction. "I question *a lot* of his data. We were running this plant successfully before he ever saw a printed circuit board. I think that he represents a lot of theoretical gobligook that next year will be discredited and replaced yet again by another fad."

Joe from marketing had been very quiet during the presentation but had a serious look on his face.

"You may not like what he said, guys, but it's something that I've wondered about for a while. We're having trouble picking up new accounts. We just don't seem competitive enough. We seem to lose at the end to someone who is cheaper, can quote a shorter leadtime, or has a quality guarantee. We still have business from our current customers, but lately some of them have been giving me some friendly feedback that we need to "pick it up," get more aggressive, and shake up the troops. They're not ready to throw us out yet, but I have to believe that they're starting to look at competitive bids. Negotiations are getting tougher.

"He was right, you know. Customers expect *a lot* more these days. And his numbers—our numbers—look scary to me."

"Rod?" Mark asked.

"It's very hard for me to say," replied his head of operations. "I want to look at his data more carefully. But if he's right, then we have to do something about it. I have a lot of pride in my organization." He paused. "I need to look at his data," was all he said.

"Bill?"

"He was good, very good. I bet he's an engineer," Bill quipped. But no one grinned. "But fixing it is going to be a lot harder than measuring it. This organization is not used to change—no offense, Mark. We move slowly and deliberately. But I can buy into it. It seemed like good, solid analysis. It's given me some new ideas for analyzing circuit problems."

"Tom?" Mark asked his I.T. guy.

"We don't have the systems to support a lot of it. I'm not sure what systems would support that type of measurement. Our systems are really corporate-based financial and planning views of manufacturing. We focus on the order and not the resources. All the data that he talked about today is on thousands of pieces of paperwork—work orders, run tickets, log sheets, and equipment records. There's no way right now for us to have visibility into what he was talking about. We don't really use that paperwork for analysis. We simply file them away unless there's a problem. I've heard that there are new systems aimed at manufacturing—I think they're called manufacturing execution systems—but we don't have one. As far as his analyses, I really can't say. But he made sense."

Bob was still visibly angry. He was used to giving the criticism, analyzing the budgets, and holding the power. This didn't sit well with him, Mark could tell.

"You know," Bob started, "we should really look at this situation another way. If the company doesn't want us anymore, we should consider doing a leveraged buyout."

Skepticism replaced concern on several faces. "This factory throws off cash. We've generated a 20% operating profit every year. If that's not good enough for them, how about for us. My brother-in-law is an investment banker. I could talk with him."

"Does he also have the money to lend us?" Joe asked with a wink. "You said he was always borrowing money from you."

Mark laughed and cut it off.

"Let me tell you where I'm coming from right now. Whether we run as a division or as Bob's Leveraged Buy-Out (LBO), we have to be competitive. We haven't kept up to date. There wasn't any pressure to change. As long as we made our numbers and the corporation credited us at our transfer prices, everything was fine. Well, that just changed.

"Right now it is still under our control. We still have the opportunity to improve and maybe even thrive as a result. Personally, I'm not feeling so good right now! Maybe I should have seen it coming. But I know how much worse I'll feel if they shut us down and I have to explain that to everyone who works here. Or if we're sold off and then 'rationalized.'

"As I see it, there's only one way for us to control our own destiny and that's to be so good at what we do that we get to compete and win any business."

"This is a big job. It is immense. And there's only one way we can do this—if each of you takes on one improvement challenge, builds a team to figure out what needs to be done, and then implements it. And by the way, keeps doing everything else you do already.

"From now on this is the stretch goals team, and we're going to go out there and make it happen.

"We need to do a lot of things in parallel. Bob—I want you to take on our costs. Bill—you take on quality. Joe—you have customer service. Rod—you get cycle time reduction. Tom—you work on systems if we need them. I'll pitch in wherever and run the overall project coordination across departments.

"I want each of you to form a team, agree on best practices measurements, benchmark where we are, set stretch goals for us, research the best practices to use, and then implement them. Your team will present each phase to us as the steering committee so that we can share ideas and progress." General Patton he was not, but everyone got his message. And so the "stretch goals" revolution began.

Appendix

Metrics of Manufacturing Performance

The technical metrics used to measure manufacturing performance of the participants (Table S.1) are summarized as follows:

1. Average cycle time per wafer layer.
2. Average line yield per ten wafer layers.

TABLE S.1 List of Companies Participating in the Main Phase of the Competitive Semiconductor Manufacturing Survey (First 18 Months)

Advanced Micro Devices, Inc. (AMD)	Nihon Semiconductor, Inc.
Cypress Semiconductor, Inc.	Nippon Electronics Corp. (NEC)
Delco Electronics Corp.	Oki Electric Industry, Ltd.
Digital Equipment Corp. (DEC)	Silicon Systems, Inc. (SSI)
Intel Corporation	Taiwan Semiconductor Manufacturing Corp. (TSMC)
International Business Machines, Inc. (IBM)	Texas Instruments, Inc.
ITT Intermetall	Toshiba Corp.
LSI Logic Corp.	

3. Die yields for major process flows, converted into defect densities using the simple Murphy model. The reported defect densities account for all yield losses, including both spot defects and parametric problems.

4. Wafer layers completed per 5X stepper per calendar day (considering only layers exposed using 5X steppers).

5. Wafer layers completed per operator per working day.

6. Wafer layers completed per working day divided by the total headcount.

7. On-time delivery (percentage of scheduled line items with 95% of scheduled volume shipped from wafer probe on time).

For all of these metrics, we encountered a wide range in scores, even though the basic process technology in use at the participants was generally similar. Table S.2 summarizes the best, average, and worst scores for each metric considering the latest data points we received from each of the sixteen participants. These data points represent measurements of manufacturing performance in some quarter between the middle of 1992 and the end of 1993, depending upon the participant.

TABLE S.2 Summary of Technical Metric Scores, Competitive Semiconductor Manufacturing Survey (First 18 Months)

Metric	Best Score	Average Score	Worst Score
Cycle time per layer (days)	1.2	2.6	3.3

The Competitive Semiconductor Manufacturing Survey: Second Report on Results of the Main Phase

September 16, 1994

Report CSM—08

Robert C. Leachman, Editor

The Competitive Semiconductor Manufacturing Program
Engineering Systems Research Center
3115 Etcheverry Hall
University of California at Berkeley
Berkeley, CA 94720

Center for Research in Management
554 Barrows Hall
University of California at Berkeley

Berkeley Roundtable on the International Economy
2234 Piedmont Avenue
University of California at Berkeley

2

Achieving Stretch Goals Through Best Practices in Quality

Current Practices—Quality

Before their TQM project in 1987, the current quality practice had been to "inspect" quality into the product at the end of the production line. The company was proud of their extensive testing practices which began with a "bed of nails" test to check for opens and shorts, then went on to a functional test, 48 hour burn-in, and a repeated functional test. Customer satisfaction on quality was high—almost all problems were caught through this extensive testing procedure. From the customer standpoint the problem with PCB Co. was not quality but cost and delivery. Shipments were frequently short boards (which were shipped a few weeks later), and prices were high compared to offshore suppliers with similar levels of quality.

Internally, the PCB Co. management assumed that the offshore suppliers had significantly lower labor and facility costs and similar quality. It was assumed that everyone had similar quality since everyone had equipment from the same equipment vendors and components from the same suppliers. Management deduced that they could improve their customer service by putting in and staffing a separate rework line that could turn around the defective boards faster (within three days) in a JIT mode since the rework rate was fairly constant. It seemed like a good application of JIT logic to them from what they had read. In addition, they would increase their quoted shipment date and internal leadtimes by three days to allow for the rework line time.

They would compete based not on manufacturing, but on their design capability that allowed denser packing of chips with a complex router and layout logic. Management was comfortable that this would solve their competitive problem which was never seen to involve quality. After all, the customer saw almost 100% quality—returned boards were a rarity. Manufacturing to top management was a fungible commodity. Anyone could buy the same equipment and components and set up an assembly operation. There was no competitive advantage to be had where there was no differentiation. Design was the key to competition.

A cross-functional TQM (total quality management) team was formed in 1987 with a goal of improving first-pass quality by at least 10%, with a goal of 90%. Members were drawn from all areas: engineering, production, equipment maintenance, and quality assurance.

There were problems early on with the team members' managers. While they all approved of the TQM concept, they had all lost valuable people who now were on the team half time. The amount of work to do in their departments hadn't changed, only the level of resources available to accomplish it. The team members felt this begrudging acceptance of their participation in remarks like: "I'll sure be glad when this TQM business is over, and you're available full-time again"; or "Sure, be on the TQM team—just as long as you can get all your other work done"; or "I've never seen anything good come out of a committee. I hope yours is the exception."

The team had management sponsorship but was left mostly on their own. They reported monthly to a management team where they reported on progress and presented their proposals for improvement.

Their first analysis of their current quality program was done after they had all attended a week-long training program—"Quality is Free"—which seemed a misnomer as the program certainly was not. However, it was an excellent program, and they came back armed with new insights, tools, and enthusiasm.

It was obvious to all of them that their current quality program was inefficient and ineffectual. They were simply inspecting quality into the manufacturing process after mistakes or problems had already occurred at incoming receiving of parts, at picking of parts for orders (called kitting), or at insertion or wave solder preparation or wave solder or degrease after wave solder. The immediate way to dramatically improve quality was to develop quality inspection procedures at each operation to ensure it was executed properly. While this was a sensible solution, it was again greeted with resistance from the production managers. To them, it meant that either additional quality inspectors had to be hired for each operation, or if the operators inspected their own quality after each operation or the quality of the batch they received from the preceding operation, it meant added labor cost—cost that would put them above their labor standards and budgets. They weren't being hurt by the current quality. It was already encompassed in the standards. What was going to hurt them was the added labor needed for the improvements to quality since the corporation set budgets and standards yearly and any changes were considered variances and affected bonuses. While it would be a win overall for the plant, and a big win for the testing operation which would show higher yields, and therefore, a positive variance to budget, each area supervisor or manager would lose.

This nearly killed the project because corporate balked at changing the labor standards midway through the year. Ultimately, Mark Ritchards had to go to corporate and get a special dispensation to make the changes. The experience left everyone somewhat annoyed—the team who felt their managers were unhappy with them for their participation, their managers who were short handed, and management who had to deal with the corporate reluctance to change their costing standards.

Ironically, the project was an overwhelming success. As the team developed operational tests for quality at each step, errors were caught before the order moved on. Once the errors were detected, the operators soon found out what caused them and rarely made that mistake twice. The attention to quality brought attention to details and fewer errors in general. After a few months, quality had improved dramatically, and productivity at each operation was nearly what it had originally been. Their goal was met and exceeded—first-pass quality at the bed of nails test had risen to 92%. The team was instantly disbanded with many congratulations and much management relief at having their resources fully available again.

Overall, the experience was called an enormous success, but no one proposed a second phase or expansion of the original scope. The company returned to a status quo and a focus on new products and process technology.

However, the group had not really disbanded. They continued to meet informally without any management encouragement. They felt that their task was not really complete. All their TQM training told them that 92% first-pass yield was not a stopping point, but management had disagreed. Management had declared a victory in the quality wars and dismissed the troops, but they hadn't disarmed them. They used their meetings to share educational materials since they were no longer encouraged to make improvements to the manufacturing process. Actually, two of them left the company, disillusioned by the return of disinterest in quality. It seemed "enough quality for now" had replaced their "quality now" slogan.

Bill was surprised to find that the TQM group still existed. He thought that it had long ago been disbanded. Given the short time frame and the fact that they still cared enough to continue, he decided to approach them and ask for their help. Right after the executive staff meeting, he had started to research the quality effort at PCB Co. and quality improvement programs in general. His first trip to the bookstore was eye opening. The majority of books were now on speed, agility, and empowerment. Quality seemed to be yesterday's news. Clearly the chic had moved on to new territory before PCB Co. had. The bookstore clerk suggested that he try the library. They didn't carry many quality titles these days.

The library was equally eye opening. There were dozens of books on quality—most from the early and mid-eighties, a few even from the seventies. Many titles covered TQM, some quality, some continuous improvement, and some kaizen. He felt like a historian here, burrowing in another decade.

He borrowed a handful and began reading them evenings and weekends. He also started calling around for the name of a consultant who could help him. He'd been impressed by the one Mark had brought in, and perhaps, not surprisingly, that consultant was the most helpful in suggesting two or three alternatives. He also read the original TQM team reports and met with the team leader. This is how he had discovered that the team still existed. He had already gotten permission to restart the team from Mark.

When he finally met with the original TQM group, he felt a bit more prepared. He had read several of the books (he stopped when they began to repeat material he'd already read) and had hired a consultant to help them (really him). He knew that *product* quality alone was not going to give them stepwise corporate improvement. If they reached perfection, it would only mean 8% higher product yield. They had to look across the board at quality in *every* department, from equipment maintenance to scheduling to facility management. He believed that meeting their new stretch goals was going to depend on moving quality to a *corporate*-wide focus from a pure product focus. That was a key component of getting twice the output from the existing equipment and staff.

The TQM group was really more of a social and debate group now. They met monthly, but with little purpose beyond information exchange. Bill immediately put a new agenda in front of them. "Last time," he explained, "you took the first steps toward product quality. You accomplished a lot, but we, the company, left the job unfinished. This time, we're going to finish improving product quality but then move on to quality in every one of your functions—in every one of your departments."

"We don't have any time to waste. I've arranged for a consultant to come in and help us. She has done a tremendous amount of work in quality, and she's going to start by interviewing each of you."

The group's reaction was mixed. There was a skepticism that anything had changed—a concern that as soon as they got started, they'd be thanked and disbanded again. But, there also was a deep belief in the value of improving quality and the pride that they'd had in their work. What convinced them to do it was Bill's promise to continue their education on the latest thinking in quality improvement.

The consultant turned out to be a unifying force for the team. Each member was both proud of the work they'd accomplished, angry at the way management had not let them complete their job, and still confused about some questions related to quality that they had never resolved. Being interviewed was a cathartic experience that put the old team to rest and restarted the new team.

The consultant unearthed many questions that gave her a good picture of where they had left off and where they needed education. She had already met with the corporate consultant and been briefed on his meticulous walk-through of the PCB manufacturing process. She was a statistician who had worked in manufacturing, then quality assurance, and finally product reliability before becoming a consultant. Their unanswered questions were her starting point, and she had tried to capture as many as possible. She would structure her training material around where they had struggled with improving quality.

Why Quality

The instructor was a medium-sized woman. She began her class.

"Quality is both detective work and police work at its best. We need to find violations, figure out 'who done it,' and strictly enforce the law—your work instructions. As we understand 'who done it,' we'll add more laws to the books to prevent it from happening again. In this way, we'll continuously improve quality until we've reached 100% customer satisfaction.

"Quality is the most basic goal a company must achieve. Quality problems in manufacturing affect each and every way you need to compete. We're going to spend most of today on the tools of continuous improvement, but first I'd like to review how critical quality is to competitive positioning, what quality is, and what determines quality—what I call the quality chain. Some of this may be review; some of it may be new. I know you've already looked at quality in your TQM work, but I want to set a powerful framework that you'll be using from now on.

"While quality is thought of as a single measure of a company's competitive positioning, it actually affects five others. Poor quality can be prevented from the customer's view by sufficient end of line testing. If we invest (or should I say waste) sufficient resources and cost in exhaustive final functional testing, we can weed out almost all product failures. Therefore, our customers see our quality as 'excellent.' However, this approach to 'eliminating' quality problems only degrades our performance on cost, technical capability, speed, customer service, and/or field reliability!

"Many companies 'mask' their quality problem by end-of-line testing, but like all problems, it only shows up somewhere else. If we scrap final product, we have higher production costs and usually will miss the customer's requested order quantity and/or date. Therefore, the most commonplace poor quality shows up is in higher costs or prices or lower margins and poor customer service. Whenever a company has a scheduling problem, I'm immediately suspicious that what they really have is a quality problem—in product quality or equipment uptime/quality.

"If a company can rework quality problems, then poor quality usually shows up as poor performance on cost and speed (or lead time). Jobs are delayed waiting for the reworked units to be repaired/reprocessed. Reworking units at an operation usually costs at least three times the normal operation cost—you have to process them, 'unprocess' them, and then re-process them. So the jobs are finished later and at a significant higher cost.

"When a manufacturing operation has significant variance or uncertainty in its process, engineers can't specify as tight an operating range as they desire. For example, if we'd like to maximize the component count on a board and minimize the trace distances and resistivity for increased speed, we'd like to use 1.5-mil traces with very precise location (± .2 mil). But if our manufacturing process can't control placement that tightly, and in fact, it varies ± .5 mil, then we either go with a different design rule (2.0 ± .5) or expect low yields. This gives a competitor with less variance (or higher quality), a technical capability advantage over us—a capability that means either lower costs or the ability to produce smaller, faster boards. Now that's a competitive advantage.

"Finally, poor quality usually translates into poor product reliability in the field. It is difficult to catch all product quality problems during testing—especially those that are seen in extended use. Therefore, the lower our quality typically the higher our field failure rate and the higher our field service and product warranty costs.

"Therefore, quality is not one competitive advantage but six! Unfortunately, many companies have not yet fully understood that point and so attack scheduling, leadtime, or cost problems independently from quality. They assume that if their costs are higher than a competitors, it must be purely a labor or material cost differential and not a quality issue. In fact, no one area has a greater impact on competitiveness than quality."

What Is Quality in Manufacturing?

"Quality has many definitions depending on whose viewpoint we're taking. Many books and articles have explained product quality from a comprehensive customer viewpoint covering form, function, reliability, ease of use, ease of learning, and so on. In our more limited scope, we're covering quality strictly from a manufacturing viewpoint or where manufacturing is the service being provided to our customer.

"Our definition of total manufacturing quality is the elimination of uncertainty or mystery in our output. This means that our plan for product manufacturing—quality, quantity, speed (or leadtime), and cost—is met without any variation. In practice, this requires the elimination of uncertainty or mystery in *all* the factory's manufacturing activities such as in product quality at each operation, but also in equipment performance, in operator availability, in vendor deliveries and in facility availability, and so on.

"Once all uncertainty is gone, we can replicate our performance indefinitely providing perfect product to schedule and cost. Obviously, this assumes that our basic manufacturing process, if done to our specifications, produces 'good' product; product that is within specified quality parameters or tolerances. Our goal, therefore, is to learn how to operate without uncertainty to eliminate any variances that affect our quality, cost, or speed."

Principles of Quality Management—A Historical Perspective

"Today, there are several well-accepted principles of quality management which have changed the way we ensure quality. Traditionally in the 1960s we measured or monitored quality at the end of the manufacturing line. We would either go through an inspection (visual test), a full functional test (where the product was tested under a range of operating conditions for some specified time), a sampled functional test (where some percentage of product was tested), a go/no-go functional test where product was tested to see if it operated within acceptable limits, or a partial functional test where a few key elements of function were tested or characterized. This approach is known today as "inspecting" quality into the product. We rely on this last inspection/test to protect the customer from any manufacturing, design, material, or handling errors. Our 'test' of the thoroughness of this final inspection is the rate of customer returns, field failures, customer complaints, or other problems at the final customer. We can look at this approach as an output test—we put a 'meter of quality' on the output of our entire manufacturing process. We do not look at quality until all the value has been added to the product, and it is completely finished and ready for shipment.

"We recognized in the '70s that this approach to quality was not efficient. It had several significant drawbacks. First, any part that was not correctly processed or had a failure early in the manufacturing process continued to have value added to it even though it would ultimately be scrapped or reworked. Second, and more importantly, the cause of the problem was not detected until much later—at final inspection—and many more jobs or parts may have similarly been affected before we recognized the problem. Third, when we finally did detect a problem, the root cause of it was more difficult to diagnose as time had gone by since it had arisen—the original person or tool, material, etc. at the operation where the problem occurred was gone or couldn't remember what had gone wrong. Fourth, now that we had a quality problem affecting a lot or work order, that order was likely to be delayed or short the quantity ordered. So we missed our shipment date or quantity, affected customer service (unless we kept large finished goods buffer inventories, or started more boards than were ordered to try to guess at the number that would pass final testing).

"For example, in your board shop, there was an occasional problem at the auto insertion equipment with bent leads. The insertion equipment seemed to bend the leads in such a way that caused shorts and opens at wave solder. The problem used to be detected at functional test and by then all the boards through insertion often had a similar problem and need to be reworked. Often, given the several-week leadtime through the shop, that amounted to hundreds of boards."What you did in your TQM program was actually apply the third principle of best practices in measurement (Table 1.3): 'Measure at the time of the activity with objective measurement tools that are replicable and accurate.' I could rephrase that to say, instead of measuring quality at the end of the manufacturing line, do it at each operation with the minimum delay from the operation itself in time and physical distance and responsibility.

"In other words, the person doing the processing, setup, or handling should inspect for their quality. This eliminates the time lag that results in additional scrap or rework and confusion over what the operating conditions were that caused the problem.

"This realization led to two changes in your basic operating procedures. Up to that point in time, operators did the work and quality inspectors at the end of the line did the final testing and inspection. In fact, there were no 'interim' quality checks at each operation! These would have cut down on the 'efficiency' of the operators who were measured on *output,* not on *quality.* Therefore, the first step in quality was to redesign the manufacturing procedure to add effective quality measures at each operation that the operators were responsible for monitoring. In Figure 2.1, we show the

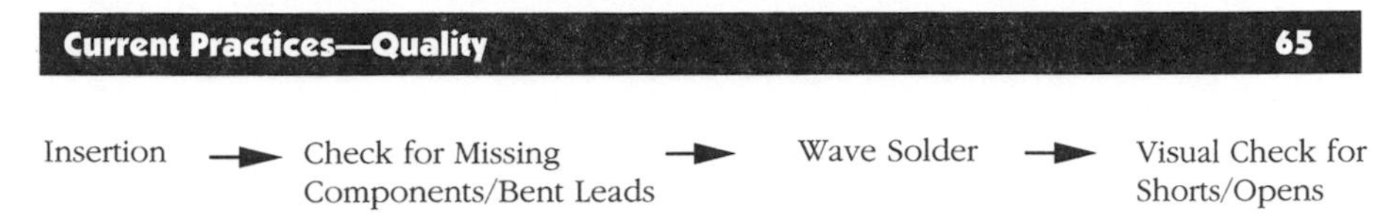

Figure 2.1 New "Quality Flow"

resulting manufacturing flow with several new quality monitors. In practice, the first set of quality monitors were soon enhanced by others. Now that the operators were responsible for measuring quality, they had ongoing suggestions on how to avoid quality problems.

"New ideas arose. It was decided that the visual check after wave solder was ineffectual. Many opens and shorts were undetected, especially on the fine line-width boards. The group decided to add a 'bed of nails' check that the operator could do faster and with less fatigue after wave solder. This was well accepted and increased first-pass yield another 3%. At insertion, the operators also found the visual inspection laborious until one of the operators suggested the use of a 'pattern mask' much like the ones used to grade multiple choice exams. This again increased accuracy and efficiency of inspections. While the inspections added total direct labor time to the boards, the dramatic decrease in scrap and yield more than offset the direct labor cost increase. And that was where you stopped. Management declared victory in the war on quality and disbanded the troops. But were we done?

"In principle, while we were closer to the source of a quality problem, we were still 'inspecting' or testing quality into our product. Therefore, we were still inefficient. We had simply shifted our "output" view of the line to an "output" view of an operation. While it was more efficient by proximity, it was still the wrong concept for improving quality. So in the '80s we moved to another new way of thinking about quality.

"What we needed to do was move from detection to prevention—or apply Principle 5 of best practices in measurement: Measure for predictive control and not for reactive postmortem (to avoid a problem or to detect and correct versus to search for the guilty).

"In other words, we would like a new 'model' of quality. Instead of measuring the output of the operation to detect a problem that has already occurred, we would like to measure the 'inputs' or 'setup' of the operation to ensure that they are *capable* of giving acceptable output quality *before* we process the lot.

"This third principle of quality management was a radical shift in thinking for most organizations, moving from detection to prevention. Unfortunately, most traditional approaches to quality had been built around catching the 'bad units,' starting with the development of sampling procedures for inspection parts.

"However, successful quality management requires a major paradigm shift—to *avoiding* errors instead of *catching* errors. This shift is a shift in focus or view from the output of the operation to a focus on the inputs and operating conditions prior to processing, from product quality to process capability and quality, or from detecting quality problems to detecting when the manufacturing process is not capable of producing quality.

"While a company can get to 99% or higher levels of quality through inspection, it cannot prevent quality problems. To get to parts per million quality, you have to shift quality measurement from detection to prevention.

"For example, let's look at wave solder. While the bed of nails test is very useful in detecting opens and shorts, by the time you detected a problem, you have already scheduled the operation, set up the machine, used raw materials, and processed the board. That 'one board' problem affected the whole shop, delayed the job, negated the setup, and created additional work for everyone.

"Is it possible to 'predict' the problem before it arises, to forecast the quality of the output as a function of something else? What could impact the output that we could (actually or potentially) measure or monitor, or even better, control? Many of you have already told me several potential root causes of quality problems at wave solder—the temperature of the wave solder, the height of the wave solder, the percentage of impurities in the wave solder, the wetting solution concentration, the humidity in the room, the tinning concentration on the components, the size of the holes on the board, and so on (Fig. 2.2).

"The basic concept is that assuming the operation is performed correctly, then the output of the operation, its statistical quality, is determined by the characteristics of the *inputs*—the characteristics of the raw materials, the facility conditions, and the settings of the equipment, for example—and if they are within 'correct' limits then the output would also be within limits. Clearly, to use this approach, we would have to have a model of how output quality related to process inputs characteristics. This also means that

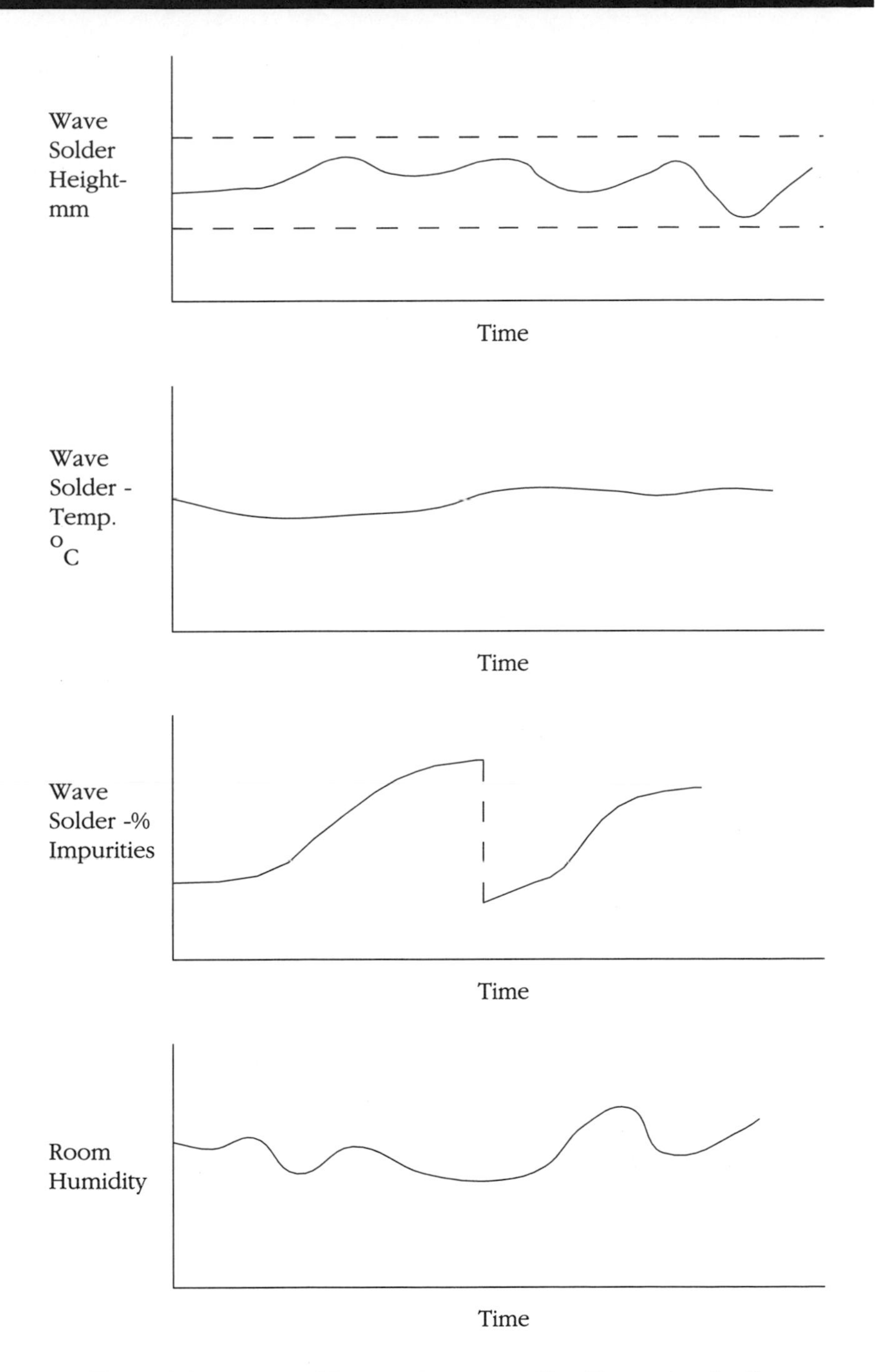

Figure 2.2 Statistical Process Parameters That Determine Quality

we would need to be able to characterize and measure the inputs exactly and be able to replicate the manufacturing process exactly. Since none of these are true, we still have output variation even when the inputs seem to be in control. However, the output quality variation becomes statistically smaller and smaller as we come closer to meeting our assumptions. *This concept leads to another principle of quality—determine, measure, and control the causal variables or controlling parameters.*

"In fact, using Taguchi design methods,[1] we could even develop 'robust' designs that would not (or rarely) be affected by the normal variation in our inputs or the uncertainty in our assumptions.

"So what 'inputs' do we need to monitor, characterize, and ultimately control? Let's define manufacturing. Manufacturing is the processing of *materials* (direct and indirect) on *equipment* (including tooling and fixtures) assisted (directly or indirectly) by *personnel* according to *work instructions/procedures* within a *facility* (including environmental conditions). We will call these the five elements of manufacturing. In our model, these are the basic inputs to each manufacturing operation (Fig. 2.3).

"In other words, how well we control the characteristics and usage of these five elements completely determines our results. They are the basic 'elements' or building blocks of manufacturing."

Our "output" quality at the operation (as measured by product quality) is determined by the input characteristics or quality of the five elements of manufacturing we utilize: if the raw materials are not within their specified characteristics or properties; if the facility conditions (electricity, water, air) aren't properly controlled; if the equipment runs outside specified ranges; if the operators do not follow their operating instructions consistently; and if the work instructions are constantly changed. All of the input variation may affect our product quality.

So what have we learned so far about the determinants of quality? We know that final product quality is determined by operational quality; and operational quality is determined by the characteristics or quality of the inputs to it, the five elements of manufacturing used there, and whether they are applied or executed correctly to your instructions or specifications.

[1] Ross PJ. *Taguchi Techniques for Quality Engineering.* New York, McGraw-Hill, 1988.

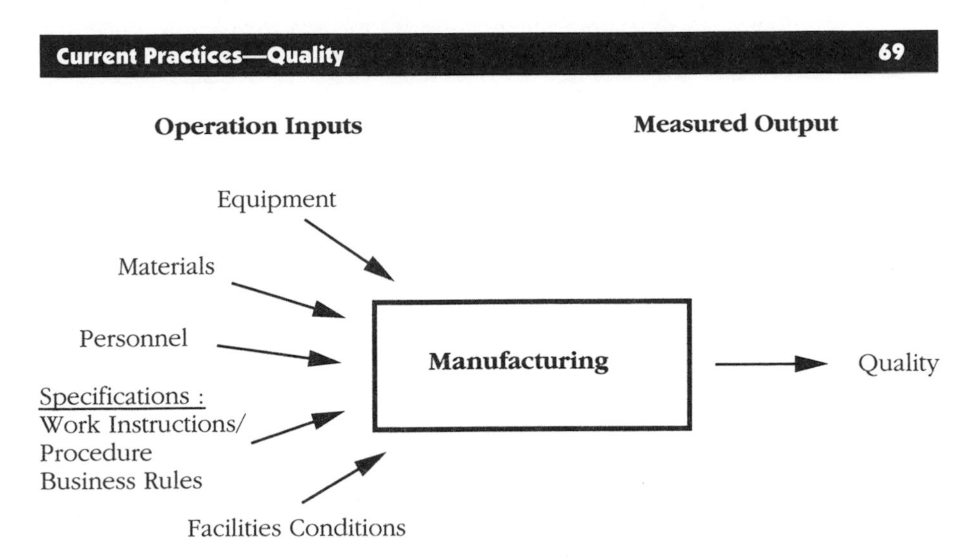

Figure 2.3 The Five Elements of Manufacturing Completely Determine Your Quality and Performance

So what determines input quality? Let's look at a chart of wave solder height. The height of the wave affects the bonds created and splattering. If the height is too high, the board is splattered, and we get solder where it is not desired leading to shorts. If the height is too low, we don't get sufficient bonding coverage, leading to opens.

From an operator's viewpoint, we can see that wave height varies over time (Fig. 2.4). We consider that normal variation and within our limits.

If we view that chart from a maintenance technician's standpoint, we may only see wave height periodically when we do calibrations, cleaning,

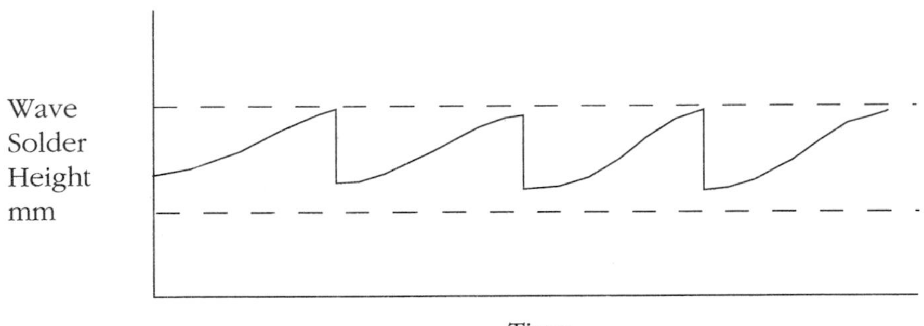

Figure 2.4 Wave Solder Height Over Time

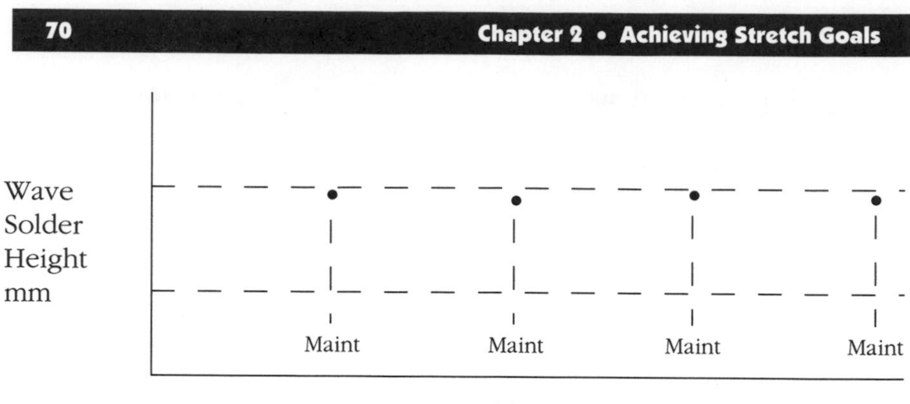

Figure 2.5 Wave Solder Height at Maintenance Time

and other maintenance functions and reset it or clean and lubricate the motors or pumps (Fig. 2.5).

It's only when we put the two views together that a pattern emerges that answers our question of what affects input quality (Fig. 2.6). What we see is that the wave solder height is a function of the maintenance and calibration program and equipment degradation over time.

So what determines the quality or characteristics of our inputs? In each case it is dependent on the quality of indirect activities of the department responsible for that input.

For example, what will determine the quality and characteristics of our raw materials? It is the quality of our raw material vendor selec-

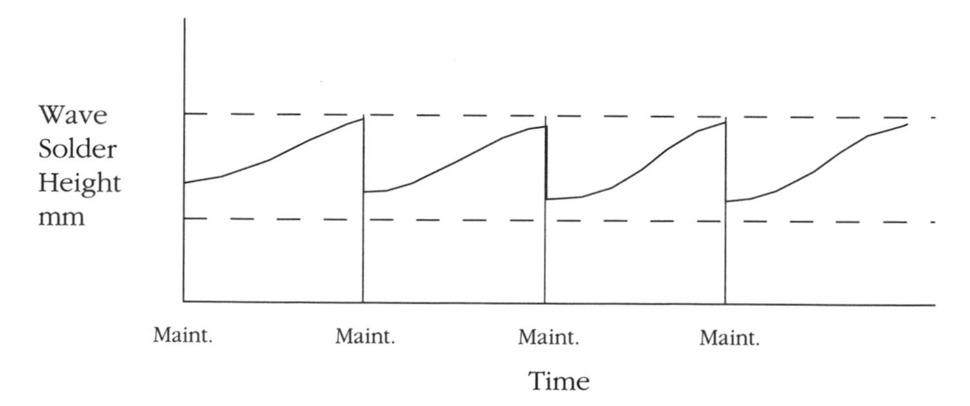

Figure 2.6 Wave Solder Height Versus Wave Solder Maintenance

tion process and of our incoming inspection or vendor certification program.

What determines the quality of our operators' performance? It is the quality of our operator training and (re)certification and hiring programs.

What determines the quality and characteristics of our equipment? It is the quality of our maintenance and calibration programs and our specification and selection of equipment and equipment vendors.

What determines the quality and characteristics of our facility conditions? It is the quality of our facility maintenance procedures and the specification and selection of facilities vendors.

What determines the quality of our specifications? It is the quality of our product and process development rules and procedures.

What we see is that the quality chain (Fig. 2.7) goes far back into our manufacturing network (back to suppliers of equipment and materials) and is all encompassing. If any department ignores quality, ultimately, the effect will likely show up in your product quality. If maintenance is not done on time and to specifications, if training programs are not carefully designed, if raw materials are not certified as within specified limits, or if facility maintenance is not enforced, product or process quality problems can arise.

The reason to trace back quality to its root cause is so that we can be sure we are treating the root cause and not the symptom. If we look at Table 2.1, we see this illustrated. The symptom may be bent wires at insertion. The apparent cause the insertion equipment, but the root cause may be our maintenance program or choice of equipment. To only fix the symptom (reworking the affected boards and resetting up the machine) only ensures it will reappear and *systematically* affect our quality.

Root causes of product quality problems are often found in indirect activities. So to assess the quality of our *manufacturing* capability, we also need to measure and monitor the quality of our indirect activities. This has been the focus of many leading companies in the last five years to institute quality programs in *all* departments. This means enlarging our focus on measuring quality to include indirect activities and departments. Again, we can both measure quality "after the fact" (ex post facto) or measure to ensure that each activity is done correctly.

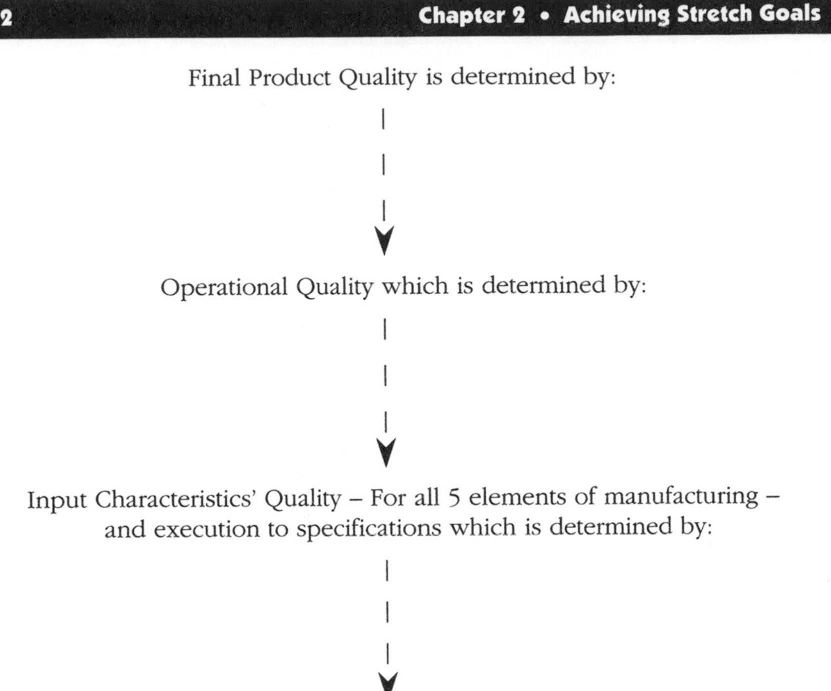

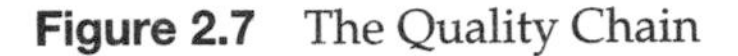
Figure 2.7 The Quality Chain

Let's look at how to measure the quality of our equipment programs, encompassing maintenance and repair, and equipment vendor improvement programs (Table 2.2). The key measure of their effectiveness is going to be measured by the percentage of time the equipment is available and operating within its specified ranges. The converse measurement would be of total downtime—when the equipment is not available—due to breakdowns or "out of operation."

Another form of this measure is the mean time between failures (MTBF) and mean time to repair (MTTR). We can also measure product losses (scrap/rework/off spec) due to equipment malfunction.

These are all ex post facto measures of equipment quality. They are taken *after* the problem has already occurred. More and more companies are also using preventative measures of the quality of indirect activities themselves. These are checks made to ensure that setups, calibrations, re-

TABLE 2.1 Wave Solder Height Over Time

Symptom	Apparent Cause	Root Cause
1. Bent wires at insertion	Insertion equipment Feeding jig	Incorrect maintenance schedule at insertion; feed pressure outside limits
2. Poor solder bond strength	Wave solder	Wrong tinning on leads; poor incoming inspection of raw materials
3. Operator misprocessing error	Operator error	Poor training and certification program
4. Operator misprocessing error	Operator error	Ambiguous specifications; poor testing and acceptance of engineering specifications

pairs, and maintenance were done correctly in the first place, at the right time, and by the correct people to the specified procedures. This is analogous to the quality checks we developed for each operation to ensure performance to specification. We can then also report on the number of setups, calibrations, maintenance, and repairs done correctly and on time. Or conversely, you could measure the number done incorrectly or late. Such measurement could be done by a quality assurance group on a sampling basis, by the department itself through a peer inspection program, or by the "receiving" department itself, production.

TABLE 2.2 Measures of Equipment Quality

- Total time available (operating within specified range)
- Total downtime:
  - (Broken/out of range)
  - (Being repaired)
  - (Maintenance)
    - Mean time between failures; number of failures
    - Mean time to repair
- Total product losses/rework/off spec
- Number of setups done incorrectly
- Number of calibrations/maintenance done incorrectly
- Number of repairs done incorrectly
- Number of calibrations/maintenance done late
- Number of broken/worn parts replaced before their specified time

Again, the actual uptime experience is a good check on the quality of the maintenance and repair program. Another check on the current scheduling of maintenance (is it frequent enough) are the number of parts replaced before the scheduled maintenance time due to equipment part failure.

Few companies have accurate data for management on indirect activity quality. Most of the information they have is either anecdotal or relates to the current cost of maintenance and repairs. They don't have visibility into the mean time between failure or total downtime or mean time to repair. They know the cost of repair but not the cost of lost production or the actual time the equipment was down and unavailable. The accounting systems tend to focus on the labor and parts costs for maintenance and repair and not the equipment availability and predictability that are more critical for customer service and quality.

Ironically, we collect *all* of this data—on manual equipment logs and maintenance records. We collect an extraordinary amount of data but the majority of it is on paper without set formats and without any summary capability. We can look at the repair or maintenance records for any piece of equipment. We simply can't (or don't) summarize its operating characteristics overtime or equipment types.

Again, our more advanced clients have moved to computerized systems which collect the same data but in standard formats that allow reporting on downtimes, mean time to failure, mean time to repair, most frequent cause of breakdown, parts replaced, and other critical quality and productivity monitors of their equipment.

What We Have Done—The Shift of Measurement to Cause and Away From Effect

What we have done throughout this exploration of equality is move our focus on quality from the product to the process to the elements of manufacturing to their management or from symptom or effect to cause. What we've done is intensify our focus on quality by moving it beyond direct production to looking at the quality of all the indirect functions that affect our product quality. By doing so, it becomes clear that quality is *everyone's* responsibility and to truly eliminate waste means every activity on the factory floor must occur without variation or uncertainty. In Tables 2-3, 2-4, and 2-5, three other indirect groups—facilities, purchasing, and manufac-

TABLE 2.3 Measures of Facility Quality

- By resource—electricity, water purity, facility cleanliness, temperature, humidity
  - ❑ Time within specified limits
  - ❑ Variance of the measured parameter
  - ❑ Number of maintenances done late
  - ❑ Number of maintenances done incorrectly

turing/engineering—were asked to develop measurements of their quality. What we see is that every group can and must measure their quality if they are to improve it. These measures must be parameters that impact our product quality or operations efficiency.

Often there are no quality measurements for indirect activities, and they have to be added, just as the production groups did at each operation. The most striking example to me is measuring the quality of work instructions. Most companies don't think of measuring the quality of work instructions. They are simply "there." But there is no reason not to measure their quality. In fact, incomplete, ambiguous, or out-of-date specifications are a common cause of operator error for new hires and trainees. Some of your work instructions have not been updated in years, and while operators know the "current" standard procedures, new operators struggle with tools and parts listed that are no longer used, safety procedures that have been augmented or replaced, ambiguous wording, and so on. In fact, this discussion led one of my other clients to consider looking at measuring the quality of their training procedures—a clear partner of quality documentation in operator and technician effectiveness.

TABLE 2.4 Meaures of Raw Material Quality

- By raw material and parameter measured
  - ❑ Percentage of time within specified limits
  - ❑ Percentage of material within specified limits
  - ❑ Variance of the parameter

TABLE 2.5 Possible Measures of Work Instruction Quality by Work Instruction

- Number of engineering change notices (ECNs)
- Number of operator errors in applying them
- Operation yield/rework rate
- Number of operator questions asked about specifications
- Variance of the time to execute a specification
- Number of discrepancies from current practices (specification not yet updated)
- Number of obsolete practices

Summary

So let's summarize where we are in our best practices in quality:

Quality Best Practices Principle #1

Final Product Quality is determined by operation quality. To ensure product quality, we need to ensure that each operation is performed exactly to specifications.

Very simply, this principle states that if we perform each individual operation properly and eliminate all uncertainty at each step, we'll have final product quality. So what determines each operation's quality? Our next concept was that each operation's quality was determined by the quality of the *inputs*—the five elements of manufacturing—and how they are applied for the execution of the manufacturing process.

Quality Best Practices Principle #2

Operation Quality is determined by input quality—for all five elements of manufacturing—and process or execution quality (to specification).

As we discussed, we can measure the quality of each of the five elements and ensure that it is operating within its specified limits. We also need to ensure that each operation is executed exactly to our instructions or specifications, or quality problems will arise. We'll be talking about how we ensure compliance to specification or "execution."

Note that if we stopped our pursuit of quality here, we would never manufacture a defective unit. We would detect a manufacturing problem with one of the five elements and shut down that operation until it was corrected. While this is certainly far superior to producing one defective unit or batch and then shutting down that operation, it is not going to be enough for world-class manufacturing in the 1990s!

In the remainder of the 1990s, companies will strive for perfection—choosing equipment, materials, and vendors; training personnel; and designing products, processes, and maintenance activities that eliminate unplanned downtime or "out of specification" conditions. Therefore, we continually need to detect and find the root causes of input and process execution problems to eliminate those as well.

So perhaps, more importantly product quality is ultimately determined by (and the responsibility of) the indirect activities of: maintenance, repair, calibration, training, incoming inspection, engineering, purchasing, planning, scheduling, and so on.

Quality Best Practices Principle #3

Input and Execution Quality is determined by the indirect activities of: Maintenance, Repair, Calibration, Engineering, Training, Incoming Inspection, etc.

While all companies measure final product quality and many companies measure operational quality, fewer companies measure input quality and fewer yet measure "indirect" activity quality. We need to think of "indirect" activity quality as quality for the '90s. It is the only way to approach perfection in quality—the elimination of all uncertainty. How do we know this? Because Principles 1 and 3 tell us that all final product variation can ultimately be traced back to our indirect activities—how well our operators are selected and trained, how precisely and clearly our work instructions are written, how carefully our raw material and equipment vendors are selected and their incoming product quality assured, how carefully our equipment is maintained and calibrated, and so on.

Indirect quality will be the playing field for quality in the '90s, the drive beyond 3 sigma and even 6 sigma to perfection.

What Determines Preventable Poor Quality—a Model for Improvement

Now that we know in general what determines quality, we can further divide all causes of preventable poor quality into two specific categories—errors that arise from not following known procedures (our direct or indirect activity work instructions) and errors that arise from "unknown causes" or mystery that are not prevented by current work instructions. In other words, quality problems can arise from what we already know but don't do properly (called random errors) or what we don't know, and therefore, don't do (called systematic errors).

Quality Best Practices Principle #4

All preventable poor quality arises from random errors or systematic errors:

- **Random Errors: Not following the (known) work instructions/specifications/procedures.**
- **Systematic Errors: Not properly controlling the inputs to the manufacturing process procedures and/or factors not covered in the existing instructions/specifications/procedures.**

Examples of random errors are: using the wrong tool, raw material, or work instruction; letting an operator who isn't certified perform the operation; missing a scheduled preventive maintenance or tool calibration; not notifying the floor of an engineering change notice just released; skipping an operation or repeating one; physically dropping a job; using a machine that is "down"; and so on. In fact, there are 28 of these that can be listed (Table 2.6).

They are termed random errors because they are usually attributable to human error (even though they may be repeated without detection and correction). We will look at an approach to eliminate all random errors called execution control or Poka Yoke, or foolproofing. Random errors are a common component of waste and normally account, in our experience, for 3 to 7% of total costs in yield losses and rework.

Systematic errors are harder to detect. They arise because the work instructions are incomplete, not covering some aspect of monitoring, and/or

TABLE 2.6 Examples of Random Errors

Sources of Error
Before processing • Wrong ingredients Status Quantity Batch number • Wrong Equipment Status ID number • Out of spec facility conditions • Wrong documents—expired/not approved yet/wrong one used Specifications/instructions • Missing custom instructions/ECNs • Wrong recipe • Trends—equipment/process capability not detected • Product quality hold exists • Next event, eg, clean—setup skipped or done incorrectly • Skipped OPN/step • Repeated OPN/step • Timing—minimum/maximum—violated between operations • Job on hold • Incorrect feed forward parameter value used
During/after processing (documentation/evaluation) • Instructions not followed • Missing information (to collect) • Missing signatures/approvals • SQC results not checked • Error in data collected—times, quantity, values • Job sent to wrong next operation • Feedback parameter values not collected and recipe reset • Job damaged in handling or transit • Setup/teardown done incorrectly

tightly specifying the inputs to the manufacturing process or the manufacturing process itself or the indirect activities that significantly affect our manufacturing quality. Examples of systematic error are: time intervals or triggering for equipment maintenance, tool calibration, or raw material changes that are incorrectly set; key facility conditions (such as temperature, humidity, particle count, electricity spikes, vibration, or electrostatic charges) that are not monitored, regulated, or specified; and acceptance levels for raw material parameters and attributes that are not specified or incorrectly set such as board flatness or trace element levels of nickel in solder paste or tin thickness on leads.

These are not easily recognizable errors, or they would have already been corrected. We call them systematic errors because they are affecting you continually as part of your currently accepted manufacturing process. They are part of your standard operating procedure by default!

We will look at a set of methods to detect systematic errors, relate them to one or more of the five elements of manufacturing, and eventually correct our work instructions to eliminate them.

Most factories are significantly impacted by systematic errors whose individual operator, tooling, and equipment have: higher downtime, scrap, or rework rates than others; vendors whose material causes higher scrap or rework rates; work instructions that cause greater variance in operator performance; maintenance technicians whose equipment suffers higher downtime or scrap rates; trainers whose operators have higher error rates; and so on. The key to any improvement program is to systematically detect these variations and search for their cause. We will begin the study of detection and improvement tools by focusing on random errors. The first step will be to eliminate all random errors to ensure that our specifications are carried out exactly as specified. The second step will be to detect, then identify the root cause of systematic errors, and then modify our specifications to eliminate them.

Human Error—The Source of Random Errors

Humans are, by nature, error prone. Undoubtedly, you've heard the expression, "We're only human; no one is perfect." It is realistic to expect an error rate of between .5% and 3% in trained personnel. This rate can be higher under extreme conditions such as toward the end of 12-hour shifts or when operating in inhospitable environments—at elevated temperatures or humidity. It can also be affected by long periods of tedium or inactivity where a person's concentration lags or periods of excessive activity where concentration is hard to maintain. None of your PCB operations fall into these categories. Errors also tend to be higher on third-shift operations or when there is a large and changing product mix (many new products or engineering change notices or customization required). This is true in your packaging area where each order may require special labeling or packaging. Empirically, we also tend to see higher operator error rates during periods of high turnover when there are many new personnel on the floor.

As discussed, we can dramatically affect our error rate by the quality of our hiring and training programs. Any time error rates rise, we need to check our training programs, especially as we add new products, operations, equipment, instructions, or processes. Better yet, many of our leading clients have regularly scheduled operator training or (re)certification programs to ensure that they are properly trained.

Finally, the quality of our instructions or specifications themselves can dramatically affect operator errors. Will an operator understand and be able to follow the specifications? Will two operators follow the instructions identically, or is there unacceptable room for interpretation or ambiguity which leads to variance in how instructions are followed? Specifications themselves must be quality controlled or "accepted" by manufacturing just as you would certify raw materials from a supplier. They must be unambiguous, lead to consistent quality, and be efficient and safe to carry out. As we discussed, not enough companies have standards for accepting specifications from process engineering. They simply put them into practice and then "tweak" them into acceptable performance. Instead, they should be Q.A.'d by operators to see if two operators interpret them the same way and then execute them consistently and easily.

But even under "perfect" operating conditions, we should expect some operator error rate due to our innate variability. So how can we eliminate these as well?

The concept we will use is "Poka Yoke" or foolproofing. This is a concept extensively developed in Japan to ensure that the physical processing of material was executed properly. Shigeo Shingo has written extensively on this subject.[2] The approach is to build a series of physical devices (jigs, fixtures, testers) that ensure or at least assist and verify that the physical steps of loading, setup, feeding, insertion, positioning, testing, and unloading are done properly. It originated in metalworking operations and was expanded to all types of physical processing.

[2] Shingo S. *Zero Quality Control: Inspection and the Poka-Yoke System.* Productivity Press, Cambridge, Massachusetts and Norwalk, Connecticut, 1986.

Today, however, we can expand this concept to include the informational checks we'd like an operator or technician to make. Are we using the right raw materials? Are we using the right batch of that raw material? Are we using the right piece of equipment? Are we using a piece of equipment in the right (legal) state—up and available? Are we using the right tooling? Has it been cleaned and calibrated? We can check for every potential random error shown in Table 2.6 to ensure 100% compliance to specifications. The key is that we make these checks prior to our processing. Many companies record some of this information after processing for use in costing or financial systems. They want to record usage—what resources were used by a work order or task and for how long—to assign a cost to that work order or task. But while this is useful cost data that your ERP or MRP system may collect, it does nothing to prevent errors. It simply records history without trying to change its direction. We want to perform our checks *prior* to processing for each step in a typical operation (Table 2.7).

So what are alternatives in implementing a Poka Yoke approach to eliminate random errors? A traditional approach, borrowed from the pharmaceutical or health care industry, is the use of a manual protocol (or batch record as it is called in that industry) where you are shown the instructions for each step and given space to fill in the actual choice you make, the equipment selected, the amount of raw material used, and the time of processing with legal values indicated alongside.

This approach has several inherent problems. First, there is no editing or checking of the data entered. If it is entered incorrectly or late after the task has been performed, mistakes can still occur. In this way the protocol is not a true Poka Yoke tool.

Second, it can't prevent errors which can't be "seen" by the operator. If the machine has missed a calibration and should not be used or the wrong recipe has been downloaded, these conditions are not visible to the operator. Yet, the equipment is not "legal" to use and will ruin the job. The manual prototype does not have the current state of the factory.

Third, the protocol doesn't see trends. It only collects the data for each job as it's done. We need a separate form (or forms) to look for trends in product quality or process conditions.

In general, the protocol is a passive tool which certainly prevents some of the human errors otherwise made but is not really foolproof. Experience in the pharmaceutical world shows that typically half of these forms

TABLE 2.7 The Nine Steps in a Manufacturing Operation

Step	Type	Description	Example
1	Selection	Select the next job/task for a machine or operator	• Job 1623A • Preventive maintenance • Engineering experience
2	Obtain resources	Get the materials, labor, tooling, job, work instructions, etc., needed to process the job/task selected	• Manually pick up the next kit • Have A.G.V. deliver job 1623A to the work center • Continue use of raw materials, tooling at your work cell as setup
3	Review work instructions (S.O.P.) and review nonstandard operating instruction (special)	Review how to set up and process the job/task Review any exceptions/deletions to the work instructions for this job/task	• Special customer processing • Special quality test • Safety warnings • Rework instructions
4	Validate legality	Check that all the resources and the job are in the correct state/legal for use	• Make sure the job hasn't skipped any operation; isn't on hold, hasn't "expired" • Make sure the machine is in the "up/available to process" state • Make sure the ECN is the latest, hasn't expired
5	Setup	Set up the equipment, handlers tooling, download any recipe, run any verification tests	• Clean out the previous color • Check out the mold on a few piece parts
6	Processing/task	Run the job or perform the task	
7	Monitor for exceptions	Monitor equipment, tool, facility, material, job parameters for warning/shutdown conditions	• Machine temperature • Part dimensions • Plastic viscosity • Facility humidity
8	Teardown	• Collect relevant data • Teardown of setup	• Collect quality parameters • Remove handler • Turn off machine • Fill out run ticket
9	Return resources	• Move job • Return the tooling, material, work instructions • Labor sent back to appropriate department	• Reverse of step 2

have errors in data collection! They may be relatively minor (missing data, transposed numbers, or data entered later), but it again shows the problem of human error even in a highly regulated industry.

So how can we prevent all errors? There is a new best practice now available where all the data is edited or validated by computer. We call this electronic Poka Yoke or computerized execution control in support of best practice Principle 5.

Quality Best Practices Principle #5

To prevent random errors, each step should be validated at the time of execution using the current state of all resources, jobs, instructions, and operators guided through its execution.

The operator or technician is guided through the protocol on a computer, and either the computer suggests each resource that is legal to use or validate the ones chosen. It automatically displays the latest instructions. It automatically tracks all trends in quality or process parameters and triggers on out of specified range conditions. It automatically changes equipment and/or tool status if scheduled calibration and maintenance are missed. It alerts operators if jobs are finished or waiting too long (these systems—called manufacturing execution systems—will be discussed in more detail in Chapter 7).

We can think of this as active execution control—an ever-present, ever-vigilant supervisor watching every task and job as it is executed to see that it is done correctly to our instructions. This starts at Step 1 in Table 2.7 in choosing the right job or task to do next and continues as we select our resources (or are directed to use the appropriate ones), display the latest instructions (in full multimedia if desired), download the correct recipe to the equipment and monitor the equipment (if automated), and update the state of all resources upon completion of the operation with instructions on teardown and where to return resources (examples given in Table 2.7).

Some clients have even used these systems to go to the ultimate step of full automation. The system selects the next job to process, moves it to the equipment, downloads the recipe, uploads any quality or process data, and monitors for alarms (exceptions) and completion. Then it unloads the machine and moves the job to the next operation or a storage area.

This is the ultimate in active execution control. It is used when automation is more cost effective than manual operations and/or manual operations cannot ensure the accuracy or consistency or human safety of processing.

One other side benefit of the electronic batch record is that we automatically collect all the data—on resources utilized, equipment status, and job location—which gives us a complete database of plant information to use in best practice measurement, process improvement, scheduling, costing and all other applications. We also end up with the certificate of analysis (C of A) for each job (certifying how it was made), the batch record for the boards you produce for medical devices, and assurance of compliance for ISO 9000 certification.

Closing the Loop—Continuous Improvement of Quality

Whatever active or passive batch record or protocol we utilize, we will continue to find random errors that the protocol does not prevent. We need to document these random errors and pinpoint their root cause. Once determined, we can if necessary revise our protocol to prevent these errors and/or revise our indirect activities (such as training or incoming inspection).

Systematic Error = Systematic Waste

Let's move on now to systematic errors, variation that arises in product or operation quality even when the specifications are followed exactly. Systematic errors seem like a mystery—a variation in operational or product quality that appears unexpectedly and seemingly random. Many companies simply accept it as part of their manufacturing process and plan for a certain rework or scrap rate. We can immediately detect systematic error by looking for variability, for example, in product scrap and rework rates, or equipment uptime, productivity, and output quality over time and across equipment (Figs. 2.8, 2.9).

In fact, this systematic error is usually traceable to a root cause—a sudden or more subtle variability in one of the five elements of manufacturing or in the execution of the step. It may be due to variation in the characteristics of the raw materials or ingredients such as the percentage of contaminants in the wave solder, the surface conditions of the leads, or the pH of the wave solder. Or it may be due to variation in the equipment itself, the motor speed of the wave solder machine, or the setup of the insertion

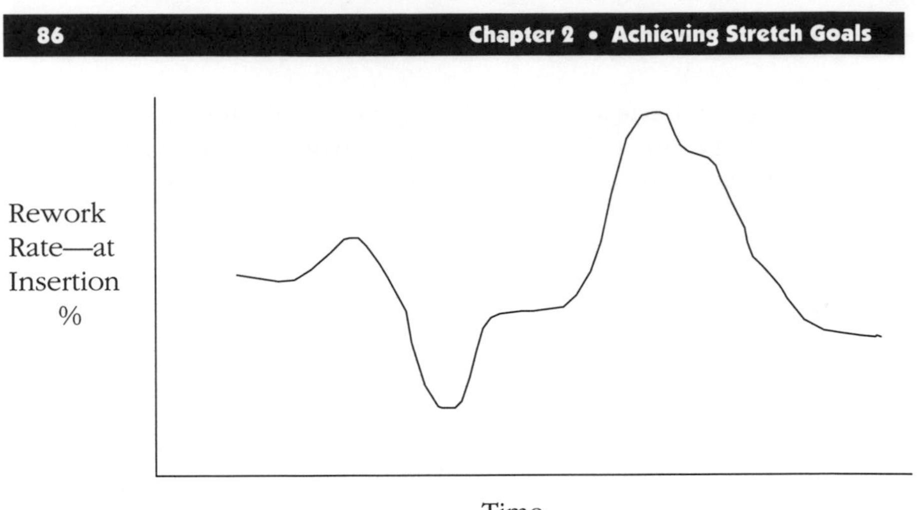

Figure 2.8 Systematic Error as Seen in Rework Rate Variation Over Time

equipment. It may be due to variations in the specifications—typically the specifications have a range of settings we can use. Each batch may see a slightly different setting; or we may have changed the specifications with an engineering change notice. It may also be due to variation in the facility

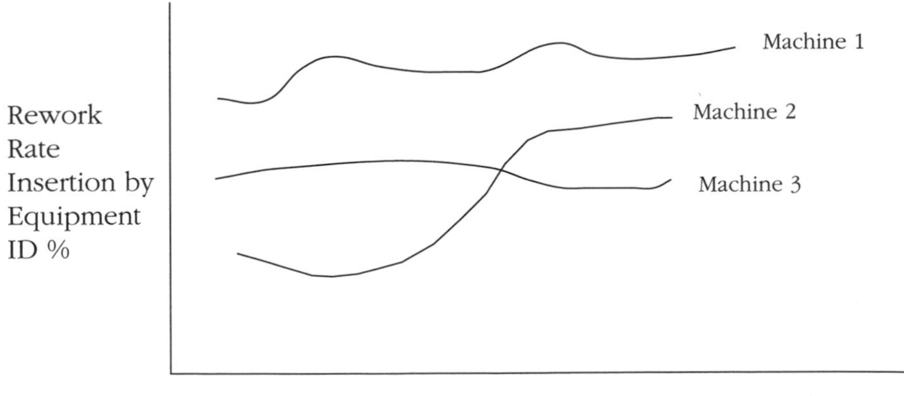

Figure 2.9 Systematic Error as Seen in Rework Rate Variation Over Equipment

conditions—the room's humidity, temperature, or cleanliness (particle counts). It may be due to variation in the way operators carry out or interpret the instructions in setup, processing or inspection, and measurement.

As we see, there is a lot of potential variation in the conditions under which we manufacture. Most of this variation does not affect our quality. The process has hopefully been designed with this variation in mind. However, some of it may impact quality. Our job is to solve the mystery, find the root cause of the variability, and then remove it from our manufacturing environment. Basically, the search for quality is the search for the causes of variability and their elimination. Once we've found the root cause, we can change the specifications to eliminate the problem.

Why do I call these systematic errors? Because they are systematically and continually affecting your quality as part of your current accepted manufacturing process. They are actually, by default, part of the standard operating procedures. Your current specifications do not prevent them. They may occur every time you execute those particular operating instructions.

We will look at a set of methods to detect systematic errors, relate them to one or more of the five elements of manufacturing, and eventually update our work instructions to eliminate them.

Continuous Improvement Tools for Quality—Systematic Errors Best Practices in Detection

The first step in eliminating systematic error is to detect it—to find which operations, equipment, operators, tools, etc. have quality variation outside our norms. Once we detect a quality problem, then we'll try to relate it to the inputs, verify the relationship observed (our hypothesis) with a carefully designed experiment, and then change the specification as necessary.

Our tool for detecting changes or unusual variation in the manufacturing process is the statistical process control or SPC chart—a plot of our product quality monitors or process parameters or manufacturing performance over time. We can use it to detect changes in both the mean (center) or variance (range) of a manufacturing process or measurement. Figure 2.10 is a plot of one process monitor from our test after wave solder—number of opens.

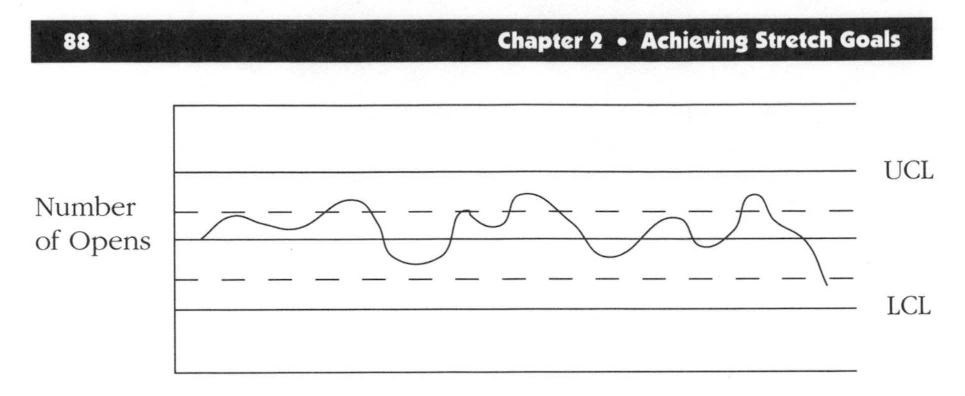

Figure 2.10 Number of Opens

A process that is under control has a stable predictable or static (unchanging) distribution of points. Over 99% of the points will continually fall between its upper and lower control limits (UCL, LCL) unless the process has changed its center mean (X) and/or range (R).

The control chart is used to detect abnormal variation or changes in the process that suggest we are no longer in statistical control. The most obvious exception would be a point outside the control limits. However, we can also detect patterns or trends that are precursors to points outside the control limits—patterns that suggest a change to the mean, range, or distribution of the underlying process. Visual examples are shown in the following charts.

Figure 2.11a shows a process is gradually drifting out of control—a higher percentage of opens are occurring over time. Figure 2.11b shows a sudden change in the mean—we have gone from an average of 1.1 opens per board to over 2.2 at a point in time. Figure 2.11c shows a sudden change in our variability—the range has dramatically dropped from 6 opens to nearly 1. Figure 2.11d shows a cyclic process—repeating with a steady pattern. Figures 2.11e and f show what appear to be two or more patterns mixed together—a mixture of data.

More Precise Detection Rules—Western Electric Rules the Quality World

While visual detection of a change in our manufacturing process is possible, changes are more accurately and consistently detected by more precise mathematical techniques. Operators or technicians may argue over whether a change has actually occurred, but mathematically, we can deter-

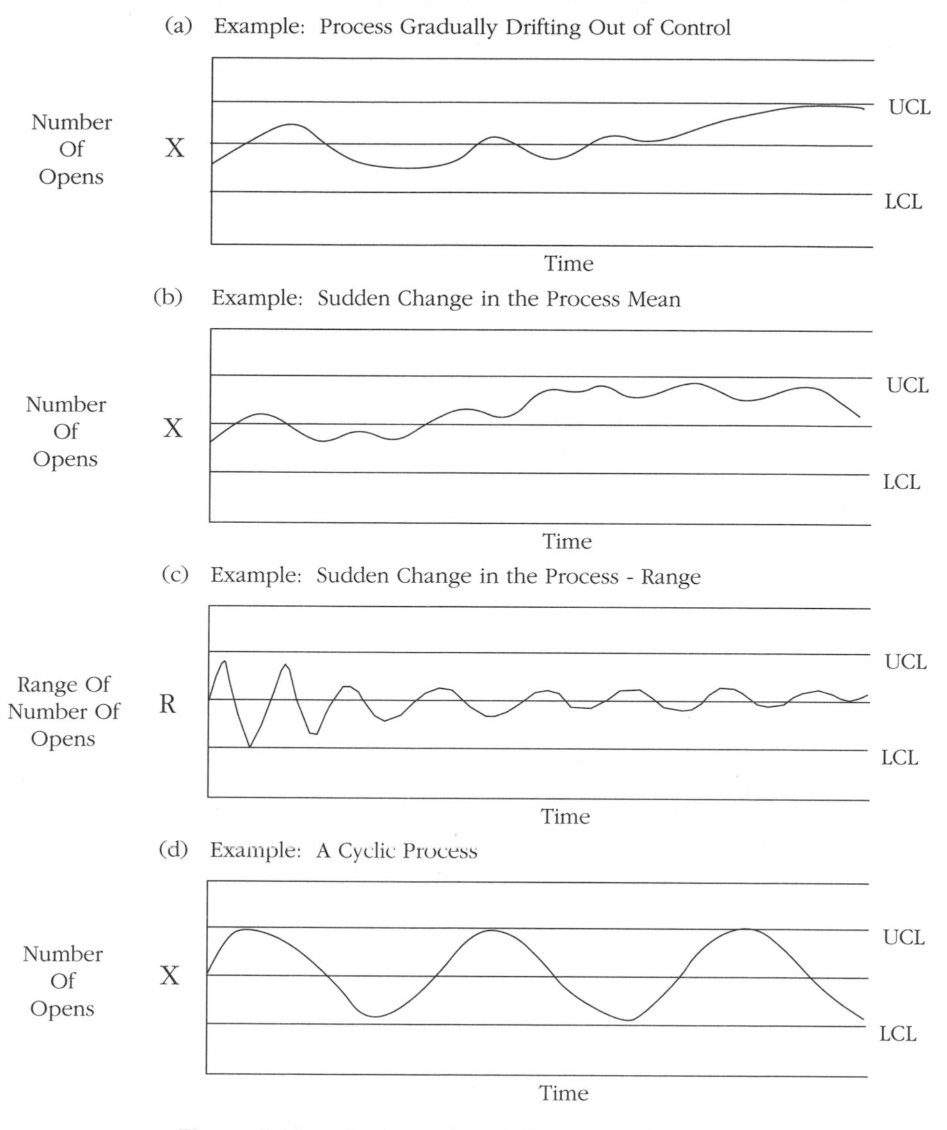

Figure 2.11, a–f Examples of Changes in the Process

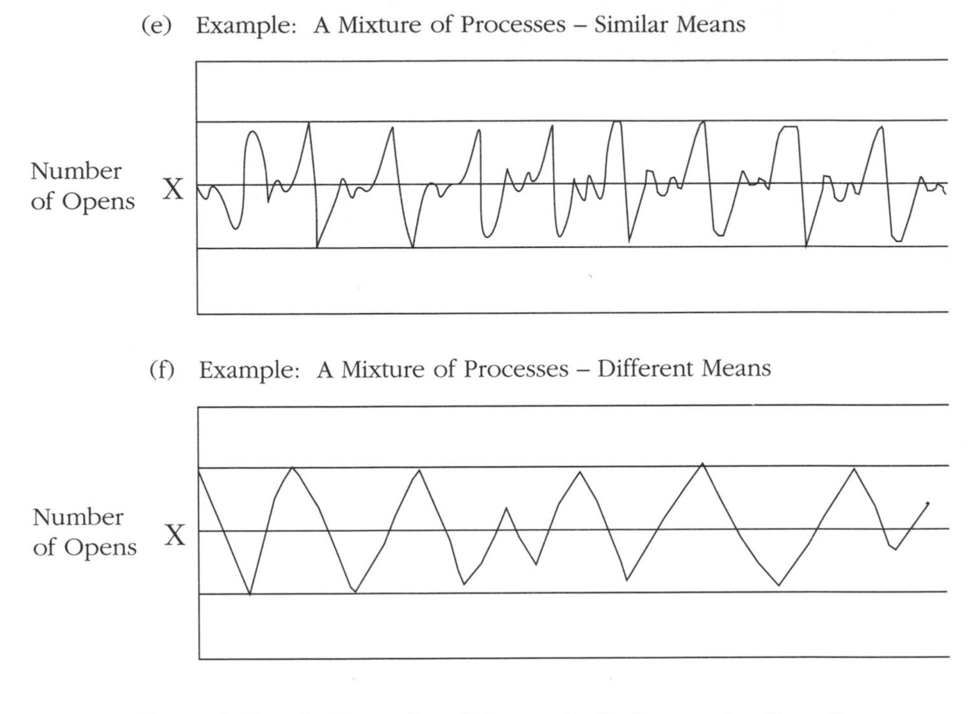

Figure 2.11, a–f Examples of Changes in the Process (continued)

mine it with a given level of probability. AT&T/Western Electric developed a set of rules that are based on the probability of a given pattern of sample points occurring to signal likely changes in a process. While there are many other detection (or "triggering") rules, Western Electric rules are perhaps the best known and illustrate the general approach.

We divide each half of the control chart (between the mean and its upper and lower control limits) into three equal zones (Fig. 2.12a). Our detection rules are then looking for an unlikely series of observations or data points or "runs" given our assumed distribution of points. Examples are:

- A single point outside the upper or lower control limits (Fig. 2.12b)
- Two out of three successive points in the outermost zones (Fig. 2.12c)
- Eight successive points on one side of the mean or center line (Fig. 2.12d)

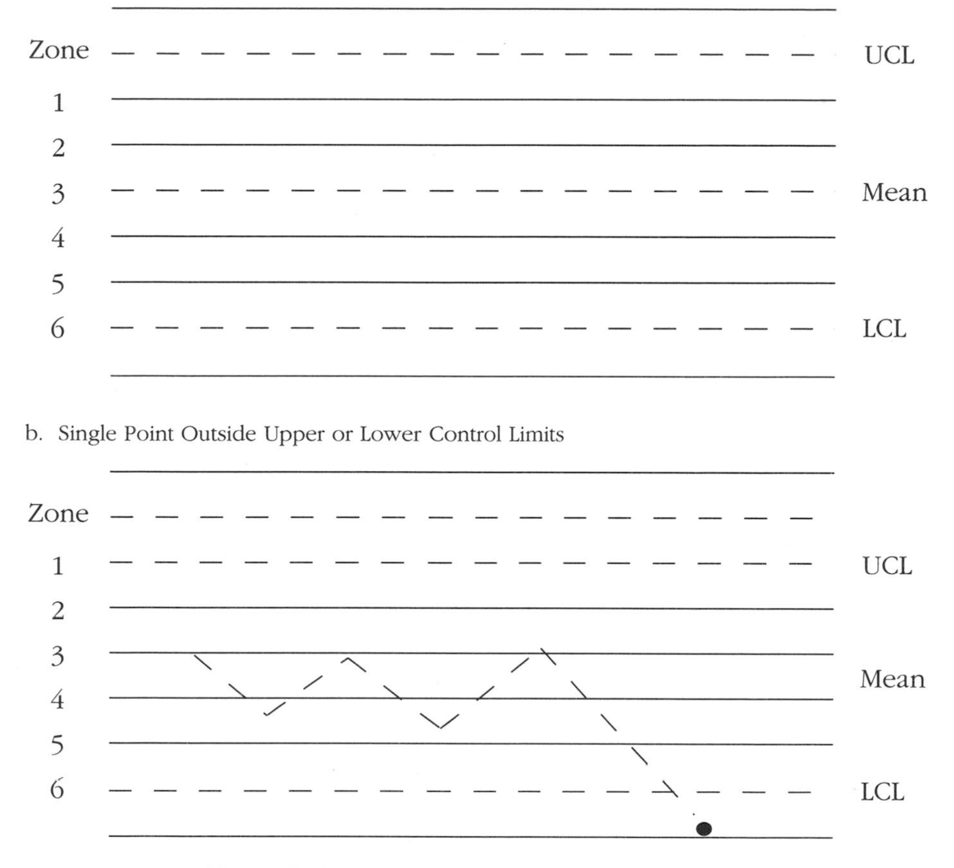

Figure 2.12 Western Electric Rules—Examples

Each of these occurrences has approximately a one-in-a-hundred chance or less *unless* the process has changed. We can use these tools, the SPC chart and detection rules, to monitor any part of our manufacturing process or performance that we can measure and detect changes in the process. We could measure final product quality at final test, operational quality at any step, process conditions like wave solder temperature or height, or raw material characteristics like acidity (pH) of our solder. We can use them to monitor our performance to schedule (percentage of jobs shipped on or before schedule) or our inventory level or cycle times.

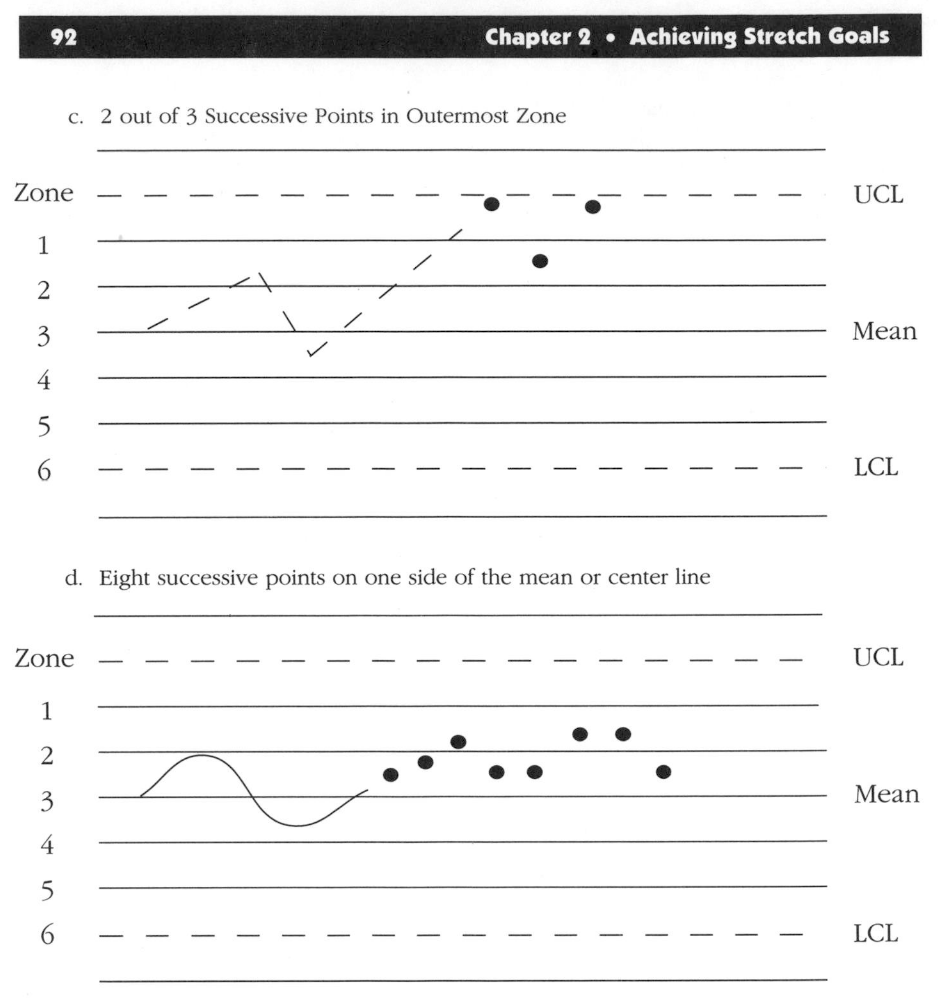

Figure 2.12 Western Electric Rules—Examples (continued)

There are many variants of our X and R chart including: using a measure of variance instead of the range; using the median instead of the average; and using individual readings (X) instead of an average of a sample. We can also plot control charts for "discrete" attributes that have two or more values (pass/fail; on/off; yes/no; done/not done; in/out; on time/late; grades 1–4). What we do is measure the *percentage* (or P) of nonconforming (failing) occurrences in the total sampled that period. For example, we could measure the number of maintenances that were not done on schedule each shift (on time/late) and divide by the total number scheduled. We could measure the number of tools that broke (not broken/broken) divided by the total number we used that shift. We could measure the number of operators who were late (on time/late) divided by the total number of operators. Now we can see if the attribute is in statistical process control—just as we did for continuous parameters—where we use p as our sample value instead of (X).

The Devil Is in the Detail

The key to detecting and solving systematic variation is to measure it at the lowest level of detail or responsibility. Many companies measure product or operational quality or productivity across all operators or equipment performing that operation. This averages out or masks systematic variation that is occurring by "hiding it" in the law of large numbers. If we look at Figure 2.13, we see a process that looks in control. However, when we monitor and display the quality and productivity by individual equipment (Fig. 2.14), the information is much more useful and interesting from an improvement viewpoint. Clearly all equipment is not equal in this simple example. However, our aggregated view gave no indication of this variation across equipment.

Many companies relying on manual data collection systems or ERP systems simply don't have the tools or data to support or allow analysis by individual equipment, operator, raw material vendor, or engineering change notice. And yet, that is the level of detail that often holds the key to detecting systematic variation. In quality analysis, the devil is in the detail. To track down the cause of variation, we'll need both to detect it and then trace it back to the differences in operating conditions experienced. Without this level of data availability, we are left with mystery and variability.

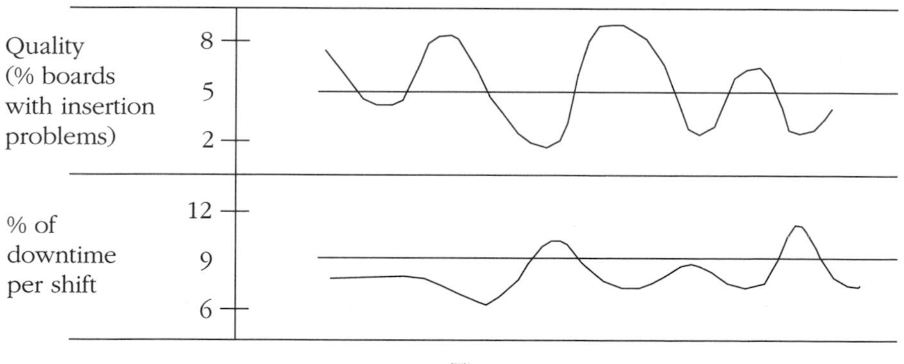

Figure 2.13 Operational Quality and Productivity—At Work Center Insertion

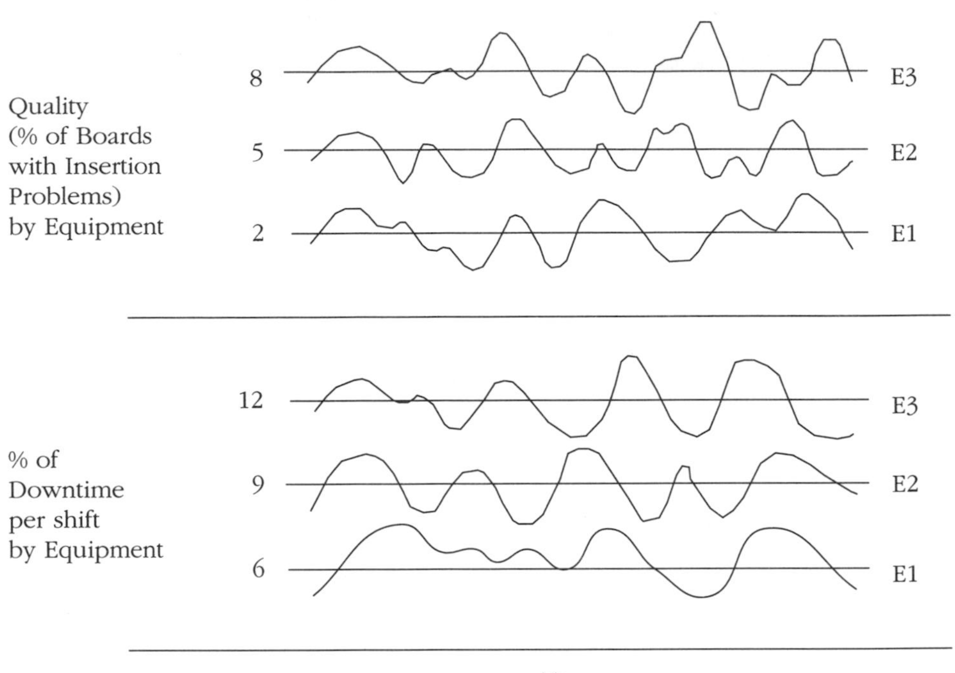

Figure 2.14 Operational Quality and Productivity—By Individual Equipment ID at Work Center Insertion

That is again why many companies are using new systems called manufacturing execution systems or MES *along* with their ERP systems. They allow them to track the detailed operating conditions during production of each batch or run so they can relate product quality to process conditions and see variation in quality, productivity, or uptime across their resources. As previously discussed, they also actively enforce compliance, eliminating most random errors.

What Is Causing the Change? Correlation or Regression Analysis—Relation to Inputs

If we do detect a systematic change in the pattern, or systematic differences in performance across resources, the key in any improvement program is to understand what caused the change or difference in performance and correct or control it. Furthermore, if the process is stable but there is "considerable" variation in the process, we would like to try to find the major causes of this variation and reduce or eliminate them. As we discussed, today many people view this variation as "mystery" and have accepted it as part of their manufacturing process and their cost of doing business. However, uncertainty is one of the greatest causes of waste.

Certainly every process has variation. Not every input to the process can be controlled (or even measured) to the same exact value. Humans cannot exactly repeat a procedure whether it directly impacts the quality of our product (as on a manual insertion and soldering of a power transformer) or indirectly impacts the quality of our product—as on a setup of the insertion equipment feeder. Equipment has variation as internal parts wear and power varies. However, much of the supposed random variation *can,* in fact, be explained by relating it back to variation in the inputs in the characteristics of the five elements of manufacturing. That is our basic model of manufacturing.

In other words, if we see changes at an operation in product quality, or differences across equipment performance, we will try to explain them by looking at differences or changes in the inputs of the five elements of manufacturing. There are three types of situations we will consider: a sudden and dramatic shift in the process, cyclic and/or gradual degradation over time, and a seemingly constant process with significant variation in results by resource. These are shown in Figure 2.15.

Without an exact model of each operation, we can't completely characterize or model how the product output is determined by the values of the operation inputs or status. If we had one, we could attempt to set each input within its specified limits and assure "perfect" quality (this approach is used in adaptive process control to build a "model" of the process). Instead, improvement programs work backwards. We detect when the output changes and try to deduce the input that could have caused that change by relating output changes to those inputs that also changed or are different. We postulate in Principle 6 for normal operative ranges that the way the input has changed will cause a similar type of change in output. Small variation in inputs cause small variation in outputs. Major variation in inputs may cause major variation in outputs. Because this may not be the case or we may find a plausible but false correlation, we will have to further test our hypothesis with a carefully designed experiment.

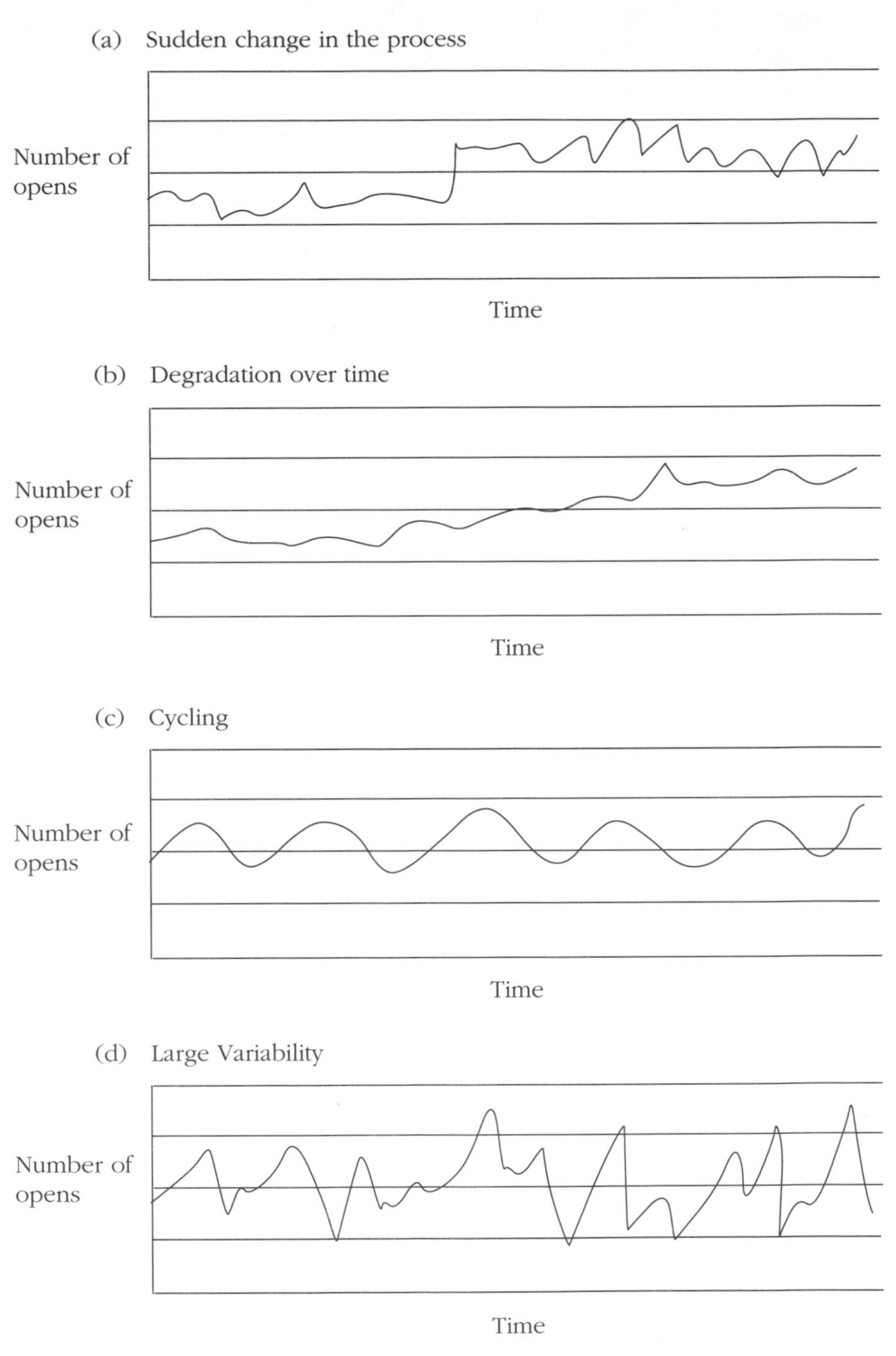

Figure 2.15 Patterns of Systematic Variation

Quality Best Practices Principle #6

Observed changes in operation output quality are caused by *analogous* changes in inputs.

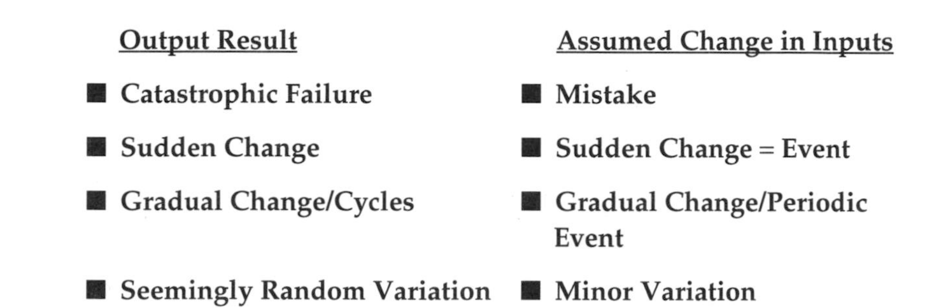

Output Result	Assumed Change in Inputs
■ Catastrophic Failure	■ Mistake
■ Sudden Change	■ Sudden Change = Event
■ Gradual Change/Cycles	■ Gradual Change/Periodic Event
■ Seemingly Random Variation	■ Minor Variation

Note that this approach to quality improvement requires *memory* or history. We need to be able to track product quality or resource productivity or uptime along with the associated characteristics of the five elements of manufacturing over time. No data means no memory, and that, as the historians have said, condemns us to relive the past! Systematic variation in output quality is a pattern that is detected over *time.* If we do not have a previous record of input values and resulting output quality accessible by a statistical package, we can't easily perform this type of improvement analysis. While it is fashionable to argue that true JIT/TQC factories run without systems, any statistician or improvement activist will clearly state the need for *data*—accurate, timely observations that can be analyzed for their patterns and meaning. Every job you run in a factory is an *experiment* with potential *value* for improving your process *if,* and only if, you have the data from that experiment (the values of the inputs and the output quality observed). This approach will be even more valuable in understanding how to achieve perfect output—what input conditions gave rise to the *best* quality output.

Best Practices in Root Cause Analysis—Explaining Sudden Changes in Performance

Using our hypothesis, in Principle 6, the cause of a sudden change in product quality at an operation—detected in either our $\bar{X}$, R or p chart—will be a *sudden change* in one (or more) of the five elements of manufacturing at that operation. For example, we may have just changed:

- a lot, batch, shipment, or vendor of one of the raw materials (the components we are inserting, the solder, or the boards themselves, for example)
- the operator at the operation
- the setup on the insertion equipment
- the calibration of the wave solder equipment
- the work instructions (or an ECN) at any of the feeding operations
- the work instructions for maintenance or setup at the operation
- the facility conditions—temperature, humidity, power surges

We can think of these changes as "events"—some indirect activity that has occurred affecting one or more of the five elements. We'll usually find the record of such events in our indirect activity logs, in our maintenance logs, specification histories, incoming receiving logs, lab results, and facility monitors logs.

To trace the root cause of a sudden change in product quality or process conditions, we use a product quality/event chart as shown in Figure 2.16 to relate the events that have occurred at that machine or work center to its quality.

On this chart we can see the timing of key or nonstandard events for the workstation/equipment. If any of the events seem to correlate in time with the sudden change in product quality, we can then look in more detail at their history. We can indicate the events with a code for the type of event (PM = preventive maintenance, S = setup, BR = breakdown, R = repair, RM = raw material change, ECN = work instruction change for example). If this is a test or quality measurement or lab operation, we probably will look

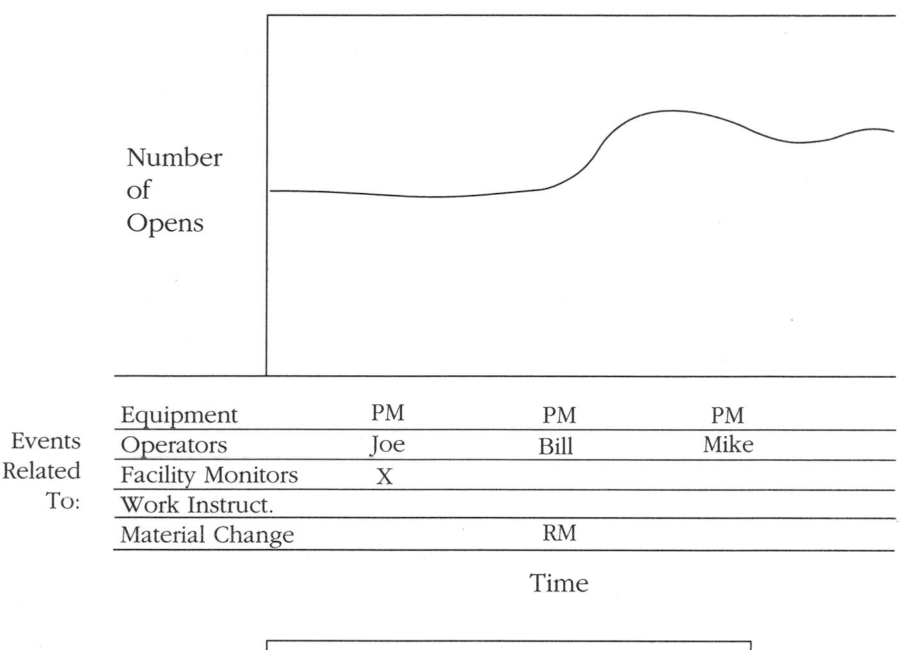

Key

PM = Preventive maintenance
RM = Raw material change
X = facility monitor alarm
BR = Breakdown
R = Repair
C = Calibration
ECN = Engineering change notice
S = Setup

Figure 2.16 Product Quality/Event Chart for Number of Opens

at the "event log" for the previous manufacturing operations and offset the time of the SQC log to when the job went through that previous manufacturing operation (to get the product/event log recalibrated onto the same time scale). For example, if we are looking at an SQC chart for opens discovered at final test and want to correlate it to wave solder events, we would recalibrate the time of the quality measurement of opens to reset it to the time that batch went through wave solder and look at a product quality/event chart for wave solder.

In the case of Figure 2.16 we can see three suspect events: the preventive maintenance, the raw material change done shortly before the number of opens jumped up, and the change in operators. We could then look at the logs for these events and recheck the state of the equipment, the raw material quality, and Bill's familiarity with the product specifications.

Similarly, sudden changes in quality at a manufacturing operation may also have to be related to events or changes at previous operations. For example, the change in the number of opens may be related to a change in the vendor of components (recorded at the incoming inspection step), the work instructions for insertion, or when a handling tool was last calibrated. While these may be less likely causes, detective work can't rule them out just because we lack easy access to the data!

In many factories today this information is spread out over many departments, stand-alone systems, and manual log sheets, and so is hidden or very hard to consolidate onto one chart. As a result, valuable time is lost in careful diagnosis, or the process is adjusted haphazardly or tweaked based on gut instinct and experience in an attempt to bring it back into control. For example, while the problem may really be in the raw material—the solder's chemical properties—the engineer may instead adjust the temperature and/or wave height in an attempt to correct a problem. In doing so, he has only brought the process into a less stable operating range! With each of the five elements under the control of a different department and tracked on a separate system or database (purchasing/inventory; maintenance/equipment; tool/tooling; production/operators; engineering/work instructions), it is hard to produce charts like these. The operator may not know what events have occurred. One manual approach is to have each department update this main chart kept at the work station. This obviously doesn't help in getting the detail of each event rapidly or in looking up previous occurrences of similar problems as shown in Figure 2.17 (these capabilities will be discussed in Chapter 7 on Manufacturing Execution Systems (MES); but briefly, they represent two of the key features still required in systems to support continuous improvement environments—even in JIT/TQC situations—"memory" (or history) linked to analysis tools.)

This type of improvement program is also dramatically assisted by eliminating the long time lags between operations by first physically laying out the manufacturing line with minimal distance between stations; second, running with minimal WIP in the line; and third, enlarging the role of the operator to be responsible for the indirect supporting tasks at that work

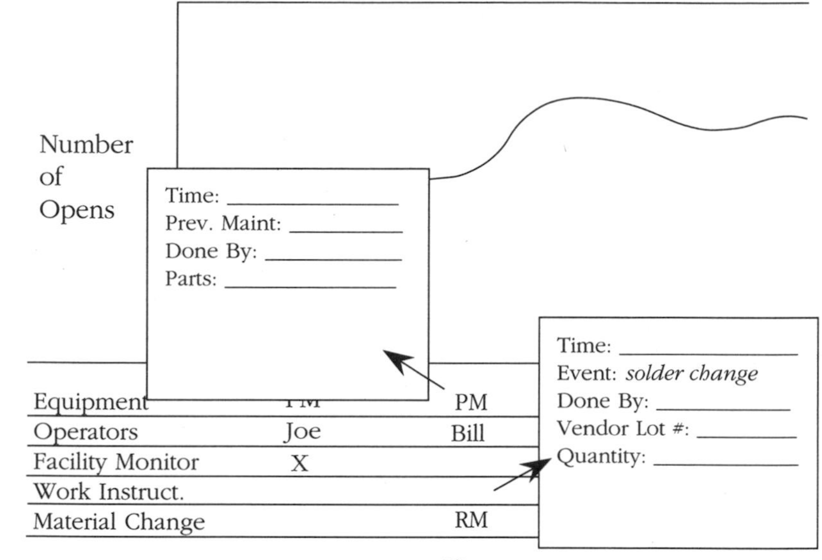

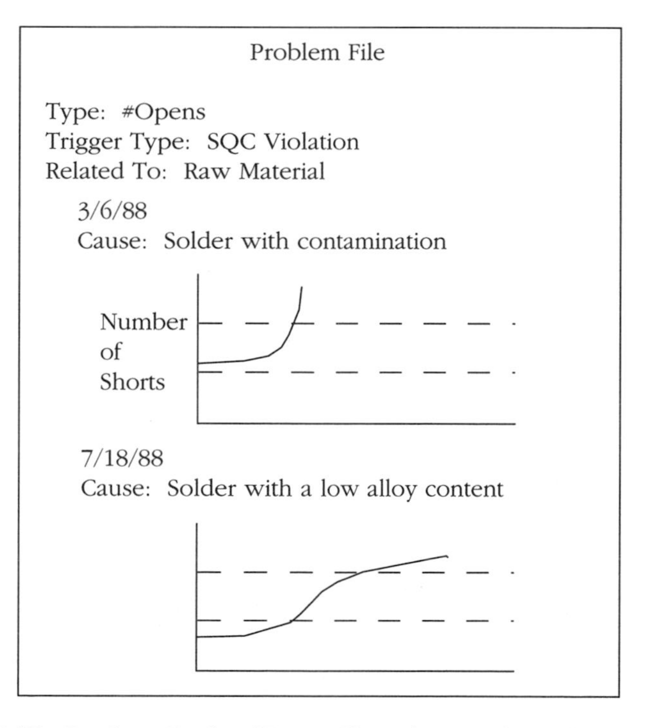

Figure 2.17 Product Quality/Event Chart for Number of Opens—Detail

center. This eliminates the lag (in time, distance, and personnel) between cause and effect. If the same person (the "operator") is responsible for the majority of the events (setup, maintenance, processing, work instruction improvement), then he is even more aware of the cause and effect relationship.

However, many relationships that are subtle or discovered only after many problems ("experiences") have occurred could easily have been seen using the tools of statistics. It is a mistake to eliminate one of the key improvement tools—statistical analyses programs tied to historical data—instead of adding it to our arsenal of improvement weapons such as layout and organizational redesign.

Best Practices in Root Cause Analysis—Explaining Gradual Changes in Performance and Cycles

The second type of systematic variation often seen is a gradual change or degradation of quality or performance over time. This pattern can usually be observed before an actual upper control limit violation occurs. Western Electric rules will trigger the steady shift in the pattern. Figure 2.18 shows an example of gradual degradation in a p chart for percentage of misinsertions on the insertion equipment.

From our theory, a gradual change in a process will be caused by a gradual change in one of the inputs—one of the five elements of manufacturing.

In general, we are looking for an input that causes a similar but less dramatic effect than in our sudden change examples, an input with some discrete event whose effect has a time factor. Examples could be:

- a setup that goes slowly out of calibration from vibration or impact
- a raw material that gradually loses strength or gets contaminated (as in a solder or etch or clean bath)
- a tool or part that wears over time
- a raw material whose chemistry changes (loses or gains moisture, changes potency, or begins to break down)

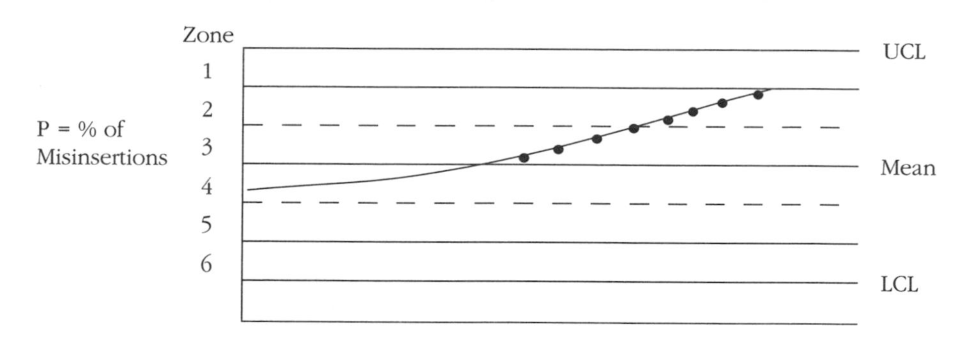

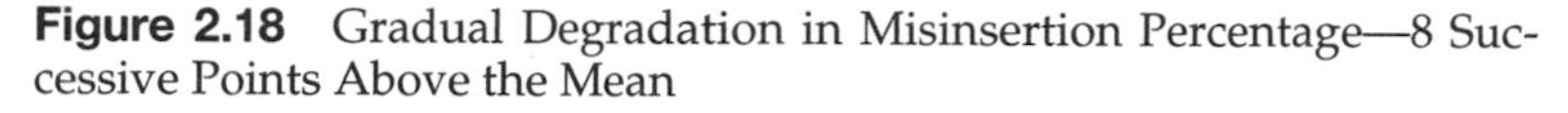
Figure 2.18 Gradual Degradation in Misinsertion Percentage—8 Successive Points Above the Mean

- a room that heats up during the day or cools down at night
- an operator who tires over the shift

A factory is a set of patterns in fixed or variable time cycles—of operators changing shifts, machines being maintained, raw materials resupplied, tools resurfaced or calibrated, facilities heating up over the day or cooling down at night. Often distinct patterns of quality changes are hidden when the affect of all these cyclic events are viewed together. Therefore, we have to attempt to pull out the affect of each cycle or suspect event in isolation.

The key to detection is to isolate the SQC chart from the time when the suspect events start and look for cycles—of degradation. Figure 2.19 shows several examples. Instead of looking at a continuous time plot (Fig. 2.19a, c, e), we reset the clock and replot each time the suspect event starts to see if there is a causal pattern (Fig. 2.19b, d, f). In the first example, there is some seemingly normal variation in the percentage of boards with misinsertion problems. Visually, however, the pattern has a potential suspect cyclic appearance (Fig. 2.19a). If we chart a quality/event time chart, we see a cyclic event of calibration of the insertion equipment. In Figure 2.19b, we replot the misinsertion chart as a percentage of misinsertion versus time since last calibration. We note that there is a strong visual pattern suggesting that our quality is degrading before we recalibrate the equipment. We could prove this precisely with statistics, but the visual data suggests we examine our calibration interval and consider shortening it to a length prior to where we see the degradation in quality starting.

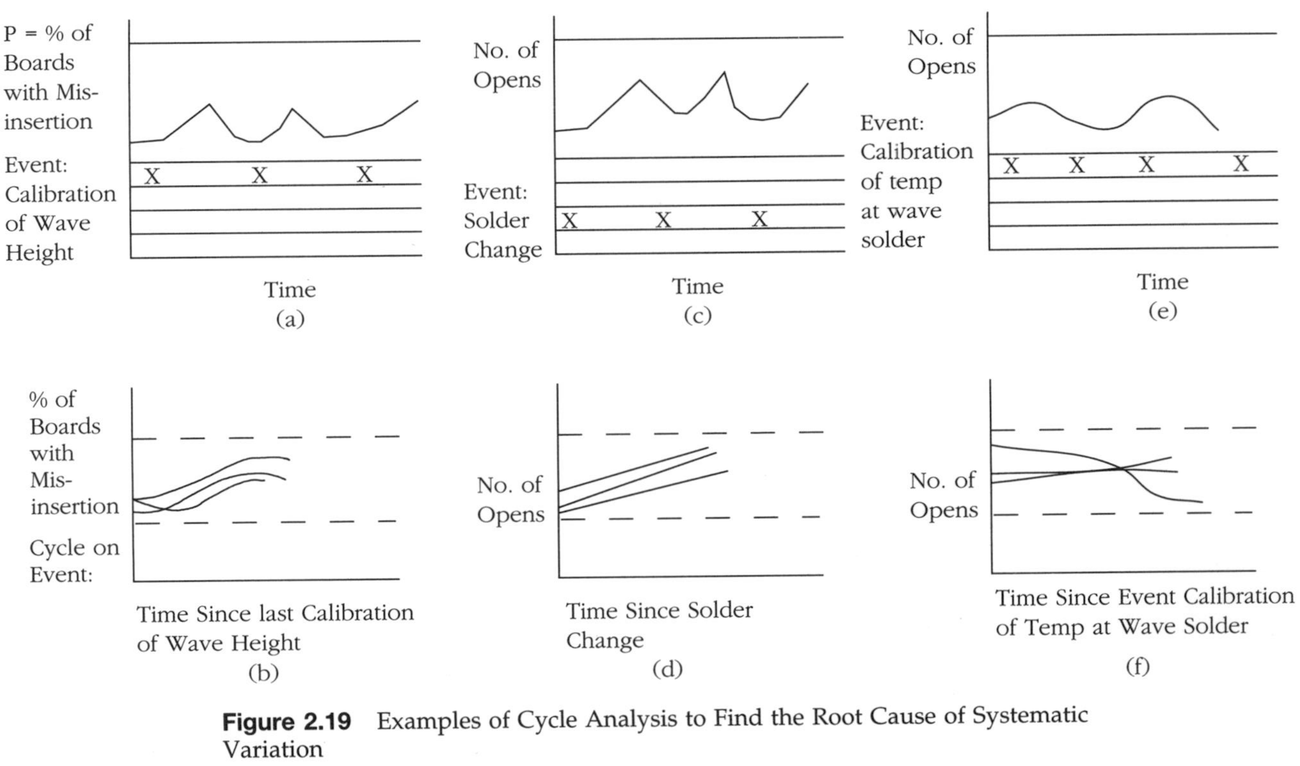

Figure 2.19 Examples of Cycle Analysis to Find the Root Cause of Systematic Variation

Similarly, in Figure 2.19c and d we examine the relationship between the number of opens found at our bed of nails test and when we change our wave solder bath. Figure 2.19c shows the pattern over time. Figure 2.19d shows the pattern since the last time we changed our wave solder. Again, the data suggests that we consider changing our bath more often. Note that this conclusion is premature without considerably more detective work. The pattern suggests a possible cause of degradation that is systematically related to how often we change the wave solder. In practice we would need to explain why waiting longer is causing more opens. Is it due to a gradual contamination of the bath, or is it due to some other recalibration or change we also make when we change the bath? We are using visual data analysis techniques that are not explanatory (explaining why the relationship may exist) but correlative (detecting possible or seeming relationships). In the third case (Figs. 2.19e and f), the cycle analysis shows no clear pattern. Once we have visually detected a suspect pattern, we can run more precise statistical measures for verifying the relationship and then an experiment to confirm it (as we discuss in the next section on design of experiments).

This form of analysis is a useful method for setting or evaluating fixed or set interval maintenance or calibration or cleaning cycles (until we go to predictive maintenance based on observed product quality, process quality monitors, or in-situ sensors). In most factories, these intervals have been set years ago and rarely changed unless clear evidence arises as to a problem. With a time cycle analysis we can pinpoint the need for more frequent maintenance (meaning we can improve quality) or less frequent maintenance (meaning lower costs—from higher utilization of equipment and lower indirect labor costs). Figure 2.20 shows a chart of percentage of misinsertions as well as when we calibrated the feeders on the insertion equipment.

Since this data is difficult to interpret, we can replot this in an event cycle plot (misinsertions versus time since last calibration) and fit the points.

As we can see from Figure 2.21, more frequent calibration would appear to improve quality. This type of analysis can potentially help set cycle periods for all cyclic events (operator retraining and recertification, vendor recertification, facility cleaning, specification reviews, etc.).

Again, these analyses are infrequently done in most factories because of the way we've organized manufacturing into separate departments, each with their own systems. As a result, the data necessary for quality improvements is spread over separate systems for maintenance, facilities, incoming

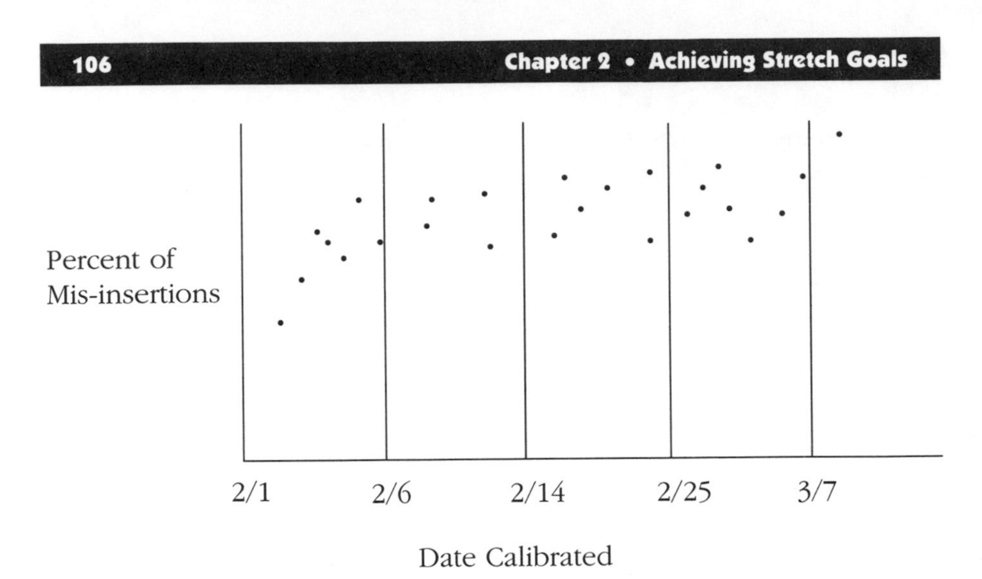

Figure 2.20 Percentage of Misinsertions Over Time

inspection, work instruction, time and attendance, tooling, cell control, quality, and shop floor control. Some of these are manual systems; some reside on personal computers, the equipment itself, or a host computer, but almost none are integrated together into a unified plant view. Few are aimed at improvement at all! Most are meant to support financial or performance reporting needs. They were not designed with any concepts of best practices in measurement, quality, or any other manufacturing goal.

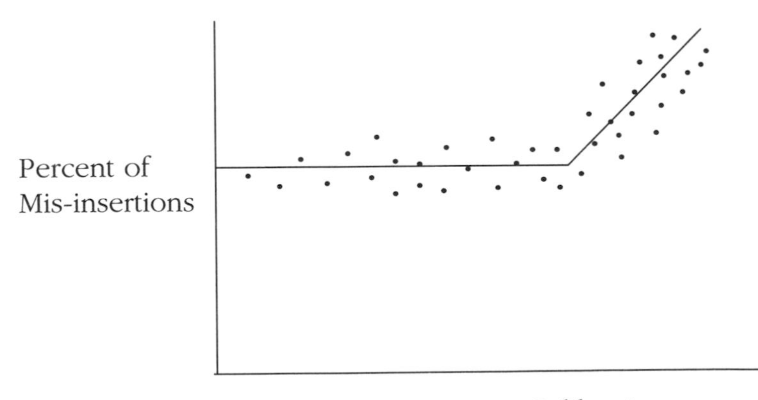

Figure 2.21 Percentage of Misinsertions Versus Time Since Last Calibrated

By cutting up manufacturing artificially into isolated fiefdoms, we've lost the *integrated* or holistic view of cause and effect of manufacturing as a team sport. "Perfect quality" manufacturing requires systems and procedures that support it (systems that provide information integration), integrating our view of manufacturing. MES (manufacturing execution systems) were invented to provide this capability.

In companies that have implemented MES systems, these analyses are done by engineers automatically.

Best Practices in Root Cause Analysis— Explaining "Controlled" Variation

The third type of systematic variation we will examine is one that displays "considerable" variation but all within set limits. In Figure 2.22 we see that our solder bond strength (measured on a pull test for the solder ball) varies considerably between its upper and lower control limits but seems to be under control with stable mean and range.

Each new generation of product seems to require finer manufacturing tolerances in line widths, bond strengths, board flatness, facility particle counts, and solder chemical characteristics to meet new product specifications. We can either continually purchase new equipment, tooling, and facilities, and/or find new vendors and operators, tolerate lower yields

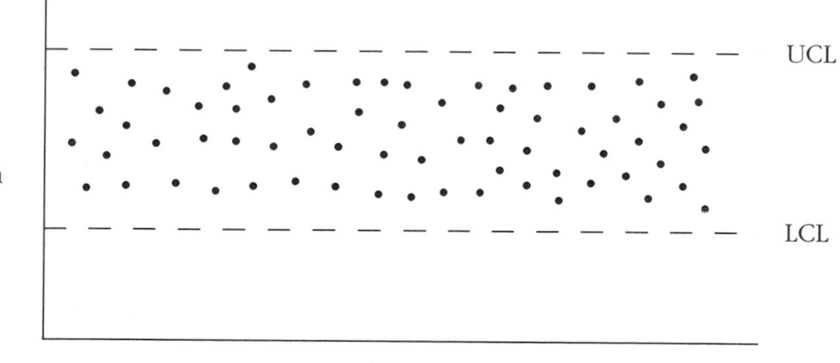

Figure 2.22 Solder Bond Strength Over Time

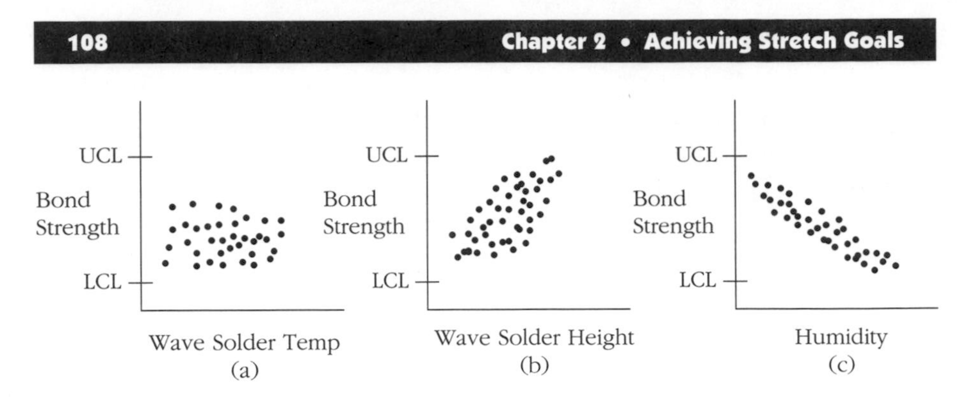

Figure 2.23 Scatter Plots of Bond Strength Versus Process Variables

and/or higher rework, or at *no* cost learn to improve the process to operate within these tighter bounds.

The goal of this root cause analysis is to learn what characteristics or settings of inputs may cause the finer variation we experience in quality and control them to eliminate wasteful variation. The principle we use is that small or seemingly random changes in product quality are likely to be related to small variations in "continuous" input values.

The pattern in Figure 2.22 shows no obvious cycling or major change in quality level or significant trend. However, it has large (if constant and predictable) variation that affects product life in the field. The product must be rated for a shorter effective life as bond strength can vary greatly, and many fail the "shake, rattle, and roll" test (vibration) for mil spec grade. We have been examining how changes in quality may be related to discrete events. Now we will try to relate the output quality directly to the inputs process values to try to better understand the details of this relationship. We do this through a *scatter diagram*—plotting the value of our product quality measure on the horizontal axis and the accompanying value of our input or process variable (one of the five elements of manufacturing) on the vertical axis (we could conceptually do this for several inputs at once, building multidimensional response surfaces).

Suppose that we have been measuring several of the "input" process conditions during manufacturing—some continuous such as wave solder temperature, wave solder height, and humidity; some discrete such as operator, equipment, vendor of the raw boards, and the ECN used. We could first look for any relationships between the "continuous" parameters and our bond strength on scatter diagrams (Fig. 2.23a–c).

We see from the first diagram (Fig. 2.23a), now that we are isolating potential individual cause and effect relationships, our specified limits for wave solder temperature seem to produce no systematic or correlated variation in bond strength. However, from the second diagram (Fig. 2.23b), the wave solder height may be causing a problem. It appears that the lower heights *may* have a negative correlation to bond strength (possibly not enough solder is deposited). This is a weak correlation. However, there seems to be a stronger correlation between bond strength and humidity, shown in Figure 2.23c. As humidity increases, our bond strength appears to fall. We have never changed the upper and lower specified humidity limits over the last six years while actually changing the wave solder composition and manufacturer. It appears that we could improve the bond strength by tightening up our humidity limits.

Again, we can be more precise in the degree of correlation by going to a statistical measure called a correlation coefficient. However, visual review of such plots is extremely important to detect what are often more complex patterns that purely calculated (linear) correlation coefficients may not represent adequately. We often see nonlinear relationships as shown in Figures 2.24a–c where the pattern degrades after a set point, varies parabolically, or has a small stable operating range.

With these visual relationships, we can either go directly to change our specified operating limits or hopefully design experiments to verify our hypotheses. We need to make sure that a spurious relationship (based on too small a sample of data or caused by a correlated but different parameter we have not postulated). In practice, each observed point during manufac-

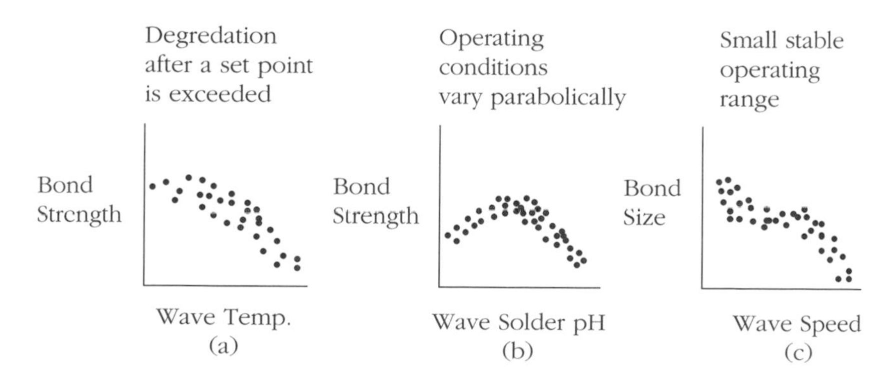

Figure 2.24 Examples of Nonlinear Relationships Between Product Quality and Process Parameters

turing has many simultaneously varying parameters. A carefully designed experiment will let us *isolate* the variable we want to test and keep the others constant.

We must also be careful before changing any specified setting to see how that would affect *other* key quality measures. For example, lowering humidity may improve adhesion bond strength but could cause a problem in the number of shorts as the solder balls are slightly larger or tend to flow more freely. Therefore, before we stop at our first conclusion and implement it, we must look at how that process parameter affects our *other* quality measures. Often what we are really doing is a balancing act.

Best Practices Correlation—to Analyze Differences Across Resources

While the scatter plot has been extremely useful for analyzing the effect of continuous process parameters, we will use a different technique to detect differences across resources used in production, such as the equipment, operators, or ECN used to detect whether they have an effect on our quality. The basic concept is that we will look to see whether the mean/average and/or variance/range of the performance differs significantly across resources.

For example, to assess the effect of equipment on quality, we have plotted the frequency of number of shorts per job (fixed number of boards) for wave solder machines (Fig. 2.25).

We need to be careful that these curves represent the same underlying "population"—meaning that they represent the same products and lot sizes done on both machines. If, for example, machine 1 was used on more difficult products (products with tighter tolerances or closer traces) we would expect that the number of shorts might be higher. However, we have used the same product for both charts (product #124) and the same job size (48 boards/job). We can see that these distributions are not equivalent—they have different averages and ranges. In practice, most companies do not collect statistics by individual element of production (equipment, tool, operator, ECN) and therefore only see the quality by product operation (as in Figure 2.26) as discussed earlier.

This distribution really represents a *mixture* of the two individual machine distributions already shown. Again, we can statistically test for signif-

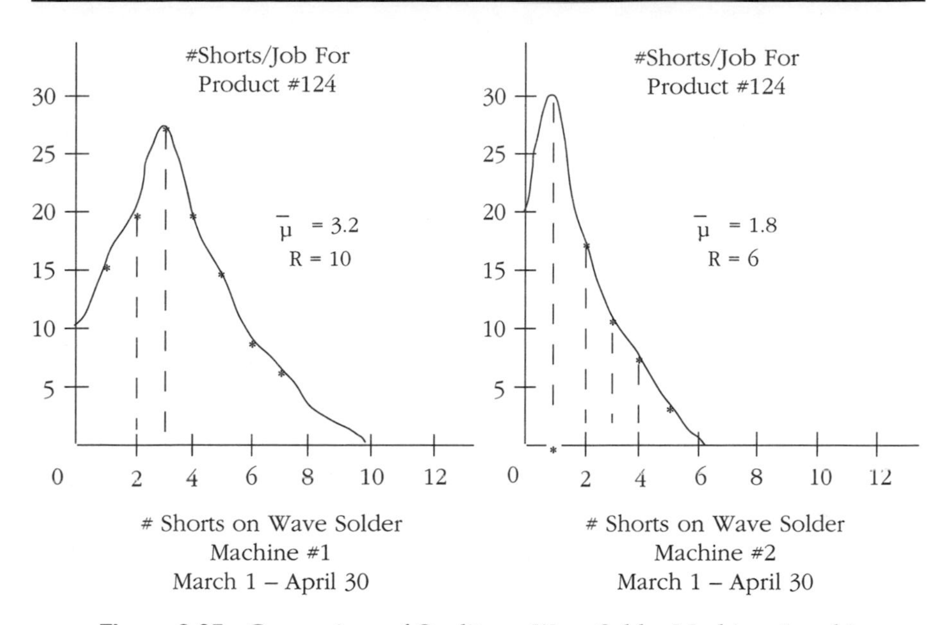

Figure 2.25 Comparison of Quality on Wave Solder Machines 1 and 2

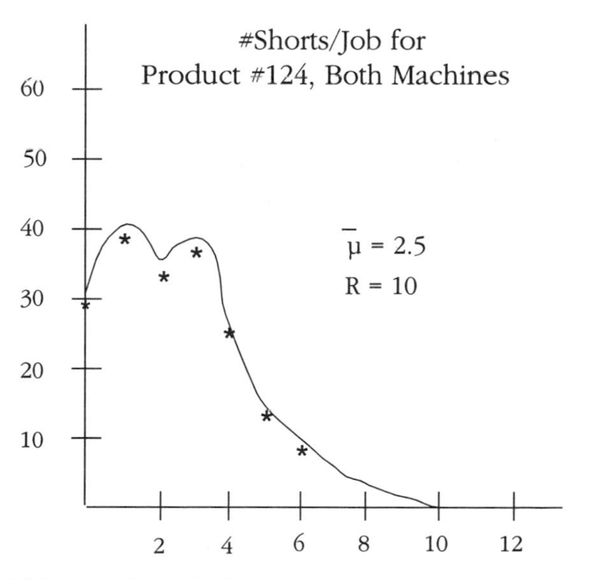

Figure 2.26 Number of Shorts/Job for Product #124 at Operation Wave Solder

icant differences in the two sample population means and ranges, but visual analysis is an important starting point.

To plot individual charts by resource would be a significant task. It is one of the reasons companies tend to collect "pooled" data and so miss key differences between equipment, tools, operators, ECN vendors, etc. that could lead to continuous improvement. The more we can isolate the potential source of variation, the more we can search for the root cause and attempt to control it. If we do find a significant difference between product quality on machines 1 and 2, then we can investigate the root cause of the superior quality on machine 2 and try to improve machine 1's quality.

The ultimate root cause may be directly related to the equipment and its characteristics (operating temperature or speed), to the raw materials used on the jobs done on that equipment, to the operator doing the processing using that equipment, or to the facility conditions around the equipment. Or it may be related to the indirect activities affecting that resource (maintenance calibration, cleaning, and setup of equipment; training, supervision, and certification of operators; cleaning and monitoring of facilities; and so on) and who performs them. But a significant difference in observed performance is the starting point for our search for the root cause. These are all potential correlations we would need to explore using the techniques we've already discussed.

The Search for the Source of the River of Quality: Back to Our Indirect Roots

Let's take a step back and use an analogy to see what we have been discussing in our search for perfect quality. If we want perfect apples, we can search through all the apples for ones that, through some seemingly random events, had the correct combination of water, wind, branch length, sun, shade, insect protection, and so on to grow to perfection. Alternatively, we can prune, water, shade, fertilize, and spray the tree so that *every* apple we grow is perfect. In one case, we intervene at the real causal point. In the other, we're at the mercy of chance and inspecting quality into the system.

In manufacturing, the care of the tree is the care of our five elements of manufacturing: the selection, training, and management of our people;

the selection, maintenance, and repair of equipment and tooling; the selection, maintenance, and repair of facilities; the selection, inspection, storage, and management of our raw materials/vendors; and the manufacturability of our work instructions/specifications. In *our* search for quality we want tools to improve the tree—our manufacturing resources—so that we can build a manufacturing operation which runs perfectly all the time. This means measuring each operational input for its quality (as seen by its effect on each operation's quality), and trying to find a "root" indirect cause of input variation to control it.

Using our tools, for example, we can measure resource quality and the individual's responsibility for that resource. We can compare the mean time to failure on a set of equivalent equipment (in terms of hours in use or number of units produced), by a maintenance/repair technician (Table 2.8). We can compare the number of product failures by vendor for raw materials. We can compare the number of errors per operator by trainer. In this analytic or quantitative approach we can search for the potential root causes of the differences seen between resources as well as finding the "best" indirect performers to try to duplicate their successes elsewhere.

While there may be other factors impacting differences found, the concept revolves around finding the "indirect" effect on quality and improving it.

Alternatively, we can use another technique for finding potential "indirect" problems or successes called similarities and differences. Table 2.9 lists insertion equipment with the highest and lowest downtime and operators with the highest and lowest quality.

We can look to see what, if any, potential indirect root causes correlate with the individual elements in each category. For example, to help explain the potential root cause of equipment downtime, we could see whether

TABLE 2.8 Measuring Indirect Quality: Mean Time to Failure (MTTF) at Insertion Equipment (Month to Date) by Maintenance Technician

Equipment Maintained By	MTTF
Maintenance technician #1	96 hours
Maintenance technician #2	112 hours
Maintenance technician #3	63 hours

TABLE 2.9 Best and Worst Performers Equipment Downtime and Operator Quality

Group 1 Equipment With Greatest Downtime	Group 2 Equipment With Lowest Downtime	Group 1 Operators With Highest Quality	Group 2 Operators With Lowest Quality
Insertion A123	Insertion A133	Joe S.	Bob W.
A161	A124	Bill L.	Tom F.
B214	B166	Mary T.	John S.

there is a high overlap/commonality/similarity among the maintenance technicians, operators, raw materials, maintenance interval, age, tooling, or other direct or indirect resources that might have affected equipment downtime. Similarly, we could look to see if the "highest quality" operator group shared characteristics of trainer, supervisor, supporting engineer, education, length of time (experience) at the company, equipment, and so on.

By searching for any common attribute, we can search for the likely root cause and, therefore, the improvement factors. For example, we see in Table 2.10 that there is an 80% "correlation" for high downtime equipment with the maintenance technician responsible for the machines. High uptime machines all share the same production supervisor on first shift. Highest quality operators also share the same supervisor as a common link. Lowest quality operators have no common link that has been found so far. These searches are helpful in bringing quality back to the root cause—our *indirect* activities in support of manufacturing. Finding the "cause" of the best and worst performance gives us the ultimate means to improvements.

TABLE 2.10 Similarities Among the Best and Worst Performers

Equipment		Operator	
Group 1 Similarities Equipment With Greatest Downtime)	Group 2 Similarities Equipment With Lowest Downtime	Group 1 Similarities Operators With Highest Quality	Group 2 Similarities Operators With Lowest Quality
Maintenance Tech.: Joe (80%)	Production Supervisor: Bill (100%)	Supervisor: Bill	None at 75%

Proving Our Hypothesis—Design of Experiments

From our data analyses, we will have formed many opinions on how our input variables affect our product quality or performance. The natural reaction is to immediately change the process—to change the allowable control limits, an equipment setting, or maintenance interval. In fact, many of our conclusions may be correct. However, the only way to be sure is through the use of a carefully designed experiment. Companies often use "seat of the pants" experimentation in which they take one job and process half of it with the current technique and half with the proposed change. While this approach is logical, it has two drawbacks. First, it may be very inefficient. A full job may not be necessary to test our hypothesis. We may be able to run several experiments on the same job or use a smaller sample size. Second, without statistical design of the experiment sample size, it may be inconclusive or even inaccurate if the new or old procedure appears slightly better! We may be led to make a change that in fact is no better or even worse.[3]

Closing the Loop—Changing the Work Instructions/ Measuring the Change/Improvement

Eliminating systematic errors is a process of continually refining our specified operating conditions for the five elements and the indirect activities that control them. As we determine the optimal operating conditions, we change our specifications accordingly. Since we are continually doing this, we automatically also see if actual improvement occurred! We are monitoring performance, so we can, in fact, assure ourselves that the changes are positive.

In simple manufacturing operations, this may seem like a trivial point, but in complex systems, changes in one area or activity may have unexpected ripple effects not anticipated here or at downstream operations.

For example, if we perform maintenance or calibrations more frequently, we may change the life or reliability of the equipment. Each time we work on it, we may be (inadvertently) damaging parts, affecting its reliability, or affecting its warranty or life. Without careful monitoring of the whole view, our "improvement" may prove transitory.

[3] Montgomery DC. *Design and Analysis of Experiments.* New York: John Wiley & Sons, 1976.

Taguchi Methods—If the Mountain Won't Come to Mohammed—Design for Manufacturability

While we have focused on improving our understanding of the "best" operating conditions, we'd be remiss in not viewing the problem from a different angle. Instead of trying to control the manufacturing process to "force" its variability into the optimal operating range, alternatively, we could design the product and its manufacturing process so that it "works" every time within the current normally expected resource process capabilities. In this approach, we measure the process capability of each resource at each operation we run. For example, we could measure the observed distribution of wave solder height and temperature and ensure that the manufacturing process as designed would produce acceptable product for all values within this range.

Obviously, this might constrain the products or processes we could employ and therefore affect our competitiveness. But so does specifying an operating condition that we cannot attain consistently.[4]

The Best Practices Approach—Joint Process Capability/Product Roadmaps

Too often product design/engineering is specifying future products, selecting parts, and even vendors without considering whether manufacturing can economically produce and procure them.

The best practices approach to ensuring quality long-term is to develop a set of matching roadmaps—one for the product requirements and the other for the manufacturing capability needed to produce them.

For example, your future products will need to shrink the trace width to less than 1 mil. That also means controlling the variance of those traces to within .05 mil. They will require placing leadless components. That means new bonding capabilities. You're designing twenty- and thirty-layer boards. That means overlay alignment capabilities in the future of less than .63 mil so that the layers all line up. These are just a few of the product requirements coming over the next three years.

[4] Ross J. *Taguchi Techniques for Quality Engineering.* New York, McGraw-Hill, 1988.

TABLE 2.11 Roadmap of Manufacturing Requirements by Capabilities

Manufacturing Capability	1997 Current Capabilities Required	1998 Required Capabilities	1999 Required Capabilities
Trace width	1 mil	.90 mil	.75 mil
Trace variance	.05 mil	.04 mil	.03 mil
No. layers alignment	20	24	30

Yet, there is no roadmap of how you are going to attain those manufacturing capabilities required to align design with manufacturing (Table 2.11). Many companies don't even know if their actual current capabilities align with the design requirements for their *current* products. This analysis is done using a measure called C_p, or the process capability. It is a measure of the actual variance observed in the process versus design requirements (Fig. 2.27). The goal is to have a process whose natural variation falls well within the specified limits at critical operations.

The "Lecture Is Over—Their Education Is Starting"

"So, now you have all the techniques and tools. The key is to apply them. I am going to recommend that PCB Co. actually hire a statistician to add to your staff just as you have maintenance technicians and engineers. Actually putting this into practice is going to require a lot of training and then actual experience. Meanwhile, I'm going to lend you an associate to get you jump started so you can try this at home."

Bill was scratching his head. Frankly, he'd never have guessed at the amount of science behind a quality program. He'd always thought of quality as poking around until you found out what had gone wrong. It was surprising and actually interesting to him to see the amount of theory in a "science of manufacturing." He'd never been taught any of this in engineering school. No one at PCB Co. had this type of training or education. But he could see that there was an enormous body of material to master. He could

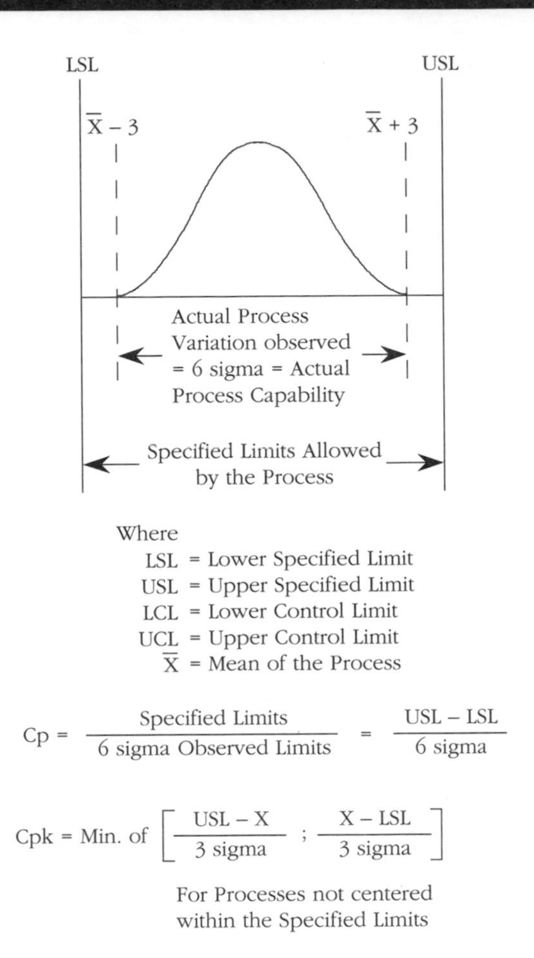

Figure 2.27 Process Capability Indices

see that he was facing a real challenge—to teach a wider group about these techniques and how to use them. Bill had no illusions that this was going to be easy. He could see all the practice one would need to really be effective with these tools and the effort to start to change their design culture to consider manufacturability.

But, on the other hand, he saw the logic, the reasoning, the approach, and the power in it.

"Okay, team," he stood up and said. "That was the easy part. Now let's go out and apply it."

Our Group on Their Field Trip

In addition to the class on best practices in quality, Bill had arranged a visit (based on the instructor's strong recommendation that seeing was believing) to a PCB supplier who was already world class. The PCB plant was a unit of their major computer vendor. The group went as a whole on the Monday right after Saturday's class. Bill wanted the group focused and motivated but wasn't sure what to expect from the field trip. He personally had never done external benchmarking or comparison visits. The group was fairly quiet. All of them had been struck by the amount of analysis possible and honestly weren't sure where to start. Bill wanted to get the team together right after the trip to discuss how to move forward. He hoped this process would help kickstart the team which was more social these days.

The guide who greeted them Monday morning in the lobby explained that this PCB facility had been totally redesigned by a work group not dissimilar from themselves—a group of operators and engineers with representatives from several other production support groups (maintenance, facilities, inventory management) as well as an industrial engineer, a JIT/TQC consultant, and one representative from their main equipment supplier. This struck the group as very large and cumbersome, probably unmanageable. The guide laughed and said that it had been a large and noisy group, but that the facility had to work for everyone. Everyone was and had to be a stakeholder in its success. What they had done was to each critique the facility from his or her own viewpoint—to both improve it and to improve communications and cross-discipline understanding simultaneously.

They had also adopted a facility slogan to show their common goal. Motorola had "six sigma," referring to the percentage of parts within tolerance; Panduit had "X not R," referring to ridding themselves of variance. They had selected—"Quality is no mystery to us." They were proud of their accomplishments and had posted their progress on wall charts. The guide pointed out how, in their PCB facility, WIP had decreased by 80% along with leadtime; rework was down 85%; first-pass yield was up to 98%; and customer service was at 99%. The charts were quite large and had not only the actual values of these measures, but also control limits to show how the variance was also decreasing. The group was amazed to see these charts posted in view of everyone—customers, workers, management, vendors, perhaps even competitors. However, the guide said that their quality goals were paramount, and everyone had to have visibility into their progress.

They were equivalent to their company's financial results in importance, and they would in the long term actually drive them! If they did not have quality, they would not have financial success.

Everywhere they toured, there were so many differences from their own company. The operators asked about labor standards at several operations to compare them to their own. They were amazed to hear that there were no individual labor standards for operators. The *group* or shift was measured for quality, output, machine uptime, safety, customer service, and so on because they were expected to focus on *facility* goals and their attainment, not on individual output. Too often, individual output meant producing boards that the next operation could not work on anyway (if its equipment was down), focusing on long runs to build up activity, or allowing marginal quality to pass. Now the team had to decide what would produce the best facility results, which was much harder but more in line with their corporate and customers' goals.

The layout and organization of equipment was very different as well. Instead of the functional layout they were accustomed to with all equipment of a common type grouped together (with areas for kitting, automated insertion, manual insertion, wave solder, and test), there were several cells in a U-shape where work flowed without WIP queues. They appeared to have purchased some dedicated wave solder machines with lower throughput rates to match the processing rate of the cell. Inventory was at a minimum everywhere—both of raw materials and work in progress. They were accustomed to shelf after shelf of raw material components, many that were now obsolete or left over in small odd quantities from bigger shipments. They saw none of these. Parts were delivered just in time and to accurate counts as needed.

One of them asked where the rework lines were. They did not see the normal piles off to one side to be reworked after test. The guide smiled proudly and said that their rework rate was now down to less than 1%—except on new products—and so needed no separate line. As long as they kept the process running within its C_P specifications they expected no rework and rarely saw any except when an unforeseen problem arose. That had certainly not been the case when they had started on this project! They had been down for hours at a time trying to get the process to stay within specified limits. For almost 16 weeks, output had been sporadic, and they had often thought of trying something far less drastic. However, they kept asking themselves why produce scrap just to make something? And since they were all in the same boat (figuratively) on their performance measurements, they'd all pitched in to help.

The group was also surprised to see desks by the side of the work area. They asked why the operators and supervisors had placed desks there. The guide explained that those desks were not for them but for the other members of the team—production control, maintenance, finance, and the other support personnel. They all were on the floor, visually in contact, and immediately ready to assist. The rest of the time they worked on improvement projects.

What especially struck everyone was how extraordinarily clean the facility was. In their own factory, the floor often was littered with scraps of paper, solder, and an odd component or two. There were boxes of boards, components, fixtures, and rework piled in various corners. Here, there were very few boxes resting outside the cells. The floor was immaculate. Somehow it all added to a feeling of quiet control and precision.

"Where is all your inventory?" someone asked. The guide explained that they had developed an entirely new relationship with their vendors. Instead of selecting them solely on price performance to their specification, they had requested new capabilities for just-in-time deliveries to the floor with certified quality. At this point, they had reduced the number of vendors from 35 to 7 for components. All of these had met their requirements for eight-hour delivery on high runner parts and two-day delivery on low runners. They had also insisted on 100% quality with delivery right onto the floor to avoid the unload, reload, and unload so common (and wasteful) in receiving parts. Now they were looking at linking directly to their vendors through EDI to send them the shift schedule each shift ahead for eight-hour delivery automatically. This had required looking for vendors who were physically close enough to deliver parts each shift, had the required quality to avoid full inspection, and were flexible enough to respond on short notice. Cost was not the primary factor in selection now as they could save large overhead costs (really waste) of storage, inventory control, inspection, obsolescence, inventory carrying costs, facilities costs, and so on.

They asked their guide for the daily output rate of the facility and were amazed to hear that it was nearly forty percent higher than their own in less than one-third the space. When the space needed for weeks of inventory, work in process, rework, and incoming inspection areas were eliminated, the floor space required for value-added manufacturing was relatively minimal.

All in all they were incredulous. They had always felt proud of their manufacturing prowess, but this was a different level of excellence. They were beginning to understand the gap between where they were today and a truly world-class operation.

When they complimented their guide on his company's progress, he smiled and said, "we *have* come a long way, but it's a never-ending race. Our competitors never stop improving and neither can we. We see it as a race. We need to get faster and faster because our competitors keep training and improving. When we stop improving, we lose. Right now, we are concentrating on increasing our flexibility—our ability to adapt to meet changes in demand, so if *our* customers need to change *their* assembly mix, we can change just as rapidly. This means that our operators, maintenance techs, and engineers need to be cross-trained on more equipment and our vendors more flexible on shipments. We are also adding the PCB designers to our team. They'll actually be coming down here for the first six to ten weeks of a product launch to see how well their designs work in manufacturing. We also participate in an evaluation of their designs in a series of design reviews for manufacturability *prior* to the design being frozen and accepted by manufacturing.

Now the group was really incredulous. Manufacturing "accepting" a design? That concept was beyond their wildest notions. In their company, the designers were seen as the "breadwinners." Whatever they designed was "law." They were seen as the corporate competitive advantage, and manufacturing's job was to produce their designs, not debate them. Everyone really began to get interested now. This was something that anyone in manufacturing could get excited about. "You mean you don't simply have to accept their designs as is?"

The guide explained that they didn't view it that way at all. They viewed that every member's role on the corporate team was to add value. Designers were there to develop competitive products and new technology. They were there to add competitive manufacturing processes, procedures, and techniques to those designs. *Both* had to add value so neither one could be an absolute. The corporation needed the highest value products for its customers which combined product features, quality, cost, delivery, reliability, and so on. Manufacturing's job was to carry its own weight as experts on manufacturing. They thought of it as a team sport in which *each* member, design, manufacturing, and field service had to have world-class performance for the corporation to win. So they did not see it as "accepting or rejecting" designs; they saw it as members of a team discussing tradeoff and cost/benefit—lowering the number of parts lowered total cost, weight, and assembly time and increased reliability; using standard higher volume parts decreased part costs and increased flexibility; using standard operations did the same. Manufacturing did not *dictate;* it just added value to the final decision. That's why they were adding designers to the manufacturing team, so that they could help get the part into full production sooner by debugging any functional problems and actually see the manufacturing tradeoff/consequences of their design.

The group was silent. This was going far beyond their immediate idea of quality tools. What they were seeing was an overwhelming change in performance measures, layout, organizational roles, and power structures that had been in place for decades. They were beginning to think that they had bitten off more than they could chew. "This is overwhelming," someone blurted out. "We could never do this."

The guide did not smile now. He was very serious. "We didn't do it all overnight," he said. "This took years. But you probably *have* to do this or something similar because your competitors have, are, or will. And you can't compete the old way. You either get started now or someone else will be manufacturing those boards for you at higher quality and lower cost. Your company still needs them; they just won't come from your factory."

"How did you get started?" was Bill's question. The guide outlined their own approach as follows.

A Program for Getting Started

"We got started three times," he explained. "First we needed a champion who made the program a priority. There is no lack of worthwhile activities or programs for a company to pursue. We could have invested in thirty areas. Someone has to take a stand and literally move the whole company along to first allow and then cause change.

"A program like this involves so many changes by the time it fans out—it ripples through every department—that without a champion it soon bogs down in all the reasons it *can't* work, must be studied more, or cannot be easily cost justified without headcount reduction. Any radical proposed change that impacts the current power bases, procedures, attitudes, and beliefs engenders more delaying tactics than you'd believe.

"In many of the more successful programs, this champion was at the top—the CEOs of Xerox, Milliken, Motorola, and other companies that not coincidentally won Malcolm Baldrige awards and regained or or grew market share. However, some companies we visited started with one general manager, division vice-president, or even one manufacturing or engineering manager who continually battled for change.

"Unfortunately, the impetus often seems to come from having no alternatives for survival, as it did at Harley-Davidson or Xerox. Certainly, it is laudable to succeed in the face of adversity. But it would obviously be better to be ahead of your competitors instead of being driven by them.

"In our case, we were somewhere in between. The writing was on the wall, but the wall hadn't fallen down on us yet.

"Our first attempt at this change was education (Bill winced). Our champion Jim—the vice-president of manufacturing—thought that if he educated everyone, they'd get on the band wagon and we'd turn into a world-class manufacturer. Somehow he thought his enthusiasm and urgency would flow right down to his managers, their direct reports, and so on right down to the direct workforce. So we sent groups out for education and brought trainers in for the direct hourlies—and nothing changed. Oh, actually, productivity went down from all the hours in the class room. And morale suffered as everyone spent too much time worried and confused about what was going on and what was going to happen to their jobs. That was our first try—not too good.

"The mistakes we made were that, first, no one was *goaled* for any change; and second, nothing had changed. Everyone had the same job, the same boss, the same responsibilities, the same attitudes, the same beliefs, and the same relationships. All that had been added were a few hours in a classroom and someone else's enthusiasm. Jim thought that his enthusiasm would rub off or transfer by osmosis.

"Take two was better. Jim started in one department—board fab—and formed a team. He took a representative from each function—production, maintenance, manufacturing, engineering, facilities, purchasing, design, finance, and production control—and told them that he wanted within six months to eliminate the rework line. How they did it was up to them. They had the authority to make whatever changes were necessary."

The guide chuckled. "I remember that first meeting. We all looked at each other and thought three things—there was no way you could eliminate the rework line, he'd never give us the authority to do what we needed to do, and why are these other people here? They have nothing to add to this process. I said take two was better—well, it was better after a while. At the initial meeting, after Jim left, it was much worse. Now we were actually supposed to do something! Half of us hadn't even taken notes from the education we'd all had, and on top of that, engineering got up and took over the meeting. They immediately started to hint that all we had to do to eliminate rework was get the operators to stop "massacreing" their designs. Maintenance and facilities never said a word. Production control said that if we eliminated the rework line we'd better be able to handle more hot lots to push through the production line. Finance said rework cost us 20 cents per board and if we eliminate it, we'd add almost three million dollars to the bottom line. Overall, most of us left the meeting angry at someone else, frustrated at engineering taking over, and certain this was going to fail miserably.

"But Jim had done one more thing. He'd given us a list of companies who had eliminated their rework lines and people to call who'd meet with us or give us a plant tour. Two were in the United States, two in Europe, and two in Japan. I think that without that list, we'd have wasted weeks or months. But we figured, if we had the responsibility, we'd see if we had the authority. So we set up a two week trip around the world to see what these companies had done. Now, we all thought of ourselves as being pretty good—in fact, very good, but in a general sense. Actually, we didn't think of ourselves in *comparison* to anyone else outside the company. Only top management did that. We thought of ourselves in comparison to *ourselves* last year or last month. So we weren't sure what to expect when we went to visit those other companies. In fact, it was the the first time most of us had ever focused on anything *outside* of our own four walls.

"That trip was an eye opener. We couldn't believe how good those other companies were. They had less floor space, fewer operators, no rework lines, and made more boards than we did. We came back from that trip changed people.

"Something else happened on that trip. We got to know each other better as people. As you travel with people, you eat with them, drink with them, wait in airports together, search for cabs together—you get to know them. When you go to foreign countries and visit other companies, you begin to think of the others as *your* team. Everyone has a competitive streak. We started seeing these other companies as the "outsiders." On that trip, we started to come together as a team—emotionally. But we just didn't know how to act and work together as a team. We saw that each of us looked at the plants we visited and saw something that the others missed. For example, I was amazed when the facilities guys noticed that they used different piping welded differently to reduce particles. The maintenance guy, Bruce, found out they did almost twice the number of maintenances we did. Each member came into their own on that trip—in the eyes of the *others.* But that somewhat evaporated once we returned. Engineering took over the meetings again. We just felt ineffectual as a team. So I went to Jim and told him about it.

"Take three was the difference. Jim sent us all to training but of a different sort; he sent us to team building and leadership training. As we discussed it at our offsites, we had two types of people on the team—people who were used to talking and people who were used to listening. Each was pretty good at one but not so good at the other. The operators were used to being told what to do but not standing up and saying how it could be improved. The engineers were used to telling everyone else how to make a board, but not to hearing what was wrong with their processes.

"Those offsites really added to forming the team that had begun on our trip. Each of us also went through a battery of tests and were sent for training to add or improve missing skills we needed—for listening, for problem solving, for communication skills, and for leadership. At that offsite we decided on two key teamwork principles—that the team chairperson would be a rotating position and not fixed, and that each person was responsible for their area of expertise. This was a team—not a democracy. For example, maintenance had to take responsibility for how to improve the equipment to eliminate rework. We'd all help them, but they were still the experts. The key was that while we'd all be judged on the team's results—eliminating the rework line—that didn't make us equally expert in all areas.

"That was two and a half years ago. Now I'm on several teams. They're a way of life around here. When they've met their goal, they're disbanded. I've worked with every function, and I know people all over the company. I visit customers on a regular basis to understand how they measure us. I look at our competitors to see how they have improved their performance."When the guide was finished, several things struck the team. They were about two and a half years behind the world-class performance they'd just seen! They hadn't been given any real authority—only responsibility. Bill thought about that a long time. He had to fix that—and soon!

Getting Started Back at the Plant

Back at their own plant the group had divided into three camps—the aggressive, the passive, and the confused. The passive were clearly overwhelmed by what they had seen on the trip and were retreating from the scope that they had heard. "It won't work here," was the gist of their message. "We're biting off more than we can chew. Let's start small with some focused improvement teams in one or two work centers and work methodically as we learn from our initial experiences." The aggressive got even more religious with each passive moan. "They started from where we were," they said. "There's no reason we can't do the same thing—or more. You heard what they said." The confused grew more confused with each volley. Both sides made some sense, but neither was totally compelling. It did seem risky on the one hand, but the reward was great on the other. The group was hopelessly deadlocked.

At the same time, this same battle was raging within Bill. He was completely overwhelmed by what he had seen and wondered if they could pull it off on that scale. Back and forth he went, pro and con. If they started and it failed, it would disrupt production, lower morale, and some people would quit over it. If they did not do something major, though, he could see the inevitable. Some production would be moved to a sister plant with lower labor costs; then a new plant would go up in Mexico, Singapore, or Thailand for the high-volume boards; and pretty soon they would spend all their time planning how to close down an area one at a time.

Finally, he picked up a phone and dialed the plant manager they had just visited. He discussed his thoughts, leaving out the part about their future if it failed, and waited for some advice. All he got back was, "if you're confused, think about your workers. Because on top of that, they don't know where *you* stand. They can't move forward without that, and I would bet they are milling around wondering what your position is. Without your full buy-in, putting your weight completely behind it, the chances of it working are not good. So, if you're not completely for a particular direction, don't let it get started. It's hard enough to succeed when they have your full support; many of them will be resistant, scared, and hostile already to the change that's coming." Not much sympathy there. Just a speedy quality message.

Bill was truly perplexed about how much to bite off in project scope. While most of the team had worked on their TQM program several years back, he, himself, had almost no experience outside of the few books he'd read and now the full day offsite and visit. He tried to be very analytical and nonemotional about his decision process and view it as an engineering problem. But he also realized that he had many emotions that kept emerging: a true sense of failure that he had not seen this coming and kept more up to date on current quality concepts; real anger that they were being threatened this way, had their very jobs at risk, and were being viewed as commodities instead of as individuals who had real value to their company; fear of what he would do if he lost his job; and intellectual curiosity of what perfect quality design and engineering would be like.

Clearly, one message he'd heard from the plant manager was that he, himself, had to decide what he was committed to and lead his team accordingly. As an engineer, he was accustomed to making rational decisions based on sound, tested engineering principles and training. This area seemed no different in that respect. There were clear principles of quality management and they needed to be adopted. They just needed to start and apply these principles step by step, methodically, continually.

So why did he feel so nervous and unprepared? He scheduled a meeting with the quality group for the next day. Another thing that was clear to him was that the team had to show some immediate successes. He didn't want the team to think in terms of months before they had to show results. He wanted improvement with every step so that the team's belief in itself would gel. He needed the disparate factions to merge into a more unified view, and honestly, he needed some short-term success for his own morale and sense of purpose. And he needed it for his peers' morale and sense of momentum. The corporate feedback had been devastating for many of them—perhaps all of them. Bill knew that some of his teammates had to be rethinking their commitment to PCB Co.—even starting to circulate resumes. The sooner he could achieve some results, the sooner the team would have evidence that they could succeed in their challenge.

He would also have to depend on the consultant for advice and experience. She seemed confident that they could achieve near perfection. His nervousness seemed most under control when he felt a sense of purpose—of leadership. He realized that he was probably not the only one on the team wondering about their potential for success. He wondered if he should also start looking for another job. The thought gnawed at him as he prepared for the next day's events.

Bill drove to work the next morning focused on the meeting. He was concerned that the message the group had received from their last TQM project had been "slightly better is good enough." They had not been given the authority to solve the real root cause problems. He needed to empower them and make them reset their expectations. Last time they had seen a partially supportive management that had eventually become the road-block.

This time, he would take responsibility for making sure they were not stopped on their road to perfection. But he would have to prove to them that he would and could play that role. If they hit an organizational barrier, he'd have to get it removed, speedily. Their project needed authority—in practice, rapid access to the top management team and rapid decision making by it! If the quality team had to petition for a hearing and wait weeks for a response every time it hit a road block, its effectiveness would dissolve in frustration and delay.

And he needed to set a goal for them. A stretch goal of their own—something difficult, challenging, even audacious but also possible. He couldn't leave that up to the team. It was too divided itself. That was his leadership responsibility.

The group meeting actually started with a short talk by the consultant on how to hold meetings. Their meetings needed to have an objective, an agenda, time allotments for each agenda item, preparation before the meeting, and a careful notation of action items that needed follow-up (with a date and assigned person or team), records of all decisions taken, and formal rules of conduct. Bill realized that he took all these guidelines for granted but this was a team reformation. He was grateful for the consultant's presence.

The rules of conduct were carefully, politely, but firmly enforced by the consultant—a neutral party designated for the meeting as its facilitator. When someone was simply repeating a previous comment, she stopped them. When someone made a personal remark or criticism, she also stopped it with a polite—"please focus on the issue and not the person."

What would have normally caused open warfare was stopped before the first verbal rock was thrown. The group accepted it from her, as an outsider, and Bill noticed that in this new, safe environment, people whom he'd never heard speak before started to raise opinions or offer ideas. And if they didn't, the consultant periodically called on them. No one was allowed to disappear.

Gradually the plan began to form of how they'd attack quality problems.

Moving Forward—Empowering Breakthroughs

The group would meet daily at the end of each shift to discuss the quality issues that had arisen during the shift. These problems would be divided into different categories: problems where an immediate solution had been found or was known and could be assigned to someone to be implemented; problems where the cause was somewhat unclear; and problems arising from another department or a cause outside their control—from a vendor or design for example. These meetings would be augmented by a weekly (at first), then bimonthly, and then monthly meeting with the consultant—who would continue to train (or retrain) them on procedures and techniques that would help them; and Bill would review where they were, check their progress against their plan, and help resolve interdepartmental problems. In that way he could assist them in moving forward in a somewhat uncertain (to them) corporate environment.

The last ingredient was a target or goal. He paused now in his meeting with them and then wrote out their goal on the white board: 100% first-pass yield for one day—perfection for twenty-four hours. He had thought a lot about this point—reading some material the consultant had given him from Motorola and Hewlett Packard. In each case the CEO had set a corporate goal that was not evolutionary but revolutionary—one that could not be met by current thinking. The point was not to frustrate people but to change their thinking, to force them to question their assumptions and find a breakthrough—to change the way they did business.

He had softened this by making it for one day. He reasoned that everyone would think this possible, but that if they could do it, they could replicate it—or nearly—every day. It was a challenge that he hoped they were ready for, and that, the consultant assured him, *others* were already achieving. The group looked at the goal and was quiet. Finally, one of the operators said, "Let's give it a try. We can do it." He hoped they were right. But he, still being honest with himself, wasn't sure if they could.

The Quality Team—Back in Action

The training had been invaluable in providing a framework for their questions. They now reviewed one of their quality problem summaries from their original TQM project (Table 2.12). Back then they had found many quality problems with an obvious cause—operator error, machine malfunction, or bad components—if not always an obvious solution. But frequently the manufacturing process seemed to produce a bad part with no "warning." Examining the part did not immediately and obviously indicate the cause. It had appeared as if there were some mystery in their manufacturing process—or at least to them. Even when the "cause" was visually obvious (a missing component causing an open, a bent wire that had shorted to another solder joint, a cracked component, or a lead not clipped short enough) or obvious upon closer examination by microscope or in the lab (poor tinning on the lead ends, a burned out component, or solder with too high a level of impurities), often the *underlying* cause of the manufacturing malfunction had remained unknown.

Now as they reviewed the list again, the group began to divide up quality issues into two classes: random mistakes and systematic errors ("mystery"). They knew how to prevent the quality problems that were caused by obvious and apparent mistakes. They could add tests or proce-

TABLE 2.12 Causes of Poor Quality

Quality Problem	Cause
• Cracked board	• Operator error—dropped board
• Missing component	• ??
• Shorts	• Operator error—wave solder height incorrectly set
• Bent wire—short	• ??
• Broken component	• Equipment malfunction—unknown cause
• Opens	• Operator error—wave solder temp incorrectly set
• Missing component	• Equipment malfunction—unknown cause
• Opens	• ??
• Wrong components	• Operator error—wrong component spool
• Open	• Bad leads (from vendor) on components

dures to avoid that error. For example, to avoid operator error in setting the wave solder height and temperature, they discussed two options: putting "short specs" or abbreviated work instructions on plasticized sheets attached to the wave soldier machine, or writing the correct specs on the run card for each job. The engineers also discussed automating the "download" of the correct instructions to the equipment using a personal computer so that all the operator had to do was enter the product number. This was tabled as an overly expensive solution to the problem until they tried the first solution.

Whereas, the first time through the TQM program, the "mystery" problems had frustrated and stymied them, now they had a new set of tools and reasoning to employ. They also decided that they really needed to measure quality by individual piece of equipment at insertion and by operator for manual insertion.

They also continued to add more prevention or detection/inspection steps to avoid errors or detect them instantly.

One of the operators, who was very practical, had two excellent suggestions the group immediately adopted. He began, "I can make sure we eliminate two of the mistakes we make now—soldering boards with missing components and using the wrong ECN. I saw how they grade multiple choice for my drivers license. They have a mask that they put over the answers to see if you filled in the right blanks. I thought that we could do that for the boards to see if all the holes are filled. I could also post the latest ECN number on the machine or put up a red flag if it had changed that week.

"They also began looking for some of the cycles the instructor had mentioned that could affect quality in their everyday jobs.

Quality as a Process

The process worked better than anyone had really thought it would. With the consultant's assistance and Bill's continual support they systematically gained quality ground. In three to four months they had:

- Added quality measures for operations that previously had no quality monitors, such as insertion and deflux.

- Separated out issues of training, prevention, and improvement. Many quality problems were solved by making sure operators were trained in current operating procedures. In fact, they initiated a scheduled retraining program for operators to ensure that everyone used the same procedures.

Operations with "tricky" procedures were examined and Poka Yoke devices added (instead of the "mask" for visually inspecting insertion, they decided on shining a light from the bottom of the board to detect missing components), or the operation procedures were redesigned.

Relationships between inputs and quality were examined to build models for improvement.

They looked at their quality problems daily, and in their weekly meetings they received any guidance or assistance they needed.

However, they still had not gotten a 100% first-pass day. They had raised first-pass yield another 4%, however, making them all pleased. They had also shown that many of their problems were related to incoming materials variation. They had begun meeting alone weekly and with the consultant and Bill bimonthly.

Bill was very pleased, but still the team hadn't met its goal. But he had stopped writing and revising his resume. He was now committed to play out the hand he held. He owed that to the team.

3

Achieving Stretch Goals Through Best Practices in Customer Service/ Scheduling for Supply Chain Management

Current Practices

As Joe thought through his new assignment—to attain a near perfect delivery to schedule—he had a strange realization. Even though on paper he was "responsible" for customer service, he didn't have a clue as to how PCB Co. did their planning and scheduling. All he knew was that they never seemed to get the right jobs out on time even though they had large finished goods inventories. He also had been seeing an enormous number of articles on "Supply Chain Management," the current hot topic in the mid-nineties. In a short cycle time, world customers expected immediate on-time deliveries of their orders.

Joe had a real feeling of trepidation. It was one thing to criticize, along with everyone else, their poor customer service. It was another thing to try to fix it. He wasn't particularly quantitative. He only knew what his customers—the marketplace—wanted. So his first step had been to do some research with the planning and scheduling guys. As best he understood, the way that PCB Co. currently scheduled manufacturing was as follows:

- Current leadtime through the facility ran 4 to 7 weeks, depending on the length of burn-in time for that product.

- Each month, the corporate production planners gave the factory a planned load of orders by part number, some made to stock (for high runners) and some made to order (for low runners and custom orders) that approximately fully utilized the factory. This utilization measure was based on the total number of boards (called the "activity") to be inserted and (wave) soldered and further broken down into a total number of "burn-in activity weeks." Since burn-in capacity was limited, the load was also analyzed this way for feasibility.

- Factory schedulers then took the next month's plan and broke it out by product by week, trying to keep each high runner produced weekly and keep each week evenly loaded by capacity—again measured by "activity."

- The factory schedulers then broke out the current week into a daily start schedule based on batching or setup rules, starting from high- to low-width boards for wave solder and trying to level the load daily.

- At the start of each shift, one-third of the jobs were released into the factory.

- Jobs were dispatched through operations by a critical ratio order, which was nearly first-in–first-out (FIFO), except when overridden by "red tag" jobs being expedited or by a supervisor trying to maximize use of a major setup at an operation (at insertion or testing, for example).

- The rework line was run autonomously without any specific scheduling approach. The operators took the boxes of work on a nearly FIFO basis, grouping similar products whenever possible.

While this procedure had been followed for years, lately their customers had started to complain about late and partial shipments. One of their largest external customers had recently given them an ultimatum—get their customer service up to 99% or be dropped as a qualified and preferred supplier. Their quality was fine, the customer had explained, but their customer service, measured as shipment to first commit date, was under 75%. If PCB Co. could attain a 99% customer service level, then they would be back to talk about price and quantity commitments; but they could not be a supplier to a JIT-based customer without near-perfect delivery and quality.

The way things currently ran, this customer's production lines would be stopped 25% of the time waiting for parts. So in six months, when the customer went to full JIT, their vendors either met JIT criteria or would be dropped.

Joe had already been concerned about their customer service performance before the "corporate fiasco," as he thought of it. While this customer was not of major financial importance, Joe felt that they were simply the first of what would be many more, trying to go JIT and needing a "JIT" supplier.

He thought he knew what to do for the high runners. He would have to dramatically increase the safety stock they held in finished goods inventory. But he wasn't sure exactly what to do about the low runners and custom products—use a longer planned leadtime than actual? But this would make customers order sooner than they wanted to, especially if they wanted to order in line with their actual demand and not forecasts made months earlier.

Joe knew what to do. He would form a team of his schedulers, production, planners, and perhaps even an engineer and empower them to start working on fixing the problem. And why not shoot high? He would set a stretch goal of 99% customer service and hope the team could figure out how to meet it.

Forming a Team—Measures at Odds

The team's first step was to get organized. They understood their charter—get customer service up to 99%—and Joe's first thoughts on how to accomplish it was by increasing the safety stock of high runners and the "safety lead time" of low runners and custom orders. However, as they brainstormed on what the barriers were to a 99% service level, or alternatively, the root causes of poor customer service, several other issues came up (Table 3.1).

Production claimed that often orders were accepted and shipment dates committed to that were less than the agreed upon leadtime. Other times custom orders were accepted before the design was completed or piece parts ordered so that by the time the order was "manufacturable" (work instructions and piece parts available), the order was already late or started well past the scheduled start date.

TABLE 3.1 Common Root Causes of Missed Schedules—Brainstorming

Customer Service Problem	Root Cause	Responsibility
Late orders	Orders accepted at less than the agreed upon leadtime	Sales
Late orders	Custom orders accepted and committed to without design specifications and parts availability	Sales coordinator with engineering/purchasing
Late orders/short quantity	Insufficient parts to complete the schedule	Purchasing
Late orders	Manufacturing worked on high runners and ignored low runners/custom jobs	Ongoing
Late engineering experiments	Manufacturing ignored engineering experiments	Ongoing
Short quantity	Operator errors	
Short quantity	Process variation causing scrap and rework	Engineering processes and designs
Late order	Equipment breakdowns	Engineering

Production, therefore, saw the majority of the problem as being poor scheduling by the corporate order takers (corporate sales). In addition, from time to time, manufacturing ran out of key parts and could not complete orders. This was clearly purchasing's fault.

Scheduling saw it differently. They had the feeling that the manufacturing supervisors tended to work on the high runners since their first measure of performance was "operator efficiency" (as measured by activity—the number of boards processed daily). A supervisor was graded first on activity, then on quality, and finally on customer service. Therefore, high runners, which did not require setup changeovers and had efficiencies of production, tended to be run first and foremost. Low runners and custom orders were "fit in" when idle periods occurred or schedulers started to put on the heat—when those jobs showed up on their daily batch status reports as static (not moving) or already late. Scheduling was not in charge of piece part ordering so they had no idea that some of the low runners or custom jobs were not run because the work instructions and/or piece parts weren't yet available.

Engineering had a more narrow view of the scheduling problem. They were not sure why they were even on this task force, but as long as they were, their concern was that their engineering runs or experiments seemed to have the lowest priority (when clearly they should have the highest). And also, the operators often made mistakes when carrying out the experiments so the engineers had to personally supervise them. As to the level of outside customer service, they couldn't really speak, but as an internal customer, they were also very dissatisfied.

This set production's teeth on edge. The production manager shot back that one of the other reasons they missed customer shipments was engineering—that they continually fiddled with the process, adding new operations or procedures that took time to train the operators on, specifying equipment that was unreliable, taking jobs off the factory floor to examine without filling out paperwork, and strutting around as if they were better than the operators. If they only designed processes that worked, there wouldn't be so much rework and scrap, was his parting shot.

At this moment Joe called time out. The conversation was deteriorating into parochial departmental concerns. No one was willing to see the big picture and drop their own narrower issues. On top of that, other groups were being blamed that were not even there—sales and parts purchasing. He had never realized how provincial each department could be and how much they seemed to blame the other departments for any problems. Could they have a whole factory of ostriches? How was he ever going to build a team to deal with cross-departmental issues? The urgency of the customer service problem called for a leadership he realized he had rarely shown. As a company, the management team had let each department move forward with small evolutionary changes: some new equipment, some modest training, or some new computers. They had not undertaken a major revolutionary project, change to their standard operating procedure, or innovation (a change to their way of *thinking* or *assumptions*) in years. Now they had to. *He* had to.

He stood up. "Clearly, one of the problems is that each of you only feels responsible for a piece of this corporate problem. We've measured each of you on different grounds, and so you are each relating to your own departmental or personal view, responsibility, or concern about scheduling. But what this company needs is a solution focused on our customers. For this project, I'm going to add another group measurement to each of your individual measurements. For the next four months, the management team is going to measure you all identically by our level of customer service. You're a team now, whether you want to be one or not. You all have a common goal—to make this *company* successful. I'm going to add a representative from sales and one from purchasing. From now on I want to hear about solving the *company's* problem, not about your own issues unless they coincide."

Joe was frustrated by this internal bickering given the urgency of the task. Literally, their jobs were at stake. However, he understood that they had let each department run as an independent group for so long that they did not see themselves as a team or a company—they saw themselves as autonomous departments. It was really their fault for creating these quasi-independent fiefdoms with their own budgets, terminology, culture, measures of performance, and systems that only met at biannual planning meetings.

He was going to have to deal with the situation they had created, but first he needed to get this team formed and moving.

Forming a Team—A Unified Goal

Before Joe left the meeting, he made two more suggestions: that they meet with the quality team to see how they were planning to run their team, given their TQM project experience, and that they get whatever education they needed or even consulting if that would help. The team was strangely quiet after Joe's departure. The truth was that now they were on the hook for achieving a 99% customer service level instead of explaining how someone else made that an impossible goal to attain. They really weren't sure why they could not achieve it; they just knew they never had. If it were that easy to attain, they assumed that they would have already done it.

They had already kicked around reasons why they did not achieve 99% customer service. They decided that maybe seeing what the quality group was doing was a good idea. They would get the two representatives from sales and purchasing on the team, then ask the quality team where they were and how they would suggest getting started.

By the next week, they had their second meeting as a full team with the quality team representative. He explained that they were working away and his team believed that their goal was possible. He outlined his suggestions for the scheduling team based on their experience to date as follows:

1. Get the education on best practices needed to achieve their goal and utilize visits to customers, vendors, or other companies who had adopted these practices and achieved this target and were willing to discuss how they had. Consultants were also useful, preferably if they could find someone who had worked in their industry. But, for sure, get outside the company's walls to see new ideas, thoughts, and approaches.

2. Be open to new ideas, principles, and suggestions, no matter what their source. Forget old biases, relationships, and prejudices. The solution was going to require changes and assistance from all the departments. Work on team building, spending work and social time together. Respect individual expertise and rotate team leadership. Get help on communication/listening/team work and problem-solving skills.

3. Develop a measurement scheme that would explain the real causes of missing a schedule and lead to both improvement and, better yet, prevention. He explained how they could detect a process going out of statistical process control *before* it resulted in scrap as an example of measurement for prevention. He went through the principles of measurement.

4. Develop a project plan so that they had a clear *joint* vision of how to proceed and could measure their progress toward their goal. The plan should include design of their new performance measures, incremental improvement targets along the way, and key tasks. For example, their project plan had the steps of:

   - Intermediate quality goals
   - Education for the team on best practices
   - Other plant visits
   - Prevention methods through Poka-Yoke techniques
   - New measures of quality added at operations
   - SQC routines installed
   - Training on design of experiments

   He explained that the project plan would be difficult to develop until the education for the team was completed. In fact, that was an ideal time to develop the plan, aided by the instructor, and then to review it with any other plant sites they visited for suggestions, cautions, and other feedback from their experience. The project plan was the first step toward team problem solving.

5. Meet weekly to work on reviewing where they were on schedule performance and eliminating the key causes of missed schedules.

6. If possible, move their offices or desks so that they were physically contiguous. This would lead to much more dialogue and daily joint problem solving than just the more formal weekly meetings.

7. Get in the mindset of best practices, continuous improvement, and change—that this was not a one-time, four-month effort but a real change in the way they had been doing business. Otherwise, he related, whatever success you have will soon be lost as the causes of schedule misses change, people get lazy again, or exceptions are made for certain orders.

When he left the meeting, the team was still very quiet. There was no real leadership on the team—Joe had not left anyone in charge purposely or by omission (they weren't sure). Everyone waited for someone else to speak. Finally, someone suggested they go around the room and see what everyone thought about this project. The first person said that he wasn't sure why he was on the team since it was scheduling's job to make up the schedule and he simply worked on the jobs he was told to. The next person was more vocal. "This is a fad," he said. "We'll work on it for a few weeks, make some suggestions, and then go back to work as usual. It's all a tempest in a teapot because the plant manager has a bee in his bonnet over some missed job." The third person picked up the same theme. "All management cares about is activity; customer service is a joke. This factory gets graded on hitting a dollar revenue target out the door. If we meet our revenue targets, we are profitable and everyone's happy. That's the way it is and that's the way it's always been."

The fourth was even more negative. "Joe just wants a promotion. That's why he's doing all this. He'll look like a hero, get out of here, and we'll be left with more work and more pressure. No one mentioned any raises in it for us, did they?"

The next guy disagreed. "Joe's not worried about a promotion; he's worried about being fired. That's why he's running scared like this. He's under pressure, and he's just passing it on to us."

This negative tone hung over the group. Finally, one of the operators had enough. "Look," he said, "why don't we see what we can do? Maybe it will make our lives easier. The quality team isn't griping. I keep reading about plant closings. Maybe it's our jobs at risk. Why don't we ask Joe to explain why we're doing this all of a sudden. We should have asked him ourselves before he left."

Forming the Team—A Unified Shared Goal

After the meeting, Joe saw one of the operators on the team in the hallway. "How did it go after I left?" he asked. The operator hesitated. Joe could see his perplexed expression. "Come in my office for a minute and let's talk about it."

Once seated, the operator recounted the way the meeting had ended. Clearly, the team hadn't understood the real urgency of the problem or felt any responsibility for fixing it. But then, he thought, why would they when he had not shared that urgency with them. He realized that changing the organization's culture was going to be a real challenge. He was asking them to ignore all the messages they had gotten since the first day on the job—that they were simply responsible for carrying out orders, that each department worked on its own, that they only needed to know enough to do their job, and that activity and dollar shipments were their most important goals. He began to realize that changing the organization was itself going to be a slower continuous process and not an immediate accomplishment. His first task was to call a meeting of the team and carefully explain why the change now without sending everyone into a panic or back into the job market. The key was to have them see what he was beginning to realize—that the world had changed around them over the last twenty years. World-class standards had been dramatically increased in quality, cost, customer service, and leadtimes. They competed not just with their "old competitors" but with new ones in countries all over the world with various advantages in labor costs, cost of capital, access to labor, protected markets, government-supported R&D, and subsidies. Customers had new expectations and needs. Their only job security was through being competitive, and management had to help them achieve that goal.

He realized that he may have given the team a unified goal but not a reason to attain it, to change. At the next meeting he would do that, and perhaps get a speaker from an outside company who had already done this. He wanted someone who would be viewed as neutral to talk about what had happened in their own company, competitively, and how they had responded. But he had to tread a very fine line between spooking the team and motivating them. And personally he was also caught between those feelings. He was also bouncing back and forth between feeling very motivated and very spooked.

A Team Forms—Competitively

At the next meeting Joe did his tightrope walk. He had met with Mark Ritchards and his peers to agree on their statement. What they decided to say was that they had a significant opportunity for expansion, but only if the division could meet some very aggressive targets for performance in all areas—quality, customer service, productivity, costs, speed, and so on. If they couldn't meet those targets, they'd be creating a competitor—another manufacturing source for Diversified. What was happening at PCB Co. was the same thing that was happening or had happened to every manufacturer worldwide: competition and customer choice. They no longer had Diversified's business by default; they had to earn it. More and more, customers were demanding extraordinary (in PCB's eyes) levels of customer service.

He had also lined up an outside speaker who was a year ahead of them on re-engineering their customer service.

The outside speaker went through how they had, one by one, addressed the issues hurting their customer service. Many of them, he related, had very little to do with scheduling but had related to quality of the product (in scrap and rework), the equipment (in breakdowns), the operators (in absenteeism or errors), or the work instructions (in frequent revisions that needed new parts or were hard to understand). That was why over time the scheduling team had become more of a quality team. Once the uncertainty and quality problems had been eliminated, scheduling had become much easier. Uncertainty was the key enemy of schedule performance.

Then he told them that he knew they could do it because one of PCB Co.'s major competitors had already done it. They had a customer service level of over 95%. That was why they used them as a supplier and not PCB Co.

The team was silent at this last statement. After the speaker left, Joe faced the team. "I didn't explain very well in our first meeting why we were doing this. But now I hope I have. He's just another example of the reason. And he's right. We *can* do it, and *better*."

The team was still silent. Finally, the operator who had claimed this was a fad in the first meeting said, "If they can do 95%, we can do better."

There was a lot of head nodding and silent agreement. Another operator, the one who had claimed revenue was the only goal, stood up. "But if we start this, we do it to win—no jerking us around. You heard what he said. A lot of this is not scheduling. We do what we have to." He wasn't addressing the group but Joe.

Joe nodded back. "We do what we have to and more."

Taking the Full Measure of the Factory: Scheduling for Customer Service/ Supply-Chain Management

The team's next step was a review of current performance to schedule (Table 3.2). They were measured on performance to volume (total number of boards shipped versus scheduled), performance to mix (percentage of orders shipped on schedule within 95% of order quantity), and leadtime (average leadtime of jobs shipped that week as measured by ship date minus start date). This was a typical week—somehow volume was attained but not in the right mix. The factory tended to work on larger, often easier, runs of high-volume product and ignore smaller jobs with difficult setups. This was due to a "hierarchy" of rewards. Not meeting volume shipment targets could lead to fairly stiff management review.

Top management graded the plant on "contribution"—revenue value of the output minus "standard cost of revenue." Missing output targets meant missing contribution targets and retribution without absolution. Missing mix was not nearly as serious (as seen by the results), especially if customer complaints did not reach top management. Leadtime was rarely discussed except at biannual financial reviews when finance made its expected "inventories are too high" speech that everyone ignored. As a result,

TABLE 3.2 Current Performance-to-Schedule Measurements

Performance to volume	101%
Performance to mix	74%
Average leadtime	6.2 weeks

these results were as predictable as mix was not. Every week volume was met, even if performance to mix had to dip even lower to do so.

The group now sat down to re-engineer their measures based on the feedback from the quality group. They wanted measures that were absolute, not relative; focused on root causes or explanations for problems, not just performance; examined variance and not just average performance; measured competitive performance and trends; and could be used for prevention and not just detection.

One of the immediate things that struck the team was that their current measures did not explain *why* the schedule was missed. When an individual order did not meet its schedule, was it due to being late or scrap losses or rework delays? When it was late, was it due to equipment breakdowns, missing parts, missing tools, or waiting for an ECN or specification? From looking at these results, they had no data to even begin improving schedule performance. So the team agreed that they first needed to change their measurement systems to give the reason for missed schedules. Similarly, the volume figure did not show how many orders were early or above the needed quantity. While this made the factory "look good," it did not really accomplish their supply chain goal: 99% on-time delivery to customer order date and quantity.

Now that they had some direction, the suggestions came faster. The performance measures did not show how leadtime varied by job; it only showed the average. The leadtime was shown as an achieved level and was not compared to the absolute performance possible (or broken down by category of leadtime). These measures did not show whether performance was getting better or worse over time or how it compared to competitors or world-class performance.

Two thoughts were brought up to which no real resolution was achieved. First someone looked at the goal of measuring factory performance. "Does that mean," he asked, "that we should measure *other* groups' schedule performance—that we should see if maintenance does their preventive maintenances on time, how long engineering takes to deliver a custom work instruction, or if parts are delivered on time?"

This seemed outside their scope of supply-chain customer service, but they couldn't quite drop the idea completely. It seemed very relevant to measure the customer service seen by *internal* customers, namely production, but how that related to final customer service wasn't immediately obvious. They should be related, but how missing a preventive maintenance affected customer schedule delivery was not clear to anyone. They decided to table it but leave it on their "more thought required" list.

The second thought was more provocative. Someone said, "I was struck by what quality did—moving their quality measurements to the operation and then using them for *prevention* by detecting a probable trend instead of detection of product already ruined. Isn't there some similar approach or analogy for scheduling?" This was such a radical change from how they measured schedule performance that everyone was silent for a moment. How could they use schedule performance measures for prevention and not simply after-the-fact measurement? Every job had a final due date, and they measured whether it was achieved or not. One of the schedulers said, "I do look at whether the jobs are running ahead or behind schedule based on a critical ratio of lead time left versus days until due. I suppose if we wanted to go to the trouble, we could put a due date on the job at each major operation."

He said this cautiously to make sure no one accused him of being biased in focusing solely on his department's interests. Someone said, "That seems like a lot of work. What good would that do us?"

Another operator ventured, "That's what the quality team did—bring the measures back to each operation." People nodded at that. It did make sense.

The scheduler gained some authority and confidence from the support. "If we developed a schedule for the job for each key operation, we could measure schedule performance and leadtime by operation. And if a job were running late, we would recognize that sooner and we could expedite it."

"But could we *prevent* jobs from being late?" The original operator who had raised the issue repeated his question. This stumped the group, so they also tabled it under "more thought required."

Meanwhile, they were pleased with their progress. They had redesigned their schedule performance reports (Table 3.3) (Fig. 3.1). Next up was a series of training classes with a scheduling expert the outside speaker had recommended.

As they looked at the new measurements they had suggested in a first attempt, another scheduler raised two questions. "Where are we ever going

TABLE 3.3 Redesigned Schedule Performance Report—Format Example

Performance to Mix: 74%	
• Jobs late (by major cause)	21% total
• Rework delay	3%
• Waiting for ECN/specification	1%
• Missing parts delay	8%
• On engineering hold	—
• Promised due date less than leadtime	6%
• Equipment breakdown	2%
• Other	1%
	21%
• Jobs with less than 95% of quantity ordered	5% total
• Missing parts	4%
• Scrap	1%
• Rework	—
• On hold	—
• Other	—
	5%
Performance to Volume 101%	
• Volume delivered early	6%
1–2 days	3%
3–5 days	3%
>5 days	—
• Volume delivered over quantity ordered	3%
Leadtime Components	
Processing time	16 hours
Burn-in time	48 hours
Setup time	8 hours
In-transit time	1 hour
Queue time	4.33 weeks
On-hold—no parts	16 hours
On-hold	7 hours
Leadtime Distribution by Number of Jobs	
<5.0 weeks	2 jobs
5.0–5.4	20
5.5–5.9	50
6.0–6.5	58
6.5–7.0	7
>7.0 weeks	13
	150 jobs

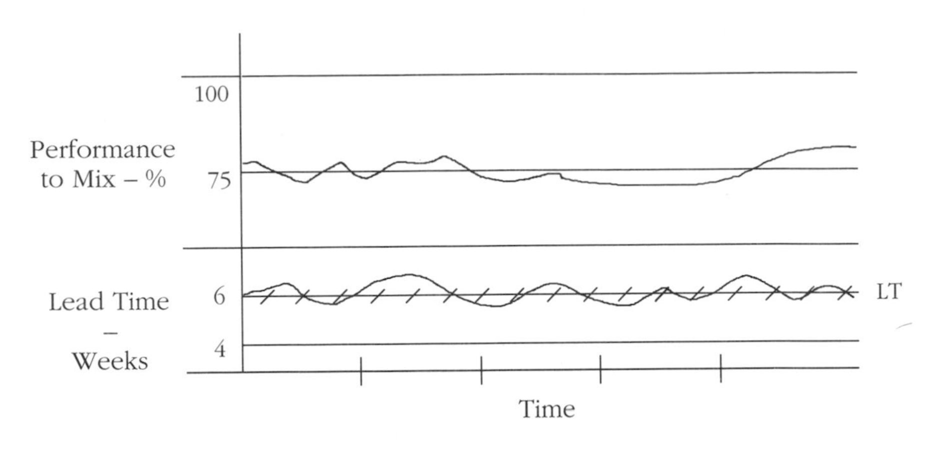

Figure 3.1 Performance to Mix and Leadtime over Time

to get the data to fill these out?" he wondered. "We don't collect most of this data today." The second question was confusing. The scheduler pointed to the performance to mix, to the leadtime component categories, and the leadtime variance and asked, "Aren't these three related? Do we need all three or do they tell us the same thing? Somehow they must be related."

By now the group had its fill of unanswered questions. They were looking forward to asking the instructor about all of them.

Determining What Is a "Good" Schedule

After the instructor had introduced himself, the class brought up their four "more thought required" questions and showed him their measurements proposal. They were very disappointed when he said, "You're putting the cart before the horse! You can't measure scheduling performance until you tell me what's *important* to your company. In other words, how do you *know* when you have a good schedule?" He went on to explain.

"In examining a schedule from a company or even a team standpoint, there are many, possibly conflicting, views of what constitutes a good schedule. Each member of the team is often judged by a measure of performance that is *directly* impacted by the scheduling of the jobs through the facility (Table 3.4). As a result, scheduling on the floor is often impacted by the 'informal' system as individuals exert pressure to adjust the schedule to

TABLE 3.4 Measures of Scheduling Quality as Seen by Individual Departments at PCB Co.

Department	Measure of the Schedule's Quality
Quality	Processing of jobs with/on "best qualified" resource
Customer service	Number of jobs completed on schedule
Production	Labor efficiency attained
Engineering	Number of experiments run
	Turnaround time on experiments
Tooling	Number of setups requiring tool changeovers
Finance	Inventory levels
	Equipment utilization
	Overtime required
Top management	Contribution ($)
	Number of customer complaints received
Facilities	Excess storage required
Customer	Leadtime
	Delivery to customer order date
Purchasing	Expediting of piece parts required
Operator	Ease of processing the jobs assigned
Supervisor	Number of complaints to be mediated on the shop floor
Maintenance	Number of conflicts over scheduled maintenance dates vs. production
Scheduling	Number of late jobs
The company	????

their *own* needs. Unfortunately, during this process, what is often lost is the *company's* goals for scheduling. What does this mean?"

Companies have various strategies for competing in their markets. These strategies can include being or having the: lowest-cost producer, most advanced product capabilities, highest quality, fastest turnaround, or highest customer service. Each of these strategies can be supported or negatively affected by the scheduling choices made. For example, if the strategy is to compete on being lowest-cost supplier, then the schedule needs to attain high levels of equipment and labor utilization at "bottlenecks" and require minimum levels of overtime, subcontracting, and inventory. If the strategy is to compete on advanced product capabilities, then the schedule needs to support rapid engineering experiments and prototyping as needed. If the strategy is to compete on highest customer service, then the schedule needs to maximize on-time delivery and flexibility of accepting orders.

Unfortunately, in all companies, most of these goals are important and usually conflict in practice. Low-cost producers who still have significant setup costs may opt for long runs and few changeovers, increasing leadtime variance and causing small orders to be late. Trying to raise the level of *customer service* may impact many of the other "hidden" measures of schedule performance and create resistance to the changes *needed* for attaining high levels of customer service. The point is that scheduling is a *team* concern and needs *team* commitment. Reaching high levels of customer service is not simply the responsibility of production scheduling or production. If the other members of the team have no incentive to reach a 99% customer service lead and, worse yet, suffer for doing so, it will be nearly impossible to attain on a consistent basis.

The organizational diversity of measures causes a major issue. It's hard for several people to evaluate a schedule and *agree* on whether it's a good schedule or not because their points of view differ. The informal system often takes care of this problem. As specific problems arise, such as a conflict between the preventive maintenance schedule and the production schedule, the customer due date and the current setup, or the engineering experiment and production jobs, they are negotiated between the "opposing" parties in real time. Any scheduling technique employed will have to deal with this reality. Therefore, it is difficult to fully evaluate a *scheduling technique* unless you can first agree on the measures of performance or the trade-offs to be considered in scheduling. Just saying that the goal of scheduling is 100% customer service is naive in most companies. You must agree as a *team* on your schedule priorities.

What Makes a Schedule Hard to Attain?

To understand how to attain the current focus on a 99% customer service level (especially as a company moves to *daily* schedules from the more common monthly or weekly schedules), we need to explore what makes a schedule hard to meet. There are two major causes and several minor ones of missed schedules.

The biggest cause of schedule performance problems is *uncertainty* or variability. If we could plan a schedule in detail and then be assured of its successful completion, then no matter how laborious the planning effort, scheduling would be relatively easy. We could argue about the schedule, but once it was agreed to we would simply execute it. However, in most factories a number of unplanned events soon make it obsolete. Common problems include equipment breakdowns, scrap, rework, forecast errors or rush (unplanned) orders, operator absenteeism, late piece part orders, process "excursions," raw material variation, human error, tool breakage, late work instructions (for custom orders), and new ECNs. While minor variations can be ignored, major changes in product quality (scrap and rework), resource availability and productivity, or demand require adjustment of the schedule. Alternatively, the schedule must already reflect this "inherent" uncertainty in the start quantity (increased to allow for scrap and/or rework), the start date (pulled earlier to allow for lead time variance), the amount of resources required (increased to allow for lower productivity and/or scrap and rework), and forecast demand (buffered by finished good inventories or higher starts to allow for forecast error variance).

In other words, we either take this existing uncertainty into account or miss our schedules.

The second major cause of scheduling difficulty or complexity is shared and/or constrained resources—equipment, tools, labor, parts, facilities, or capital. When jobs share setups or can be batched into a fixed equipment run size, they must be planned simultaneously. Similarly, when a resource is constrained, we must simultaneously schedule all jobs using that resource to be sure of a capacity-feasible schedule. This greatly increases scheduling complexity since we can't treat each job independently; we must fit them together as we do a jigsaw puzzle to make sure they all fit together *and* make their promised due dates. This introduces a concept of scheduling "complexity." Certainly, to meet high levels of customer service, we must keep our scheduled load on the facility within its capacity constraints.

Other more minor causes of scheduling difficulty are the number of products to be scheduled, the length of the leadtime, the number of parts used in assembly, the number of routings potentially used, the number of different labor classes, and the number of different tools. These cause "combinatoric" scheduling complexity and are a function of process, product or facility *design* complexity. The greater the number of unique parts used, the greater the number of individual raw materials needed to be checked for availability and reordered if needed, the greater the scheduling effort. Every time we add a new process, part, tool, or equipment, we add to our scheduling task.

When we have uncertainty and complexity (in shared and/or constrained resources or numbers), the harder it becomes to quickly *evaluate* the quality of a schedule or scheduling technique. Uncertainty means the schedule is rarely accurate for any length of time and soon needs to be replaced or adjusted. Who can tell if one schedule was better than another under those circumstances? With complexity, there are so many possible schedules and trade-offs, again it's hard to evaluate the quality of one over another. Taken together, as seen in so many American factories in the 1970s and 1980s, scheduling became a nightmare that soon deteriorated into long leadtimes and massive expediting, along with poor customer service.

Best-Practice Principles of Scheduling for Customer Service/Supply-Chain Management

From our review of what makes a schedule hard to attain, as well as your list of root causes of missed schedules (Table 3.1), we can construct our best-practices principles for achieving any level of customer service in supply-chain management:

1. The schedule must be resource feasible in plan and during execution in raw materials or parts, equipment capacity, labor capacity, storage requirements, facilities, or any other relevant resources utilized.
2. The schedule must be leadtime feasible such as when ordered parts, engineering specifications and other dependent outside activities are available, and when orders are committed.
3. The schedule must consider leadtime variance (itself caused by other variances such as rework, equipment downtime, etc.).
4. The schedule must consider product yield variance (the quantity actually produced after all yield losses).
5. The schedule should allow the user to trade off capacity utilization, customer service, cost and leadtime/speed (work-in-process levels) in prioritizing each or setting minimal or maximal levels for each (cost includes use of most productive equipment, materials, setups, etc.).
6. The schedule must use accurate and up-to-date information on parts availability, equipment capacity and production rates, yields, routings equipment and labor status, and so on.

7. The schedule must be followed!

8. The scheduler should notify the user (predict) when the schedule can no longer be met due to changes in factory conditions (equipment breakdowns, yields below normal ranges, operators' absence, etc.).

9. The scheduling technique should be understandable to and modifiable by the user. This means either training the user in the disciplines needed to understand how a schedule is generated (such as the appropriate operations research concepts) or having a technique that "explains" how it generated a schedule (such as the rule-based tree structures of artificial intelligence systems). Users should be able to override the schedule generated, if needed, and generate a new one.

Best Practices in Scheduling
Principle 1
Resource Feasibility

Most factories have some key bottleneck operations or resources. These are operations that determine the maximum output of the facility. We can think of them as the "narrowest" pipes through which the output of the factory flows. Their output rate limits or sets the maximum factory output rate. These operations are typically bottlenecks because they either require expensive equipment that cannot be purchased in small increments, and/or utilize expensive equipment with high uncertainty in operating uptime, and/or require key skilled labor in short supply, or so on. In each case, the bottleneck utilizes some resource that cannot be economically increased to eliminate the bottleneck.

The basic rule of 100% customer service scheduling is that these bottlenecks limit the output of our facility. Therefore, most of our scheduling effort must be directed at ensuring that we have planned them within their capacity and that they actually process at least the amount scheduled that period (shift, day, week, or month). If we miss the schedule at that bottleneck, it will be nearly impossible to meet the facility schedule without overtime, subcontracting, or postponing scheduled downtime for maintenance, experiments, and other nonproduction needs.

Operationally, this first principle is addressed in three ways: best practices require that our production planning or release rules must *release* an amount of work each period (as needed by customers). Second, the

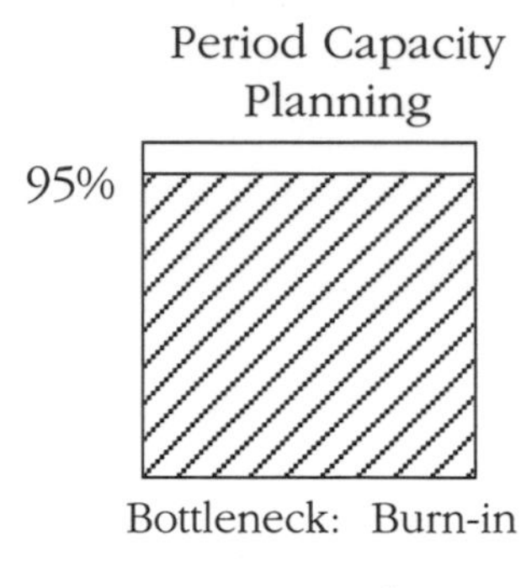

Must:

Keep the bottleneck fully loaded (within capacity) in the period plan.

Period
Dispatching

Keep jobs flowing to the bottleneck in an even load during the period to meet the required utilization.

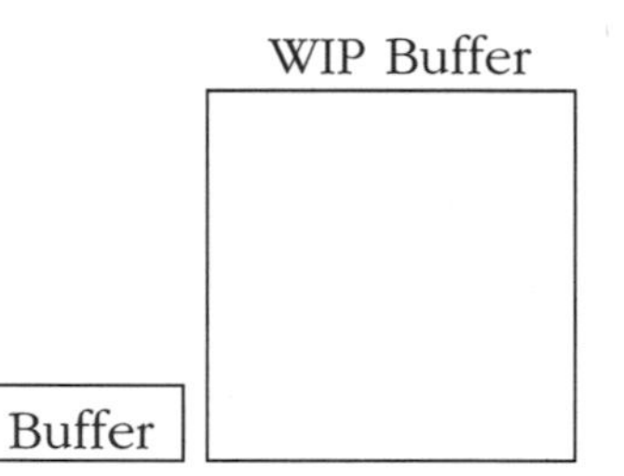

Plan a buffer to accommodate variation in the incoming job stream/rate to the bottleneck to keep it at the required utilization. Prioritize repairs at bottlenecks as top priority.

Figure 3.2 Best Practices in Meeting Customer Service—Resource Feasibility

detailed scheduling or dispatching rule must keep the released work smoothly flowing to the bottleneck during the period so that it does not run out of work. Third, the work-in-process level held as a buffer in front of the bottleneck must be sufficient to keep the bottleneck occupied given the uncertainty in the operations *feeding it* (in equipment downtime, product quality variation, jobs placed on hold or scrapped, operator productivity variation, etc.) (see Fig. 3.2).

To explain this last point in more detail, as we've said, most schedules simply ignore variation in product quality, leadtime, and equipment uptime. We'll discuss how to handle leadtime and quality variation later. But let's look at the effect of equipment uptime variation.

If we plan a steady stream of work to burn-in (which is a bottleneck), we can load it to near capacity. We've planned in that number for the average or expected planned and unplanned downtime (maintenances, breakdowns and repair, setups, cleaning, calibration, etc.) at burn-in. But, what happens if wave solder unexpectedly is shut down by a problem, such as contamination. Then eventually burn-in would be "starved" and would miss its required processing schedule. So, interestingly, we need to consider the variance in uptime of the *feeding* operations or stations to our bottlenecks in our schedule. We do this by setting a "buffer" of work—a WIP level—in front of the bottlenecks that buffers us against unexpected variation in output (typically breakdowns) at the upstream feeding operations (Fig. 3.3). The calculation of the required buffer WIP is somewhat complex because it must take into account the distribution of the downtime or varia-

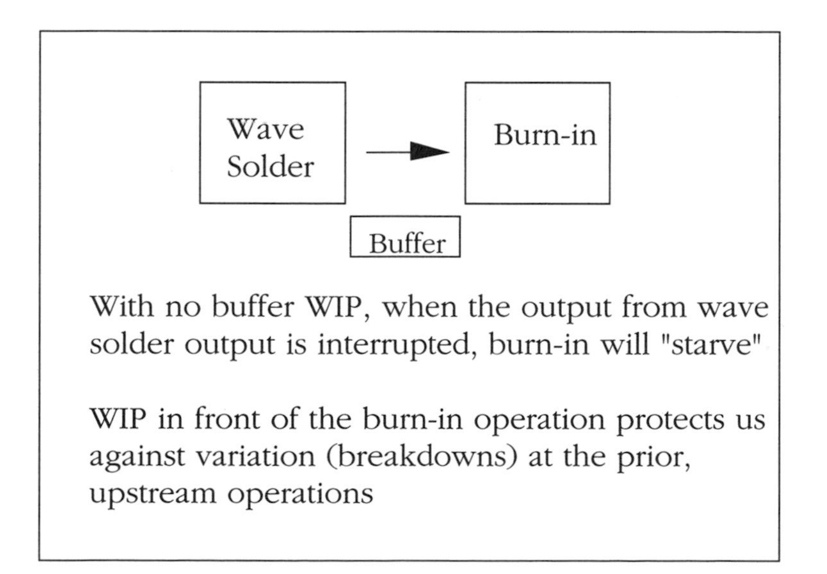

Figure 3.3 Use of Buffer WIP to Protect the Bottleneck from "Starvation"

tion at the feeding operation(s) and the length of time the bottleneck can be idle given the schedule. There are MES vendors who have developed systems to calculate the minimum required buffer inventory to assure high utilization. What this also points to is the effect of equipment downtime distributions on schedule performance. To improve schedule performance, we need to either buffer ourselves against equipment variation or improve that variation. (Hopefully the latter.)

Use of Buffer WIP to Protect the Bottleneck from "Starvation"

Focusing on the bottleneck as the key to meeting customer service also means setting our repair priorities correctly. Realistically, it means putting highest priority on dispatching repair personnel to the bottlenecks. Similarly, it would mean focusing our equipment improvement programs on bottlenecks as well.

If the bottleneck is overloaded in our production plan, the plan cannot be met without overtime. If the bottleneck loading is planned correctly but not kept continually utilized to the level planned (by the detailed schedule) the result is the same—we either utilize overtime or miss our schedule.

This brings us to another key problem in scheduling: What *is* the capacity of the bottleneck? Most companies have a "standard" they use in scheduling equipment (or labor), which has a factor for planned and unplanned downtime, idle time, and productivity variation (production rate). These standards are updated yearly at most and are often historical in nature based on either an industrial engineering study or measured actual output. However, these standards are often self-perpetuating. We plan for them, and that's all we achieve. All best practices in scheduling today focus not only on scheduling rules but also on the factors that impact schedule performance such as variation and lost capacity. As I've heard from Joe, you've already looked at *actual* equipment utilization and seen that on a 7-day week, 24-hour day basis, it's under 50%. This has enormous implications for improving schedule performance in terms of both added capacity and flexibility—at no capital cost! We need to monitor the actual (7 × 24) value-added utilization at our bottlenecks and see if our scheduling rules have caused capacity losses due to starvation caused by the feeding operations, poor release, or dispatch rules (no work at the bottleneck with feeding operations up and available).

These points are true for any resource constraint—equipment, tooling, labor, facilities, capital, materials, etc. (Note: In planning feasible parts availability, the issues are more straightforward. All schedules must be feasible given the availability of required parts on hand and on order. Substitution of parts may be allowed.)

Best Practices in Scheduling
Principle 2
Leadtime Feasibility

A key issue in scheduling performance is ensuring leadtime feasibility—not committing to order due dates without sufficient leadtime for manufacturing. This situation typically arises when manufacturing has dependencies that have *not* been considered when the order was accepted. These dependencies may be developing process plans or specifications (for custom orders). They may involve checking on parts availability, special tooling or test jigs, or test program development. They may involve a credit hold (that the salesperson may not have been aware of when quoting the commit date) that delays release of the order.

These are examples of the situations that can cause orders to actually be due before they are even released to manufacturing. To prevent this from occurring, we need to either not consider an order as available for scheduling until we have a confirmed earliest start date possible, based on a checklist of needed parts, specifications, tooling, etc., as well as any administrative tasks (eg, credit hold) needed to be completed before manufacture can begin or add these steps as operations in the schedule with estimated times for their completion.

Best Practices in Scheduling
Principle 3
Leadtime Variance

A resource- and leadtime-feasible schedule, however, still does not assure 100% customer service. The next consideration is leadtime variance. If the average leadtime through the facility is 6.2 weeks and we plan our production using it, we may only see a 50% customer service level (see Fig. 3.4)!

If we assume a nearly symmetric distribution of leadtime, then half our jobs will take *longer* than 6.2 weeks (since 6.2 weeks is the mean or aver-

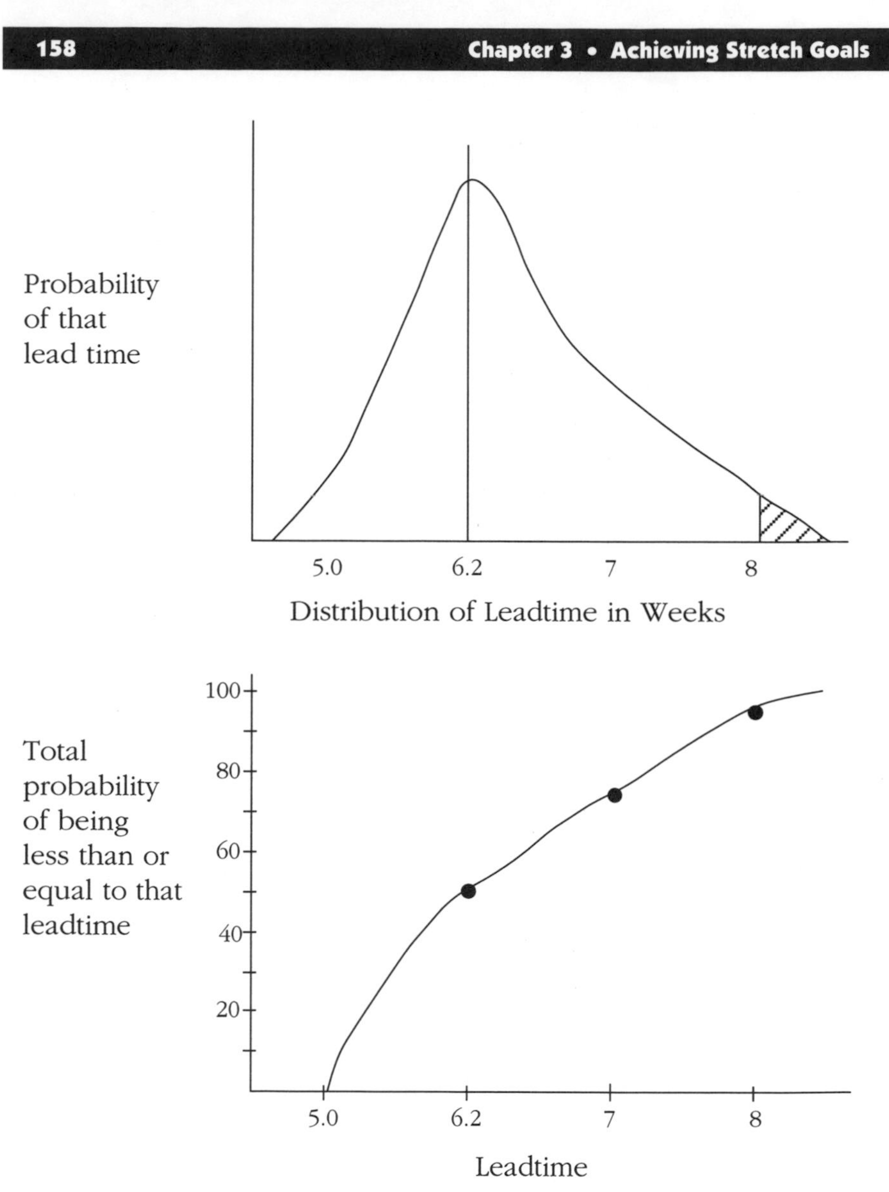

Figure 3.4 Leadtime Distributions

age) and be late! If we want a 95% customer service level, we need to plan job release using a leadtime we fall within 95% of the time, or in this case, 8 weeks. This is straightforward calculation and reasoning. However, most companies tend to ignore the *variation* in their key measures of performance, and instead, focus on averages and typically don't even measure the variance. However, as we see in customer service, it is uncertainty or variation we must "plan" for. This is another reason we see so much expediting in our factories as schedulers try to recover from a schedule that was unrealistic from day one! We plan to use an average leadtime but there is no average job—only those that make up the distribution!

Best Practices in Scheduling
Principle 4
Quality Variance

The next consideration in meeting high levels of customer service is product quality variance. Let's look at the problem first for custom or made-to-order (low-runner) products where we do not hold finished good stocks. Again, as for leadtimes, planners usually assume an average yield (or scrap) rate for the product. Let's assume our average yield for these multilevel boards is 92%. If we plan the start quantities accordingly (the customer order quantity required divided by .92), then again, on average, we may only see a 50% customer service level! The reasoning is the same as before for leadtime variance—that half the time the yield may be below .92 and we do not produce the quantity ordered. What we must worry about or anticipate and therefore schedule for is not the *average* yield but the *job-to-job* yield variation—the variance of the yield. As shown in Figure 3.5, 50% of the time your yield is less than .92. Or, as shown, to meet a 95% service level you need to plan the required start quantity using a yield you *exceed* 95% of the time (in this case .80). Of course, when you use .80 as a planning yield, you *exceed* that yield nearly 95% of the time so that you are continually overproducing most orders. This causes large excess inventories of finished goods that may, in fact, need to be written off as obsolete product (unless a customer can be convinced to accept the overage or the product can be used as a substitute for another more frequently ordered product).

We can, in fact, calculate the relationship between customer service and the expected overage for any given yield distribution (Fig. 3.6). This graph shows that in our case to guarantee a 95% customer service level, we would have an expected overage of nearly 15%. In other words, if the average order size were 200, we would be producing 230 boards *on average* to guarantee that 95% of the time we could ship the 200 boards ordered. As we increase our desired customer service level beyond 95%, we see that the expected overage increases sharply.

However, this approach is the only way we can really guarantee a 95% level of customer service. If we plan using the average yield, we end

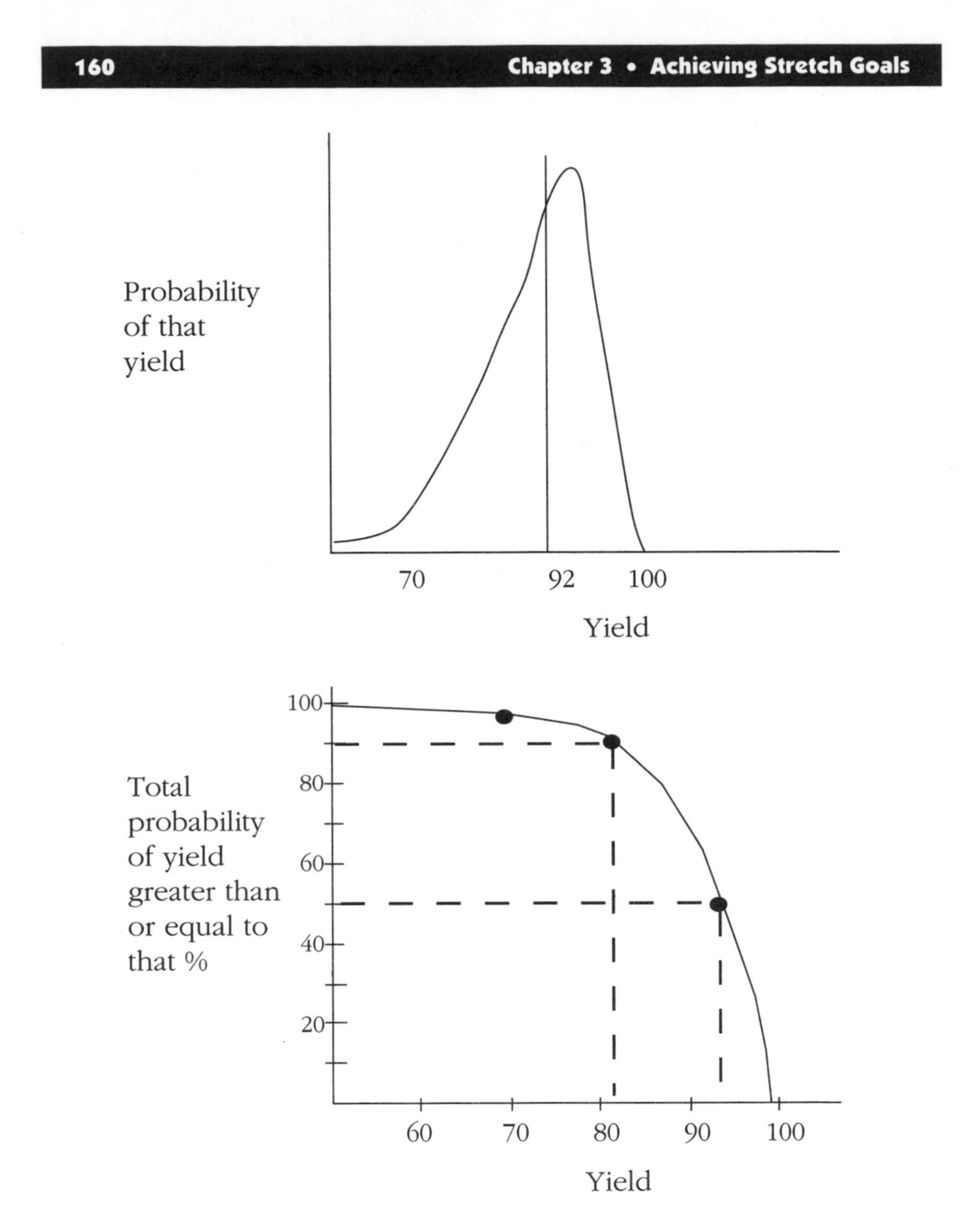

Figure 3.5 Yield Distributions

up in a cycle of plan and *react,* as yield variation occurs. Our best practices approach can be thought of as plan and *anticipate*. Obviously, the *real* solution is to eliminate (or certainly reduce) yield or quality variation. In this case, the root cause of this "scheduling" problem is not under scheduling's control. It is a quality issue, but the customer (and sometimes management) perceives it as a scheduling issue.

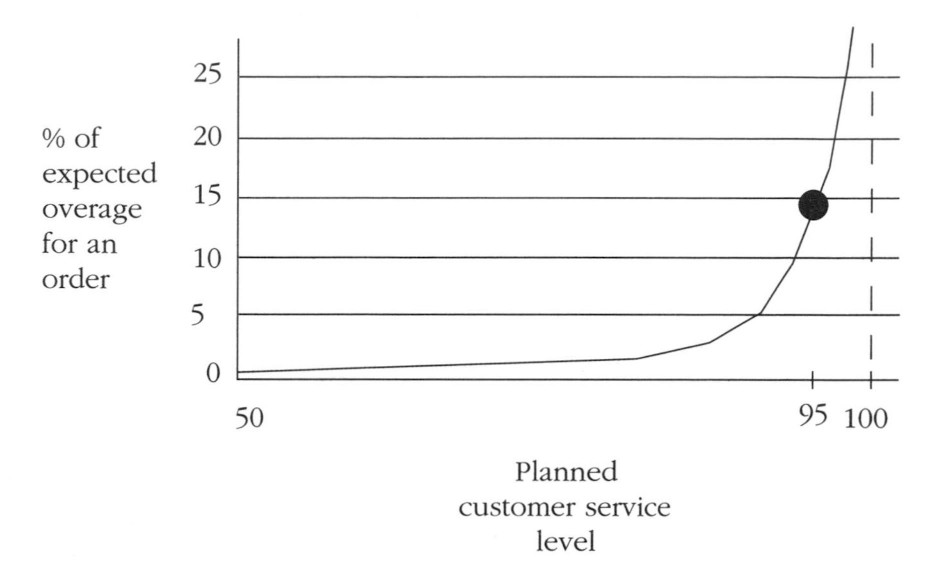

Figure 3.6 Example: Relationship of Percentage of Expected Overage to Customer Service for a Given Yield Distribution and Average Order Size

For products that are scheduled to forecast and then shipped from stock, we need to buffer yield variation using safety stocks of semifinished or finished goods. This means that safety stock must be calculated not only to cover *forecast* uncertainty but also replenishment uncertainty. For example, if we set a forecast demand of 1,000 boards per week for a production schedule, and again have a yield distribution as shown in Figure 3.5, then we'll need a safety stock of at least 1,000 × (.92 – .80) = 120 to cover the 5% case where yield falls to 80% instead of 92% that week. (In fact, we might require higher levels if we take more complex situations into account where yield in two or more successive weeks is below the average.)

Most companies set finished goods safety stock levels based on forecast error variance. They totally ignore the issue of their internal forecasting variances—the variance of yields in production. We need to anticipate yield variance to eliminate a root cause of poor customer service (until we can eliminate yield variance).

Taking principles 3 and 4 together, we can see that customer service is *not* free; it is only obtained at a cost of excess inventory until we can eliminate the uncertainty or variance of leadtime and yield.

Best Practices in Scheduling
Principle 5
Trade-offs

As we discussed at the onset, a schedule's beauty is in the eye of its beholder. In our scheduling approach, we need to allow the users to prioritize and trade off their business goals. These typically include high levels of customer service, high capacity utilization (maximizing the possible output—especially when demand exceeds our capacity), minimum cost (use of the most efficient resources, minimum setup times and cost, utilizing the lowest cost raw materials), and minimum leadtimes (minimum work-in-process levels). We may also want to set minimum or maximum target levels for one measure and "optimize" the others. For example, we may want to set a minimum customer service level at 90% and maximize utilization (output). Or we may want to set a minimum utilization level of 90% and maximize customer service.

Our scheduling rules or approaches must recognize that scheduling often is a trade-off (especially when there are capacity constraints) and guide us through our options. The desired trade-off may even vary during the month as we watch our actual performance and need to reprioritize for the remainder of the month. For example, if we have had poor customer performance early in the month, we may need to reset our rules to emphasize customer service in the second half. The scheduling practice utilized must allow this flexibility.

Best Practices in Scheduling
Principle 6
GIGO (Garbage In–Garbage Out)

Principle 6 may seem theoretically unrelated to our scheduling best practices, but is at the heart of practical results and success. Our scheduling algorithms or practices must utilize accurate up-to-date information on resource availability (status) and productivity, demand, yields, routes, and so on. In practice, few scheduling packages have been successful until they have been integrated to a real-time tracking system or manufacturing execution system (MES) covering the current status of the factory and its operating conditions. If we do not know that a machine is down, we will continue to schedule jobs on it. If we do not know that yields have improved, we will continue to overstart material. If we do not know a route has been changed, we will continue to schedule along the old route. A scheduling system without data is like a car without gas—interesting but not useful.

Therefore, we cannot evaluate a scheduling approach without evaluating its data sources. Schedules based on yesterday's factory status are as useful as yesterday's weather report.

Best Practices in Scheduling
Principle 7
Discipline

If a tree falls in the forest and no one hears it, did it fall? If we publish a schedule and no one follows it, did it exist? It's not enough to have a schedule—it must be used! It can't be ignored because the supervisor wants to push high runners through the line to maximize productivity. It can't be ignored because an engineer asked a favor. Every station "downstream" and every customer is basing their schedule on yours. If your published schedule is unreliable, due to "freedom of choice," they will buffer theirs with more work in process or leadtime.

This also means that everyone must have easy access to the current schedule. If we develop a schedule and it takes four hours to distribute it, it's already out of date. We need immediate access to the accurate current scheduling.

Best Practices in Scheduling
Principle 8
Vigilant/Current

A schedule, unlike a diamond, is not forever. A schedule is as good as the world is predictable. As soon as the factory state changes from plan (equipment breaks down, parts do not arrive on schedule, an operator is out sick) the scheduler needs to revise the schedule and notify the factory of any changes to the planned schedule as well as customer shipments, utilization, revenue shipped, and so on. The scheduler should do this as soon as the condition arises—an ever-vigilant, predictive scheduler. This is also the function of a manufacturing execution system—to alert us to deviations the instant they arise.

Best Practices in Scheduling
Principle 9
Usability

From practical experience, the phrase "trust me, you'll like it" doesn't apply to schedules. The underlying problem is that many factories are complex and cannot be fully represented in a mathematical model. There are too many "special" conditions, exceptions, or problems that arise that the user must understand whether the schedule took them into consideration or not. Also, the user must be able to revise a schedule, if necessary, to handle these exceptions or special conditions.

If the schedule can't be modified, then it has no flexibility. If a scheduling system is not easily understood, no one trusts it. In either case it usually falls into disuse.

I have summarized our best-practice guidelines in Table 3.5. Next we will look at four approaches to scheduling—the Japanese approach of eliminating the need for scheduling, optimization, heuristics—with simulation and hierarchical planning. We will examine each for its concepts, advantages, and disadvantages, and then compare it to our best-practices guidelines.

This completed the first lecture. The group left reasonably pleased. For the first time, they had a framework for understanding scheduling. Many things they had done for years, such as using the average yield and cycle time for planning purposes were now seen as questionable practices. They were looking forward to seeing what solutions were available to meet 100% customer service.

Empowered with Knowledge

The second training class was actually eagerly awaited. Schedulers and planners who for years had developed plans and schedules that were instantly out of date finally had some new insights into how to achieve their stretch goals and reduce everyone's level of frustration. They had for years heard how their plans were impossible and, in return, argued that the plant was "sandbagging" in what they would commit to deliver. Now they understood why the schedules had been so hard to achieve. Now they knew how to fix the problem! They had finally been given some basic education on the theory of scheduling. They realized that no one had ever trained them—they'd only been shown the company's current scheduling process. They finally felt empowered with knowledge. Engineers, managers, all were given training in their disciplines. Now, finally, planners and schedulers would be trained as professionals as well.

The lecturer next started on the solutions available to the scheduling problem.

TABLE 3.5 Best Practices in Customer Scheduling for Supply Chain Management

Best Practices Scheduling Approach generates schedules and alarms that:

1. Are resource feasible in plan and execution
   a. Capacity feasible—equipment, labor, storage, facilities, tooling
      1. Loads bottlenecks (equipment, labor, storage) to less than 100% utilization
      2. Dispatches jobs to continually keep bottlenecks utilized to planned levels
      3. Sets WIP buffers in front of labor and/or equipment bottlenecks to anticipate the effect of equipment downtime at feeding operations
      4. Prioritizes equipment repair at bottlenecks
   b. Are parts feasible:
      1. Plans timed parts requirements that are within the available on hand and on order quantities
2. Are leadtime feasible:
   a. Sets start dates by operation that are consistent with the availability of all required parts, engineering specifications, tooling or other resources or outside tasks needed for production at that operation
   b. Doesn't accept orders due in less than the minimum leadtime possible
3. Consider leadtime variance:
   a. Sets start dates based on the leadtime variance distribution and the desired customer service level
   b. Measures the leadtime variance distribution
4. Consider product quality or output variance:
   a. Sets start quantities (for made to order) and/or finished goods safety stock (for made to for stock) based on the yield variance and customer service level desired
   b. Measures the yield distribution
5. Trade-off customer service, cost, capacity utilization and leadtime (speed) to user priorities and/or minimum or maximum levels:
   a. During planning (in the release plan or schedule)
   b. During execution (in the dispatching rules)
6. Use up-to-date and accurate information on resource availability and productivity, WIP levels and job locations, yields, demand, and so on.
   a. Is integrated with real time tracking or MES that feeds it the latest/current status of resources
   b. Is tied to the order entry/forecast system to get the latest orders or demand forecast
7. Are followed by production:
   a. Schedules are available in real time to all personnel
   b. Any deviations made from the schedule are noted and reported on with explanations
8. Notifies the user/predicts when the schedule can no longer be met:
   a. The scheduler is notified of changes to factory operating conditions (yield, resource status, productivity, yield, demand, etc) and the schedule is rerun automatically as necessary
   b. Any customer orders that can no longer be met are indicated automatically and customer service notified
9. Are easily understood and modifiable by the user:
   a. The schedule as generated can be "explained" to the user by the scheduling system or is readily understood
   b. The schedule as generated can be overridden (with proper authority) if necessary, to handle special conditions not taken into account by the scheduling approach

Approach 1: Scheduling as Quality—The Japanese Approach

"If we look at what made scheduling difficult, we enumerated uncertainty, resource constraints, shared setups, and process complexity. Instead of trying to schedule in this environment, the Japanese approach has been to *eliminate* the factors that complicated scheduling so as to *eliminate* the need for scheduling. This can be thought of as CIMPLICITY or NOH scheduling. The elements of this approach include: SMED (single minute exchange of dies—an approach to eliminating setup times), preset and dedicated factory periods used solely for preventative maintenance, TQM (total quality management), focused factories and calls (dedicated to one process or product family), labor flexibility, and ensuring that there are no capacity constrained resources. Together, they are often called JIT/TQC (just in time/total quality control).

The greatest enemy of reliable, achievable schedules is uncertainty. As we have seen, it can appear in many guises as variance in: forecasts, productivity, quality, equipment availability, operator attendance, raw material deliveries, and so on. The Japanese have applied statistical process control to every aspect of manufacturing. They have continually investigated every cause of process variation to systematically stamp it out. They have done this not only for the quality of the product but also the quality of the five elements of manufacturing. The "quality" (or time in a state of conformance) of the five elements affects not only product quality but also schedule performance. When equipment breaks down, schedule quality and cost, speed, and flexibility are clearly affected. The same thing is true for all five elements of manufacturing. The Japanese realized early on that the key to schedule quality is manufacturing quality.

The second greatest difficulty in scheduling arises from capacity constraints. They force concurrent scheduling of all products using those constrained resources to ensure that schedules are met. Poor ordering or dispatching of jobs may cause unexpected idle time at the bottlenecks so that planned schedules cannot be made. The symptom of poor scheduling in a constrained environment is a cycle of idle time followed by overload. Usually this means that work is not being pulled smoothly (at an even rate) or released smoothly to the bottleneck resource. This is the job of the release rules and/or dispatching system. The Japanese have, in general, eliminated this problem by keeping their production lines scheduled below capacity. The philosophy is to deliberately underschedule the theoretically attainable utilization. This increases schedule reliability and also gives operators more time to resolve quality problems.

A third cause of scheduling complexity is large setup times. This forces longer runs of a product to minimize total setup change-over times at bottlenecks. It also forces coordinating schedules of other products that can share the setup, even if that product is not required at that moment. Both practices lead to excess inventory that affects cost, flexibility, and speed. The Japanese have been leading practitioners of SMED—finding ways to decrease setup times instead of "scheduling" around them. SMED is a philosophy of reducing setup times to a minute or less so that they can be ignored as a scheduling factor. Its father, Shigeo Shingo, has described many techniques he developed in the metals industry.[1] When setups can be ignored, then production runs can be scheduled to meet the exact demand without trading off early production (inventory) with setup or equipment utilization. This leads theoretically to the "lot size of one," where each order is made individually for a customer as it is received.

The fourth cause of scheduling complexity is product complexity or diversity. The more products or product variants to be scheduled, the more scheduling effort required. The approach the Japanese have taken (that also greatly aids in quality improvement) is factory or manufacturing line specialization—the focused factory or the dedicated work cell. Instead of running all products and processes on one line, dedicated lines are built for major runners, and/or major product families, and/or major processes. These can be kept set up and "tuned" for those products. These lines or cells are then run at a fixed rate (paced line), set slightly under its full capacity.

Uncertainty in product mix is handled by labor flexibility. Labor is cross-trained so that it can be moved to where the demand exists. Then the production floor can respond to shifts in product mix by "reconfiguring" itself to that resulting load mix. This means that scheduling does not have to be as precise. The factory can *adapt* to schedule variation automatically much as a thermostat can control a room temperature within certain ranges without intervention.

A last example of the Japanese eliminating the need for scheduling is their approach to prevention maintenance. In the United States, mainte-

[1] Shigeo Shingo, *Non-Stock Production: The Shingo System for Continuous Improvement.* Productivity Press, 1988. Cambridge, Mass.

nance is scheduled during production hours and melded with the production schedule. This greatly increases the complexity of scheduling. In Japan, rather than schedule within this "mixed" environment, all maintenance is often moved to a "maintenance" shift set between production shifts. For example, we may run production for 10 hours and then maintenance (with no production) for 2 hours. Or we may run production for 20 hours and then maintenance for 4 hours.

What we have seen so far is that the Japanese approach to scheduling is no scheduling—to eliminate all the factors that require complex scheduling rules. What remains, then, is the Kanban approach.

Scheduling on the "Assembly Line"—Kanban or Fixed-Rate Scheduling

Now that we have removed uncertainty, setups, capacity constraints, *individually* scheduled maintenance, and utilized focused manufacturing cells, what remains in principle is a highly reliable *assembly* line—a line that runs at a fixed rate in unison. The basic principle of the assembly line, started at Ford Motor Company by Henry Ford, is the movement of one unit from each operation group to the next in near perfect synchronization so that there is no excess inventory—the line moves continuously with minimum leadtime at a set production rate.

Over time, as American manufacturers broke up the assembly line into functional areas (Fig. 3.7) (to allow greater product diversity and variation), they also used inventory to buffer each operation and/or functional area from uncertainty in the operations of its feeders. Japan, initially, did not rely on product diversity as a competitive strategy. Therefore, it did not have to organize into functional areas; it could use "dedicated" lines for most major products.

The original assembly lines used mechanized transport to move product down the line at a fixed rate. The Japanese approach to this was to replicate the *concept* of the assembly line with a manual implementation called Kanban or JIT. The basic principle is that each operation feeds its "customer," the succeeding operation, on demand. When no problems arise, this means that all operations run synchronously at a rate set by the slowest operation. When any operation stops (due to a problem in quality or lack of a needed resource), all feeding operations shut down temporarily as well (since there is no demand placed on them from their "customer"). The very

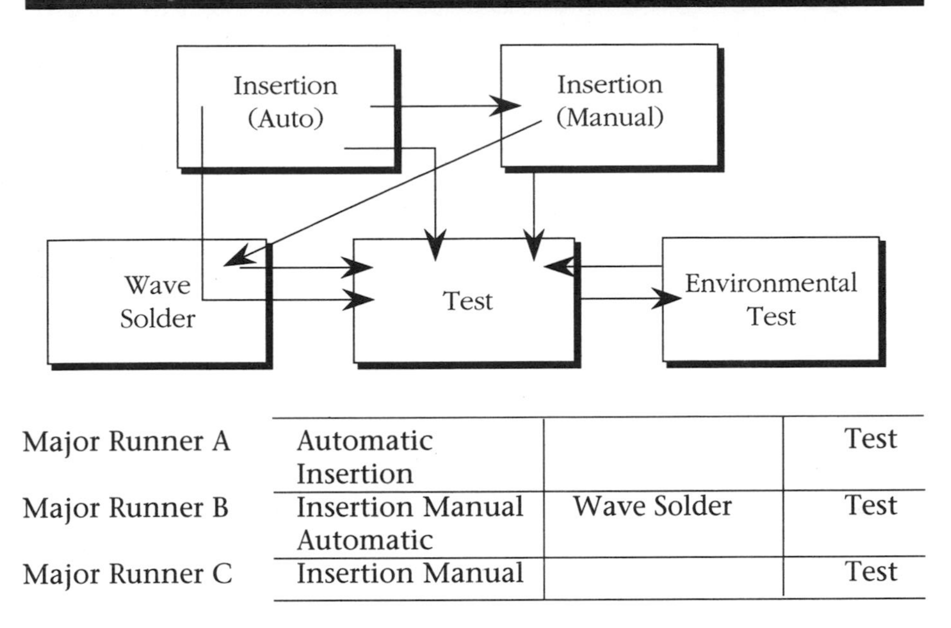

Major Runner A	Automatic Insertion		Test
Major Runner B	Insertion Manual Automatic	Wave Solder	Test
Major Runner C	Insertion Manual		Test

Figure 3.7 Functional Areas

last operation's customer is, in fact, the real customer in its fullest implementation (where we time or schedule production to the customers' orders for that day instead of to replenish finished goods stock).

For example, in Figure 3.8 when Operation 3 starts work on job A, it sends the reorder ticket (or Kanban) back to operation 2 to produce another job for it. This triggers production at operation 2, sending a Kanban ticket back to operation 1. In this way, jobs are "pulled" forward and the line is kept in perfect balance at all times. Jobs move through in first-in–first-out (FIFO) order. The amount of WIP in the line is determined by the number of jobs (or Kanban cards) allowed at each operation.

Alternatively, instead of sending back a card, the operation can turn on a light, indicating that it is ordering another job.

Pros and cons. The obvious advantage of this approach is that it eliminates most sources of scheduling complexity and waste *prior* to scheduling. Instead of "scheduling around" the problems, it drives to eradicate the need for scheduling. These principles of simplification and improvement should be followed by *every* manufacturer whichever scheduling mechanism is followed.

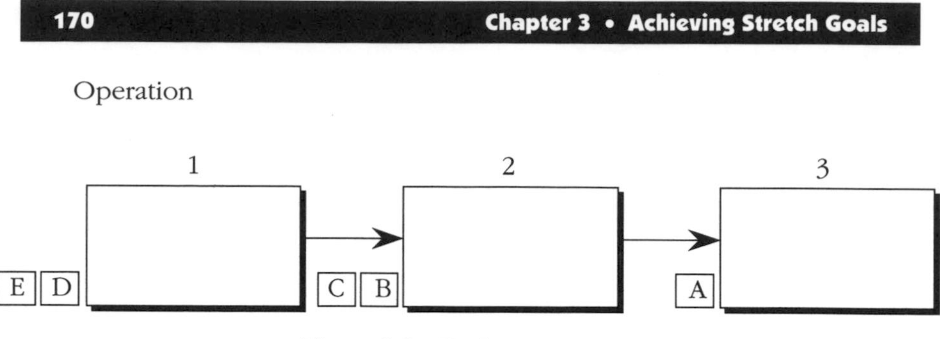

Figure 3.8 Kanban in Practice

However, the drive to eliminate the need for scheduling through simplicity also may lead to lower utilization. By forcing all equipment to be idle during a plant-wide maintenance period, we idle equipment that may not need that level of maintenance. In a strange way, it is the *opposite* of the philosophy of SMED. We force a fixed "setup" time on the shift! Similarly, by forcing the line to shut down when a "customer" (the next operation) does not demand a job, asynchronous operation is not allowed. Also, potentially lowering utilization and output, similarly keeping the line scheduled below its capacity keeps "smoother" operation with lower leadtimes but at a cost of lower utilization. It is important to note that not all industries in Japan follow the Kanban system invented at Toyota and made famous in the automotive industry. The Japanese semiconductor industry trades off higher utilization for longer cycle times to keep its billions of dollars of investment in equipment and facilities fully utilized. The trade-off is between speed and utilization. By *not* employing scheduling as a tool, we are actually depriving ourselves of a competitive weapon. In industries with high "fixed" costs of facilities, equipment, and indirect/supporting labor, higher utilization is a critical competitive advantage in reducing unit costs. In industries with low "fixed" costs and high percentages of direct labor and materials (such as in manual assembly), reducing inventory (work-in-process) cost is a competitive advantage.

The reality of scheduling is that usually we are trading off criteria—between speed (leadtime), utilization (cost), and customer service. The drive to eliminate the need for scheduling is often trading off speed for utilization. To see the problems inherent in a pure Kanban approach, let's examine Figure 3.9.

In this ideal Kanban situation, there is only one process/product type flow in this example set at 10 units per hour, which is the rate of the bottleneck (operation 2). From an output viewpoint, operation 2 is the bottleneck

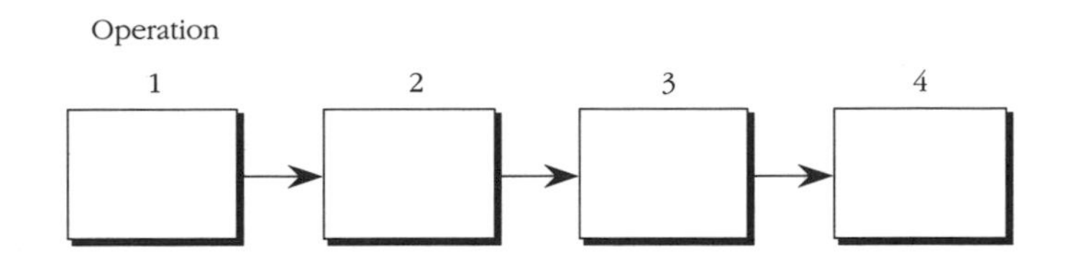

OPN	1	2	3	4
Run Rate Per Hour	10	10	10	10
Possible Rate Per Hour	12	10	15	12

Figure 3.9 "Pure" Single Process Flow

that paces the line. Using our Kanban approach, when operation 3 stops for any reason, we also stop work at operations 2 and 1. However, when we shut down operation 2, we lose the production there "forever."

However, as an alternative approach, we could allow the operations to run asynchronously; when operation 3 goes down, we continue to run operation 2. When operation 3 comes back up, we run it at a higher rate (it can run at a rate of up to 15 units/hour) until the line is rebalanced.

For example, as shown in Table 3.6, if operation 3 is down during hour 1, we continue to run operation 2. When operation 3 comes back up,

TABLE 3.6 A Synchronous Operation of the Production Line Operation Production Rate

	1	2	3	4
Hour				
1	10	10	0	10
2	10	10	15	10
3	10	10	15	10
4	10	10	10	10
Hour		Operation	WIP	
1	0	0	0	10
2	0	0	10	0
3	0	0	5	5
4	0	0	0	10

we run it for the next 2 hours at its maximum rate. We would rebalance the line by the end of hour 3 after the breakdown and not lose any possible output (given that we had "buffered" operation 4 with 10 units of work). We have kept the bottleneck fully utilized and maximized output/utilization, getting 10 extra units.

In other words, by "loosely" coupling the line (letting operations run asynchronously and keeping carefully calculated levels of buffer work in process) we can keep *output* constant by allowing WIP to become imbalanced temporarily as necessary. This approach does not sanction allowing equipment reliability problems to continue—they must be measured and continuously improved. It simply uses scheduling as a weapon to gain maximum performance *while* reliability improves. In a factory with high capital investment running near capacity with any equipment reliability problems, this may provide a competitive advantage; higher output from that capital investment.

In many lines, the rate at an operation is not fixed but can be adjusted by changing the staffing level assigned. This means that we can easily adjust rates at many operations while the problem is corrected at the errant operation. Shutting down the entire line seems like a dramatic exercise, but in practice, a costly and unnecessary one if we are willing to use scheduling as a tool. Obviously, if there are no asynchronous downtimes, there is no need for this capability (ie, if equipment is perfectly reliable).

So let's summarize the JIT or simplicity approach to scheduling versus our best practices model (Table 3.7).

1. In terms of resource feasibility, the JIT approach is resource feasible (by definition) as we schedule the line to the bottleneck's rate. However, because the line runs at the rate of the slowest bottleneck currently *in operation* (that is up and running), this approach doesn't maximize our output. Therefore, given significant levels of equipment downtime, utilization will be low.

2. Leadtime feasibility is not explicitly considered. It must be covered by the due date assignment process used by sales or customer service. The goal of a JIT line is to reduce the leadtime so significantly that customer orders can be produced in near real time (meaning that they are *all* leadtime feasible). We trade off capacity utilization for some flexibility to respond to orders in real time.

TABLE 3.7 Summary: Japanese Approach Versus Best Practices Guidelines

Best Practice Area	Evaluation
1. Resource feasibility in plan and execution (equipment/labor/parts/storage)	• Achieved by removing bottlenecks • Line runs at the rate of the slowest bottleneck in operation, so use at the bottleneck given equipment downtime is not maximized
2. Leadtime feasibility	• Not covered explicitly—must be covered in due date commitment policy • Leadtime assumed to be very short
3. Considers leadtime variance	• No. Goal is to eliminate variance, not cater to it • Leadtime is designed to be too short to need to worry about it
4. Considers product quality variance	• No. Goal is to eliminate variance, not plan for it
5. Trades off capacity utilization, cost, customer service, speed	• No. Line set at a "pull rate" with fixed speed capacity utilization and cost
6. Uses up-to-date, accurate information	• Not really applicable—no scheduling really done
7. Is followed	• Yes. Line runs FIFO and go/no go
8. Notifies user/predicts when the schedule can no longer be met	• Knowledge is whether the line is up or down
9. Easily understood and modifiable	• Yes. Simplest system of all to understand

3. Leadtime variance is also not explicitly considered. Under a JIT approach, the leadtime is (hopefully) so short that the variance is irrelevant to customer service.

4. Quality variance is also not explicitly considered. Almost all JIT programs rely on parallel TQM programs that eliminate quality problems. It is assumed that during operations, product quality is extremely high and output extremely predictable. In general, Kanban requires, or is predicated, on very reliable manufacturing processes.

5. This approach does not explicitly trade off utilization, cost, speed, and customer service. It is focused on running at a fixed speed with high reliability to achieve a high level of customer service. The cost is fixed by our line design (we do not usually consider alternate routes, materials, equipment, etc.). Utilization is deliberately allowed to fall below full capacity to give a more reliable output rate and some flexibility.

6. This approach does not need much data. It relies on knowing when the next operation needs another unit produced for replenishment of its stock (Kanban). We simply need to know, visually or by signal, that the next operation needs another unit produced.

7. JIT usually ensures that the "right" job is worked on because work is pulled forward in FIFO order.

8. Users are not specifically notified of a problem. Traditionally, however, there is a visual or electronic signal whenever a station stops work. However, as an alternative approach, we could allow the operations to run due to a problem (a red flag may be raised at that work center). Predictions of the "new" schedule are not generated automatically.

9. Of all scheduling systems, none could be much more logical to understand. Modification is usually unnecessary—jobs flow rapidly through a FIFO order. Since there is no real schedule visibility, there is no schedule to modify. Schedule modification is done "physically" by reordering the jobs in the queue.

In summary, Kanban works well when there is very little uncertainty of any kind. As uncertainty rises, utilization falls, and schedules are missed.

Sidebar: Scheduling as Science—Setting the Correct WIP Levels

The correct setting of WIP levels is one of the most misunderstood areas in scheduling. Many practitioners now speak in terms of setting cycle times in multiples of processing time. If the actual total processing time is, for example, 18 hours, experts will talk of a goal of 1.5 to 2 times cycle time or 27 to 36 hours. While this is appealing emotionally, it is not very helpful in indicating *where* to actually hold the work in process (how to distribute it through the line) or *why* we need 1.5 to 2 times the theoretical amount at all.

Given that most companies in the past ran 10 to 20 times their theoretical cycle time, attaining 2 times it is a major victory. However, in this section, we discuss a more scientific way to set WIP levels.

There are four components of total cycle or throughput time, processing time (the theoretical minimum cycle time), time associated with setup or batching (running multiple units through the equipment to utilize a setup and/or the setup time), in-transit time (to move the job between operations), and queue time. While our immediate goal should be to eliminate the last three categories of cycle time, realistically that is not always possible initially.

Therefore, while over time we re-layout the factory to minimize distance between operations (to eliminate in-transit WIP) and utilize SMED concepts (to eliminate batching and/or setup time), our goal is to reduce WIP and cycle time. Is there any reason, however, to keep WIP?

As we saw in the previous section, the reason to keep some level of WIP is to increase utilization by buffering an operation from breakdowns at its *feeding* operations. For example, in Figure 3.10, when no WIP is kept in front of operation 2 and machine 1 breaks down, machine 2 "starves" and is shut down. If machine 2 is not a bottleneck, "temporary starvation" does not affect its meeting its schedule. However, if it is a bottleneck, this starvation shutdown will affect line output. Therefore, to ensure we can meet our *output* schedule, we need to keep operation 2 running by buffering it with WIP.

How much WIP do we need? What we are "protecting" operation 2 from is a breakdown or interruption at operation 1. Therefore, we need to protect it from the maximum "likely" breakdown time at operation 1. Say that 90% of breakdowns at operation 1 occur for one hour or less (there are always unusual occurrences that could last hours or even days—when a key broken part is not in stock or the repairman/technician is not scheduled that shift or day. But we will only try to protect ourselves against reasonably likely problems).

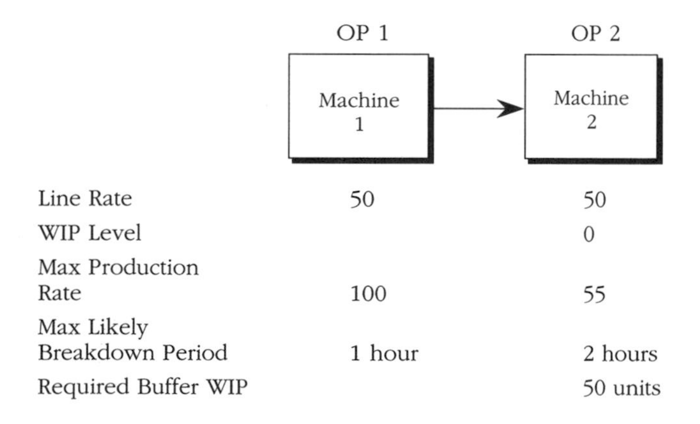

Figure 3.10 Setting WIP Levels

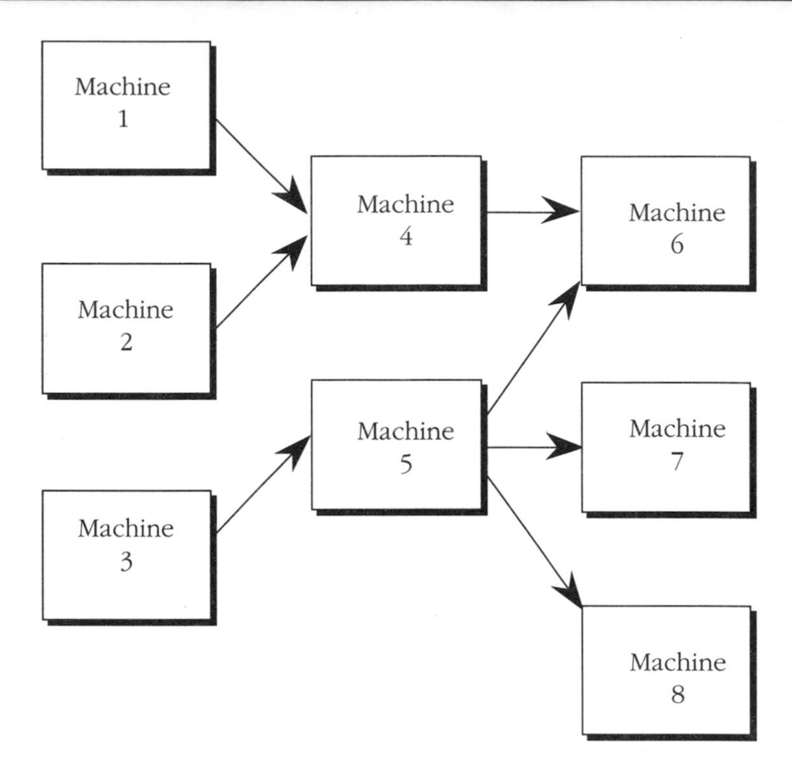

Figure 3.11 More Complicated Scenarios

Then the WIP required in front of operation 2 is one hour's worth of production at operation 2 given that its production rate is 50 units/hour, that is, 50 units of WIP.

In practice we may need to calculate solutions for much more complicated situations, as shown in Figure 3.11, where an operation can have multiple feeding operations or machines, each with different breakdown distributions; breakdowns can occur several times in a period (successive breakdowns) before the buffer is replenished; or the feeding operation may not break down but actually be starved by *its* feeding operation breaking down (chained starvation). However, as a starting point, we can calculate the required buffer WIP in this simplistic fashion.

$$\begin{array}{c}\text{Required WIP}\\ \text{at operation } n\\ \text{(bottleneck)}\end{array} = \begin{array}{c}\text{max likely}\\ \text{breakdown time}\\ \text{at operation } n-1\end{array} * \begin{array}{c}\text{Required production rate}\\ \text{at operation } n\end{array}$$

We can also see that we only need this buffer WIP in front of our bottleneck operations. Other operations can have unplanned downtime within the allowed slack or idle time in their schedule. Or, in summary, we only need to keep WIP in front of bottlenecks fed by *unreliable* operations.

This discussion should in no way be taken as a vote of *approval* for poor equipment reliability. It is simply a vote of approval for using scheduling to provide every competitive advantage (while you work to improve equipment reliability as rapidly as possible).

The pay-off—higher utilization. Looking at our simple case of operation 1 feeding operation 2 (Fig. 3.10), let's calculate the advantage of using buffer WIP versus pure unbuffered Kanban. If we assume that the average breakdown at operation 1 is 20 minutes once each shift, then by placing an hour's buffer WIP in front of operation 2 (to protect against the worst likely breakdown) we add nearly 50 units/hour × 1/3 hour = 16.6 additional units of output each shift over a pure Kanban approach. (Some breakdowns will exceed our one-hour buffer even if the average is 20 minutes, so we won't get the full benefit of 16.6 units.) This means that we will have nearly 50 extra units a day, equivalent to one extra hour of production or slightly over four percent additional output per day just by buffering WIP properly. The cost to us is an additional hour of cycle time. If we add asynchronous processing and keep bottlenecks running when their succeeding operation stops, we may stretch this gain to eight to ten percent until most of the equipment reliability problems and quality problems are eliminated. In some industries where equipment and process reliability are very hard to achieve, this difference is a sustainable competitive advantage available just through proper scheduling!

Clearly, as we reduce the *variance* of the equipment downtime (the "width" of its distribution), we reduce the amount of the buffer WIP required. This is another example of how uncertainty is the major cause of scheduling problems and waste.

In industries where WIP is not "perishable," we can keep buffer WIP for high runners that are run daily "on the shelf" so as not to impact the actual cycle time seen by the "active jobs" running through the line. This minimizes the impact of buffer WIP on speed/leadtimes.

Now that we know how to set buffer WIP levels, let's look at other approaches to scheduling.

Approach 2: Scheduling as Quantity—The Mathematical Approach of Optimization

At the other extreme from the Japanese approach of eliminating the need for scheduling is the mathematical approach of finding the "best" schedule, or optimization. Optimization involves explicitly writing down the objectives for scheduling (how we will evaluate each schedule numerically), and the constraints (the resource limitations within which we schedule) and finding the best schedule that meets the given constraints. Depending on the structure of the scheduling problem (whether there are setups or not, whether the scheduling problem is identical every time period or not, whether there is only one constrained resource or not), there are specific optimization techniques we can use (Table 3.8). The basic concept, however, remains the same—to "abstract" the scheduling problem into a mathematical representation and then apply mathematical techniques to the representation to "solve" it and find the "optimal" solution.

We can mathematically represent the scheduling problem as follows as shown in Figure 3.12. This representation says that we want to find the "best" (lowest-cost) production schedule at each operation for each product (X_{ijt}) that stays within each resource constraints at each operation each period (R_{jkt}) and meets demand for each product (D_{it}) each period. Best means the schedule with the lowest total cost of inventory, overtime, and operational production costs.

(Undoubtedly, many of you have already tested the next to last assumption when you looked at the mathematical formulation in Figure 3.12 and were too impatient or puzzled to work through the notation.)

The history of success in applying optimization to factory floor scheduling problems is fairly spotty. While there are many anecdotal (single-

TABLE 3.8 Examples of Optimization Techniques

Problem Structure	Optimization Technique
Single constrained resource; demand varies/period	Dynamic programming
No setups; demand varies/period	Linear programming
Setups; demand varies/period	Integer programming
Same problem every period; single constrained resource	Knapsack algorithms

company) examples of use reported through the 1960s and 1970s, the first large-scale or mass-market commercial thrust was provided by a software house, Creative Output, in the late 1970s. Their product, OPT, used proprietary algorithms to solve the integer programming problem based on their characterization of an optimal solution. They enumerated a set of principles (based on optimization theory) (Figure 3.13) that characterized the solution.

There are some very important assumptions made in using an optimization approach. We are assuming that:

- Uncertainty is being handled "correctly" either inside or outside the model (typically outside by setting safety or buffer stocks).
- There is a single person who has the responsibility to generate a schedule.
- We can agree on an objective function and therefore clearly and objectively evaluate one schedule against another.
- We have accurate data as to the resource requirements per unit of production and the resource availability (capacity).
- We have accurate up-to-date data as to the current inventory of work in progress of each product, and we can access it automatically to feed the scheduler.
- It is worthwhile to find the "best" solution—meaning that the best solution is far superior to the solution we get using current methods.
- Users will accept, and therefore use, a detailed schedule generated *centrally* using a mathematical approach they may not understand completely.
- The solution technique is rapid (and cost-effective) enough to implement this technique and respond to whatever level of changes that occur on the shop floor as they arise.

They achieved a great level of notoriety and near cult status but ultimately failed as a commercial venture and were restructured by a new owner.

$$\text{min.} \sum_{i=1}^{I} \sum_{k=1}^{K} \sum_{t=1}^{T} (s_{ik} \int (x_{i,k,t}) + h_{i,k} I_{i,k,t} + v_{i,k} x_{i,k,t}) +$$

Min Cost = cost of setups + holding cost of inventory + production cost

$$\sum_{h=1}^{H} \sum_{t=1}^{T} c_h \; O_{ht}$$

for: i = product 1 to I
k = operations 1 to K
h = resources 1 to H
t = time periods 1 to T

+ (overtime) cost of additional resources

Subject to: $I_{i,k,t-1} + X_{i,k-1,t} - X_{i,k,t} = I_{i,k,t}$

inventory balance at each operation:
ending inventory last period + production at feeding opn k-1
- production at this opn = ending inv. this period

last stage $I_{i,k+1,t-1} + X_{i,k,t} - I_{i,k+1,t} = D_{i,t}$

k + 1 = finished goods inv. / Meeting customer demand / ending finished goods inv last period + product at last opn - ending inv. this period = demand this period

for all t and h

$$\sum_{K=1}^{K} \sum_{i=1}^{I} b_{i,k,h} \, x_{i,k,t} \leq R_{ht} + O_{ht}$$

Total resource within capacity: use of resource h over all production is less than regularly available resource plus overtime

subject to $X_{i,k,t} \; I_{i,k,t} \geq 0$

Production and inventory must be greater than or equal to zero

$$\int (x_{i,k,t}) = \begin{cases} 0 \text{ if } x = 0 & \text{(if we don't produce)} \\ 1 \text{ if } x_{1,k,t} > 0 & \text{(if we do produce)} \end{cases}$$

Setup indicator is 0 if we don't produce and 1 if we do produce

Figure 3.12 Mathematical Optimization Formulation of the Scheduling Problem

Today, as computing power increases linearly each year, there are a growing number of successors, including a restructured OPT (called OPT 21) from Scheduling Technology Group Limited. Several have actually shifted their focus to supply chain management from factory level scheduling in pursuit of the current hot market.

Pros and Cons. Historically, optimization has not been exceedingly successful. While individuals may argue this point based on any single company's experiences, in general, there are few implementations that have succeeded over a multi-year period. In fact, they are rare compared to more generally accepted approaches such as simulation, dispatching rules, or rough-cut capacity planning/MRP II. At the height of its success, OPT had 100 implemented and operating installations worldwide (as opposed to licenses sold).

Obviously, optimization "works" (at least theoretically). So the question is, why hasn't it worked empirically? The answer goes back to the assumptions built into optimization. While many of them can be met by a careful implementation, several of them have proven difficult to achieve such as easily understood and modifiable. In scheduling problems where the assumptions have been met, users have found great success. What are those classes of scheduling problems?

The problems on which optimization has succeeded best are those that are static, deterministic, and fairly small in complexity (number of variables), and with a clear objective function. This means that the scheduling problem is *identical* every time (static), there is no (or little) uncertainty, the

where:

h_{ik} = holding cost per period per unit of product i at operation k
R_{ht} = units of resource h available in time period t
O_{ht} = hours of overtime of resource h used in time period t
X_{ikt} = production of product i at operation k in period t
S_{ik} = setup cost of product i at operation k
V_{ik} = production cost per unit of product i at operation k
C_h = cost per overtime unit of resource h
D_{it} = demand for product i in period t
R_{ht} = units of resource h available in period t
I_{ikt} = inventory of product i at operation k in period t
b_{ikh} = units of resource h used per unit of product i produced at operation k

1. Balance the flow, not capacity.
2. Constraints determine non-bottleneck utilization.
3. Activation (of resource) is not always equal to utilization.
4. An hour lost at a bottleneck is an hour lost for the entire system.
5. An hour saved at a non-bottleneck is a mirage.
6. Bottlenecks govern throughput and inventory.
7. The transfer batch should not always equal a process batch.
8. Process batches should be variable, not fixed.
9. Set the schedule by examining all the constraints simultaneously.

Figure 3.13 Principles of OPT[2]

solution can be easily reviewed, and everyone can agree on which of two solutions is better. A simple example is packing tissue boxes into a shipping carton. It is easy to compare solutions (how many tissue boxes fit into the shipping carton without being crushed), there is little uncertainty (the size of the tissue box and shipping cartons vary only slightly), and it's the same problem to be solved every time. Once a solution is developed it's easy to evaluate, no matter how it was developed. Since it's easy to evaluate, the key isn't how it was generated, it's what the solution is. This is a critical point—the harder it is to *evaluate* how good the actual schedule is, the more importance people place on *how* it was developed—what factors it considered and how it was derived.

Let's compare this simple problem to a larger problem. Every week we have to schedule production of hundreds of products at hundreds of operations with changing and/or uncertain demand, equipment availability, labor productivity, equipment quality (how well jobs are being processed on a given piece of equipment), and so on. What becomes very obvious is that it is *very* difficult to evaluate one schedule against another because of the *size* or complexity of the scheduling solution and the uncertainty of the scheduling environment. The current schedule, no matter how good, will have to be revised within the next hour, shift, or day. So how can you easily measure how good the schedule was when it will never be fully implemented.

In addition, there are many groups affected by the schedule (production, maintenance, purchasing, finance, etc.) who all need to *agree* on an

[2] Robert Lundrigan, "What Is This Thing Called OPT?" Production and Inventory Management, second quarter 1986.

evaluation method. While this is clearly critical, it is empirically difficult to achieve.

What we see is that many of our assumptions are not easy to meet in practice. In fact, most of our assumptions fall apart in complex environments. Until we eliminate the complexity and uncertainty, we can't easily use optimization. Ironically, that is exactly where it was typically applied as having the most advantage over non-optimal techniques.

In addition, most scheduling systems or packages were sold as stand-alone systems not integrated to the *data* they required: actual production to date, equipment availability, labor availability, job location and status, and raw materials availability—which had to be entered manually before each scheduling run. Without automatic linkage to the factory floor data, this manual updating was laborious and time consuming, so schedules generated were either incorrect (based on "average" data) or late (run long after the shift or day had started). Most scheduling packages still can't be used in real time without a manufacturing execution system that feeds it this data.

Most difficult, however, was the "black box" nature of the solution. Most users couldn't clearly understand *how* the solution was generated well enough to be able to evaluate its "quality" over their existing procedures. After the third or fourth time, some major problem arose that it didn't handle correctly (it didn't get updated soon enough after a major change in customer orders, it didn't get updated on operator vacation schedules, it didn't get updated on new product routes or process changes, it never was set up to understand alternate routes or material substitution; it didn't handle equipment breakdowns during a shift, . . .) it became discredited and fell into disuse. It wasn't that it didn't "work"; it was a problem of being too complicated or hard to understand, so a simpler approach was preferred.

No matter how fast computers become, these problems with optimization will not be eliminated until the assumptions needed for their success are met!

To summarize the optimization approach to scheduling versus our best practices model (Table 3.9):

1. Resource feasibility is a strength of the optimization approach. As long as the resources are explicitly and accurately modeled (in terms of capacity, productivity, and availability), the resulting schedule will be feasible and "optimal."

TABLE 3.9 Summary: Optimization Versus Best Practices Guidelines

Best Practice Area	Evaluation
1. Resource feasibility in plan and execution (equipment/ labor/parts/storage)	Strength of the approach—keeps resource feasibility and goes beyond it to optimize utilization.
2. Leadtime feasibility	Should guarantee feasibility as long as we schedule the job with all the resources required before we've committed to a due date.
3. Considers leadtime variance	Typically does not consider any uncertainty in the optimization model (OPT 21 has the concept of time buffers to handle this, but doesn't calculate them for the user).
4. Considers product quality variance	Typically does not consider any uncertainty in the optimization model.
5. Trades off capacity utilization, cost, customer service, speed	Strength of the approach, but the user must model this carefully to avoid driving the solution to an extreme. May involve several runs to explore trade-offs or reset constraints.
6. Uses up-to-date, accurate information	Typically requires a lot of data that is not available in real time without integration to MES.
7. Is followed	Typically no assurance that the schedule will be followed.
8. Notifies user/predicts when the schedule can no longer be met	Can be set up to rerun and check deviance from last run, but no triggering mechanism.
9. Easily understood and modifiable	Very hard to explain or modify.

2. Leadtime feasibility should also be guaranteed as long as we schedule the job *prior* to committing a due date. Organizationally, this means that we need to schedule prior to a salesperson accepting an order with a committed date. Many companies first accept an order (with a date) and then schedule it. Optimally, the salesperson could get a scheduled commit date while at the customer's office.

3. Optimization techniques typically do not consider variance or uncertainty. The model needs fixed (deterministic) values for yields, capacities, productivity, and demands. Therefore, in general, leadtime variance caused by unrelated events is not considered explicitly. (One exception is OPT 21, which allows the concept of time buffers to handle this but doesn't calculate them for the user.)

4. Similarly, product quality variance (yield) is not explicitly considered. The user uses a given yield or rework value to use in the model by product operation.

5. Trading off cost, leadtime, utilization, and customer service is the strength of the approach. However, doing this satisfactorily *in practice* may be more difficult than users expect. Each model is optimized *if* a solution exists. If there is not a feasible solution, the scheduler may suggest the closest solution it can achieve. The user may have to make several runs, changing the trade-offs or the constraints (on minimum or maximum: customer service, leadtimes, number of late jobs, utilization, revenue, etc.) to find a desirable and feasible solution. Depending on the system, this can be laborious and difficult.

6. One of the key requirements for scheduling success is access to accurate, up-to-date information on resource availability, capacity and productivity, production routes, product definitions, and so on. A common problem in practice is that there is no factory floor system or MES at the user's site that has this data available to feed to the scheduler. To enter this manually prior to each scheduling run is not practical in most factories with large numbers of products, operations, equipment parts, and operators. That is why many schedulers are used strictly for planning and not execution—they lack the required data to support scheduling optimization. This is not the "fault" of the scheduling approach, but will make the scheduling "solution" as a *whole* fail.

7. Typically, the results of the scheduling system are available to the schedulers and operators. However, the actual order of the jobs run is not tracked (or enforced) so there is no knowledge of whether the schedule was actually followed. In practice, if the schedule is not followed at each operation, it is almost immediately irrelevant at all downstream operations.

8. New schedule runs can be generated on user request. However, unless the issue of access to up-to-the-minute accurate data is solved, the scheduler doesn't *know* that there has been a major change in the factory environment (equipment up and running, yields, parts availability, etc.). The scheduling system needs to be notified of such changes automatically so it can then rerun the schedule.

9. The real weakness of optimization systems (and it is often fatal) is the difficulty in understanding exactly how the schedule was generated and what it has considered (as we previously discussed). In addition, most optimization packages do not allow manual overrides so that the user can't modify the schedule generated. This also can be a major problem.

Approach 3: Scheduling as Common Sense—The Tried and True Method

Obviously, people have been running factories for decades before the advent of JIT/TQC programs and optimization techniques. For most real life situations, companies have found simple approaches to addressing the need for capacity planning, lot sizing, and detailed scheduling.

The most common approach to capacity planning is to use a simple measure of factory capacity per unit processed, usually based on one key resource constraint, to limit the planned weekly (daily/monthly) start or out schedule. Examples are the total number of: units processed, hours of work, hours of work at a key bottleneck operation, weighted units processed (where some products get a higher weighting per unit due to difficulty), pounds processed, batches started, and so on. The choice is usually related to the simplest measure of the key bottleneck's capacity given the manufacturing process (Table 3.10).

This measure is used to load the factory to ensure feasible and level capacity utilization. In practice, orders (or forecast demand for products) by

TABLE 3.10 Rough Capacity Measures

Manufacturing Process Characteristic	Capacity Measure of the Production Schedule
All units take roughly the same capacity	Total number of units
Units differ in capacity required but labor can be moved to adjust capacity	Total man hours of work
There is a key bottleneck limiting production	Total hours at the key bottleneck
Units differ in capacity required	Total weighted unit of man hours
Capacity is related to a product's characteristics (eg, weight)	Total pounds processed
Capacity is related to equipment characteristics (eg, batch size)	Total number of batches processed

period are totaled, multiplied by the resource capacity required per unit, and compared to the factory's capacity until it is just exceeded. Additional orders are then placed in the next period's planning bucket or scheduled using overtime. This process "assures" that we have adequate capacity to process all the jobs released to the factory that period.

Scheduling the Capacity Plan: Product Family Characteristics

This committed plan is usually developed by a central production planning group/person and then given to the factory schedulers. The factory schedulers then further refine the plan by grouping products for release to the factory that share major setups or batching (a fixed equipment run size such as in a curing furnace), or product cycles based on a product family. They also check for detailed resource availability of tooling, parts, work instructions, or special operator qualifications required, and so on, for scheduling order releases. The orders are then released shiftly or daily into the factory.

Once launched, jobs are ordered at each operation based on either local dispatching rules or completion of an equipment product cycle at the operation (eg, going from highest to lowest temperature, highest to lowest width, or lightest to darkest color) if the processing is setup/sequence dependent. There are many dispatching rules covered in the literature, but most companies use rules to order or prioritize jobs at each operation related to meeting customer due dates, as measured by whether the job is ahead or behind schedule. Several examples of dispatching rules are given in Table 3.11.

The rules listed are all called *local* rules because they can be calculated at each operation or work center without any additional information required from *other* work centers. That was highly desirable (actually mandatory) before the advent of computerized manufacturing execution systems that could provide up-to-date information on the status of the whole factory. With local rules, each operation could run in a fully decentralized mode using these rules adjusted for local conditions that the supervisor or lead operator saw.

In this way, the scheduling problem was *distributed* among many players, as opposed to the optimization approach where *one* centralized

TABLE 3.11 Dispatching Rules

Rule	Calculation	Performance Characteristics
Critical ratio	Time remaining until this job is due divided by remaining leadtime	Same as dynamic slack
Dynamic slack	Time remaining until this job is due minus the expected remaining leadtime (how much extra time or slack exists in the schedule; how far are we behind schedule?)	Keeps jobs ordered for meeting due dates
First-in–first-out (FIFO) or first come, first served (FCFS)	Process the jobs in the order that they arrived to the operation	Keeps jobs from being very late *if* they were started in due date order
		Doesn't expedite late jobs (jobs that get delayed for any reason)
Shortest processing time (SPT)	Order jobs in increasing order of processing time at this operation	Completes the largest number of jobs, but longer jobs are delayed

Note: All of these ignore setups (sequencing and batching), and bottleneck conditions downstream.

"expert" gives the "perfect" schedule to everyone. Here each person is given an overall output target and then develops a more detailed schedule with their "local" expertise. We can think of this as decentralized scheduling. Unfortunately, it's not scheduling as a team sport because, as we will see, there is no real team coordination ensuring that *team* goals are met.

In reality, this approach was a manual solution to a complex problem developed before better tools and technology existed.

Pros and cons. The obvious advantage of this approach is its simplicity—it does not require computers (though it can be assisted by their use), and it is easy to understand and therefore to implement. Another major advantage is its consistency with the typical organizational structure.

Each organization is given "local" authority to make decisions within an output schedule or target they're given. In that way they can adapt the schedule to the "detailed reality" at their operation (or facility or work center). This means that one person does not need to understand the detailed real time constraints, operating conditions, or objectives of every work center simultaneously to schedule the factory.

The disadvantage of this approach is that it does not ensure any level of scheduling performance—in speed, utilization, or customer service. As a result, schedules must also be continually corrected (or overridden) by the informal system (supervisors calling feeding work centers to send them work or stop sending them work) and the "formal" informal system (expediters who rush around finding jobs that are late or missing and working them through the system). In addition, since the work centers are no longer "coupled" by any scheduling logic, there is no guarantee that a bottleneck work center will be kept fully utilized with a steady stream of work. To help ensure higher utilization, most factories operating with this simple approach buffer their operations with high WIP to guarantee that everyone has plenty of work (and therefore leadtime).

The end result is that customer service and utilization are obtained at the cost of speed and indirect labor cost—expediters. In addition, the expedited runs often break into more efficient setup cycle runs, hurting efficiency. These factories often become the proverbial "black holes" that jobs go into and eventually emerge, without predictability and speed.

The reason for this "chaos" can be easily understood by re-examining the principles of scheduling we enumerated: To make our schedule, we need to fully utilize the bottlenecks and carefully plan our setups there (until we have eliminated setups using SMED methods). In the "common sense" approach to scheduling we are assuming that if we plan enough capacity to meet the schedule, the detailed scheduling rules will sequence the jobs through the facility to meet their due dates *and* fully utilize the capacity planned. We can see the problem with this assumption in Figure 3.14. If operation 2 is the bottleneck and both operations 2 and 3 are fed by operation 1, our dispatching rule must not only worry about meeting customer due dates, it must also keep operation 2 steadily fed with work. In the case shown, operation 2 will "run dry" for 10 hours until a job is dispatched to it. All the "local" dispatching rules we have looked at *ignore* the need for keeping the bottlenecks steadily utilized. They order jobs at an operation based on local conditions *there*. They do *not* look *downstream* at the status of bottleneck operations in setting their priorities. This is one of the reasons companies have gone to a *pull* system—to keep bottlenecks fully and steadily utilized without allowing WIP to pile up at already busy operations.

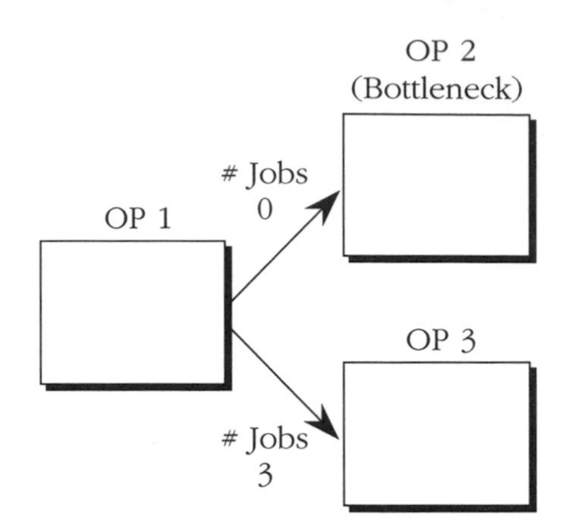

Dispatch List at Operation 1

Job #	Critical Ratio	Next Operation Job Goes To	Hours of Processing Time at Operation
16	1.0	3	4
17	.9	3	4
18	.8	3	2
19	.7	2	4
20	.6	2	3
21	.5	3	2
22	.4	3	4

Figure 3.14 Local Scheduling Rules Ignore Capacity Utilization and Downstream Conditions, Typically Focusing on Due Dates

So, we can see that for a dispatching system to work, we must carefully plan job *releases* to the factory each shift that keep our capacity evenly utilized and extend our dispatching rules to recognize the state of downstream operations. In the next section we will look at more "intelligent" dispatching rule sets that combine the principles of optimization with the ease of understanding the common-sense approach.

One of the growing trends is for companies to use simulation packages to develop schedules. But note that merely simulating these local rules does not improve schedule performance. It simply shows the problems that will result, and then forces the schedulers to *manually* fix all these problems at each operation. The solution is better rules, not simply faster or more simulation.

TABLE 3.12 Summary: Common Sense Versus Best Practices Guidelines

Best Practice Area	Evaluation
1. Resource feasibility in plan and execution (equipment/labor/parts/storage)	While the plan should be approximately (or on "average") resource feasible, there is no release and dispatching system that ensures steady flow of work to the bottlenecks. This failing is usually compensated for by large amounts of WIP.
2. Leadtime feasibility	Typically the common sense approach accepts orders and sets a due date based on standard leadtimes. Infeasible due dates are common.
3. Considers leadtime variance	Not usually considered. We assume an average leadtime and then expedite like crazy.
4. Considers product quality variance	Not usually considered. We assume an average yield or rework rate and use safety stock to fill in as needed.
5. Trades off capacity utilization, cost, customer service, speed	No real scheduling trade-off is being done. We load up to capacity and release jobs against that capacity measure in due date order.
6. Uses up-to-date accurate information	Usually uses historical data on productivities for loading and standard leadtimes for setting due dates. Then dispatches to local conditions and expedites as jobs are late.
7. Is followed	Notorious for not being followed as problems arise.
8. Notifies user/predicts when the schedule can no longer be met	No real prediction except capability.
9. Is easily understood and modifiable	Yes. Just not effective or accurate in practice.

To summarize the common-sense approach to scheduling versus our best practices guidelines (Table 3.12):

1. Resource feasibility is rarely guaranteed if there are bottlenecks. This approach works on the theory of averages. We schedule, on average, enough work into the facility with some form of rough-cut capacity planning. However, the simple release and dispatch rules do not guarantee (or even relate to) the steady flow of work to the bottlenecks. In practice, this failing is often compensated for with large amounts of work in process.

2. Leadtime feasibility is rarely guaranteed. This is again, in practice, a very decentralized approach to scheduling. Typically one group accepts orders, one group does capacity planning, and one group does detailed scheduling. Jobs are accepted using a standard leadtime

against remaining available rough-cut capacity without any knowledge of the actual release schedule. Therefore, it is common to find jobs that are released and are already late waiting for resources not considered in the rough-cut capacity.

3. Leadtime variance is rarely considered. In simple approaches, planning and scheduling is done using the average or standard leadtime. Leadtime variance is not considered. Expediting is used instead.

4. Again, product quality variance is rarely considered in simple approaches. The average or planned yield is used for planning purposes (and rarely updated more than quarterly). As yield losses occur, many scheduling systems don't even predict whether the order will still be met.

5. This approach does not really trade off scheduling alternatives. We load up to capacity and release by due date against that capacity.

6. These systems rarely run off of real time MES data. They are appealing because they do not require a lot of data. We need some estimates of capacity and productivities and then we dispatch to due date.

7. This approach, since it is relatively inaccurate, is notorious for being overridden in practice. We expedite late jobs. We shift operators to where bottlenecks appear. We ignore job priorities to run efficient setups. Since there is no fixed schedule, no one knows when it is not being followed. If there is a simulated schedule, unpredicted events soon make it obsolete.

8. There is no real prediction involved. We have plans based on averages (leadtimes, capacities, and yields), and then we expedite like crazy as problems arise.

9. These systems are easily understood. What is difficult for many users to understand is why they produce such poor results. Hopefully, now we know.

Approach 4: Scheduling as a Hierarchy of Intelligent Rule-Based Systems

If we look at the "tasks" of scheduling, we can divide up the scheduling problem into a hierarchical series of decisions (Table 3.13). In a rule-based or artificial intelligence approach, we will use scheduling rules which

TABLE 3.13 The Hierarchy of Scheduling Decisions

Question	Decision
Do I have enough capacity to meet demand for my planning horizon?	Capacity-feasible out schedule for the next month/week, or production rate for the next month/week
How can we buffer ourselves against uncertainty?	WIP levels needed for buffer/safety stock Planning yields and cycle times to use
What jobs do I release this shift/day into the factory?	Release schedule into the factory (starts)
What is the schedule at each key work center (job/setups) for this shift/day?	Shift schedule at each operation (including setup changes)
What job do I do next at each work center given current conditions?	Next job (setup/maintenance) to do now

incorporate the best practices principles we've enumerated to provide an initial solution to each problem. These rules do not guarantee "optimal" solutions, but our experience shows that there are many near-optimal solutions in real life, and it is more important to have an "implementable" approach than a "theoretically" perfect one that no one uses. Again, we are assuming that this scheduling is used to *complement* the improvement and simplification approach of the Japanese and not in its place.

Our first best-practice principle is that a schedule must be capacity feasible. No amount of expediting will get more work through a bottleneck than its capacity allows. If we are scheduling beyond a week (say for a month or several months), we'll need a rough-cut capacity scheduling system that translates an output schedule by product by day (or week) into an expected load at each key work center, as well as a cost and revenue plan (Figure 3.15). Since we are *not* trying to develop a *detailed* schedule to be implemented but instead assure capacity feasibility, this plan is approximate. It is used as a guide to match demand to capacity—of equipment, raw materials, tooling, labor, and storage—using *linear* relationships (ie, ignoring setups). If there are major setups involved, then we may have to, in fact, schedule *major* cycles of runs within the planning period.

Once we have a capacity-feasible schedule, then we need to develop both shiftly and daily release schedules (jobs to start into the factory) and work center schedules that consider detailed setup changeovers, preventive maintenance schedules, equipment batch sizes, and so on. To do this, we

Out Schedule By Product

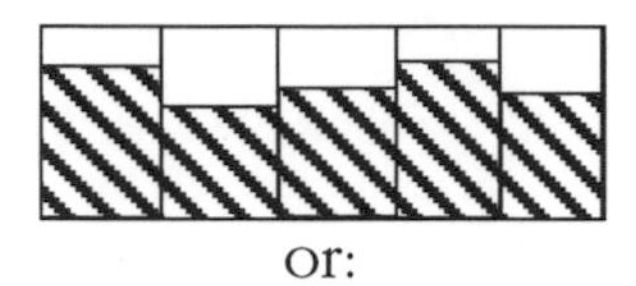

or:

Production Rate Schedule By Product

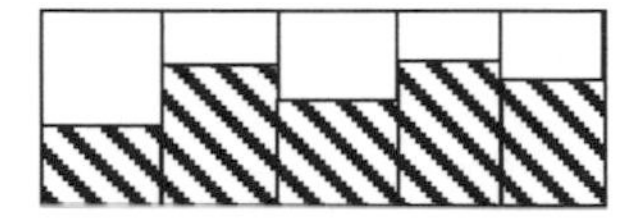

Capacity Load By Equipment Type

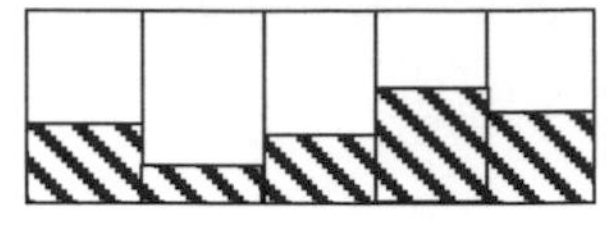

Capacity Load By Labor Type

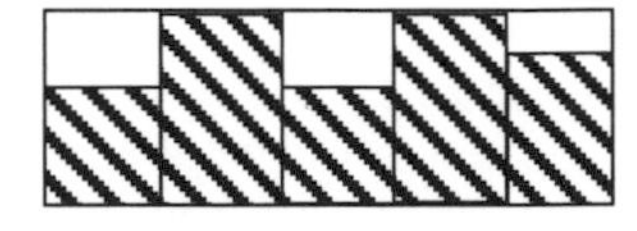

Raw Material Requirement By Part

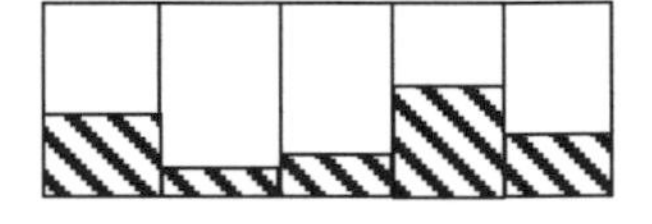

Revenue/Margin

Figure 3.15 Translating an Out Schedule into a Loading, Cost, Revenue Profile

again use our best-practice principles of scheduling. Our approach is to build a set of rules for scheduling that keeps the bottlenecks occupied, maximizes use of setups at bottlenecks, stays within minimum and maximum inventory levels at operations based on our buffering bottlenecks, and allows asynchronous operation production rates. An example is shown in Table 3.14.

Our rules first use the concept of a time-period-sized bucket of capacity that we are filling with jobs to release into the factory for each time period in our horizon. We set the rules for what goes into the bucket based on our *priority* or key measures of performance.

If we examine the release rules shown in Table 3.14 for bucketing (step A), some of them are based on releasing an *even* load of jobs into the factory

TABLE 3.14 Release Rules Set for Shift Scheduling—Examples

A. Global/facility rules—bucketing size:
- Total WIP in line must be less than 6400 units
- Start each high runner daily
- Linear output—daily starts/out must be between 17 and 21% of WTD remaining
- Total daily moves within capacity
- No job starts/moves without all parts available or if the job is on hold

B. Packetize for joint release into a bucket:
- Shared setup on wave solder (width)
- Uses same test handler
- Requires same burn-in operation

C. Order the packets:
- Go from highest to lowest temperature at wave solder
- Go from lowest to highest width board
- By average due date
- By user defined rule

D. Order within packets:
- By rule set—due date, setup sequence
- Schedule

E. Revise packets:
- Move any job more than 2 days late to the next earlier day

to keep the bottlenecks running smoothly. For example, load each day's release bucket with at least one seventh of the planned capacity utilization of each key bottleneck in order of job due date (assuming there are seven scheduled days per week.

Others keep the line balanced so as not to allow WIP to build up beyond allowable levels: Load the bucket with no more than the total allowable WIP in the facility (current WIP in the facility).

Other rules can throw out jobs waiting for raw materials, tooling, or work instructions, and not yet available for release: Don't load (release) any jobs still on hold for incomplete materials, tooling, or work instructions.

Once we have calculated the desired size of the bucket, we can "packetize" jobs together that share major setups or cycles (Step B). This will physically group jobs together for assignment into release buckets that takes advantage of setups. Packetizing rules could look like:

- Group all jobs in a packet sharing the same width (or setup) on the wave solder.
- Group all jobs sharing the same due date.
- Group all jobs using the same handler on the tester.
- Group all jobs sharing the same formula/recipe/color or other processing characteristic.

This rule batches jobs that should move together through the facility into the same packet.

Now we can order the packets (Step C) to take advantage of "cycles" (from high to low temperature, light to dark color, narrow to wide width, low to fine tolerance, and so on) or of similar due dates to fill the buckets.

- Order the packets by increasing or decreasing of temperature, width, tolerance, etc.
- Order the packets by average due date.

The order of the rules we use to fill the buckets must reflect the priorities of the facility. If we are most concerned about speed first and utilization second, we will first limit the bucket size up to the WIP level allowed in the facility, and then fill it to try to keep bottlenecks busy secondarily. If we are more concerned about customer service, we will first load in the orders due out in the normal cycle time and then worry about equipment utilization. Now we have our planned job releases into the facility.

Within packets, we can order jobs by due date, sequence, dependencies, or any other user rule (Step D). Finally, when we evaluate the release plan against our goals (speed, utilization, customer service, etc.), we can revise assignment of jobs to packets and/or buckets (Step E).

For example, if utilization is fine but customer service will be too low, we can adjust the "bucket" rule to force in jobs that are late into an earlier bucket. We can think of "parameterizing" our rules. To change the tradeoffs, we can adjust parameter settings (such as how late a job must be to be forced into an earlier bucket).

The next step is to define the dispatching rules at each key bottleneck for ordering jobs during processing.

The goal of these rules is to move jobs through the facility based on management objectives. Again, we use best-practice rules that implement the scheduling philosophy of the facility—JIT, high utilization, and/or customer service—in the trade-off order chosen.

For example, to keep high utilization, we would use a rule that put the highest priority on pulling jobs to bottleneck work centers that are low on work and maximizing the use of setups. For high customer service, we would use rules that moved jobs that were falling behind their operational schedules. For JIT, we would pull jobs as our WIP level at an operation fell below its Kanban level.

An example of these rules is shown in Table 3.15 for implementing JIT. Table 3.16 shows an implementation for ensuring high utilization. With these rule sets, the facility can, as a *team,* develop their operating rules and trade-offs.

To develop shift schedules, the intelligent rules can be simulated forward given the current job locations, equipment status, labor availability, and the results evaluated for performance—in customer service, utilization, number of setups (changeover), and cycle time. Parameters or rule sets can then be adjusted to change the trade-offs.

TABLE 3.15 "JIT" Rule Set

Priority	Issue	Rule
1	Starvation	If the next OPN group this job goes to has a WIP level less than its critical level (Kanban), then its priority is 1.
2.	Next WIP level too high—don't move	If the WIP in the next OPN group this job goes to is too high (greater than its Kanban), then its priority is 2; don't move.
3.	Next machine down—don't move	If the next OPN group this job goes to is down (preventative maintenance/broken/testing), then its priority is 3; don't move.
4.	Current operation down—don't move	If the current OPN group for this job is down, then its priority is 4; don't move.

TABLE 3.16 "High Utilization" Rule Set

Job Priority	Issue	Rule
1	Next bottleneck starvation	If the next OPN group that this job goes to is a bottleneck and has a WIP level less than, this job has critical level priority 1.
2	Next setup waiting	If the next OPN group that this job goes to is a major setup station and a setup is waiting that matches this job, then its priority is 2.
3	On schedule/FIFO	Otherwise, run jobs in FIFO order as priority 3.
4	Next WIP level too high	If the WIP at the next OPN group that this job goes to is too high, then its priority is 4.
5	Next machine down	If the next OPN group that this job goes to is down (preventative maintenance/broken/testing) then its priority is 5
6	Don't start, can't be run	If the current OPN group for this job is down, then its priority is 6.

Hierarchical Intelligent Rule-Based Scheduling Versus Simulation

Many commercial systems are now available for simulation. The question is whether a hierarchical rule-based scheduling system as we've discussed is simply a simulation tool. It is and it isn't. It is, in that it has a representation of the factory (jobs, routes, equipment capacities, operators, setup times, etc.) and simulates a set of release and dispatching rules through this factory model to develop schedules and projected performance.

Where it differs from most "standard" simulations is in the best-practice principles of scheduling represented in the rules. Many simulation packages use very simplistic rules for scheduling much like the rules discussed in the common-sense approach. Jobs are scheduled based on either dynamic slack or critical ratio. The scheduling "model" does not support sophisticated rules for batching, looking ahead, leveling utilization, and WIP buffering that are needed for predictive, anticipatory, near-optimal performance. As a result, the simplistic simulations give misleading pictures of the true capacity and performance *possible* in the factory.

As a simple analogy, think of simulating the traffic patterns in your city with rules that only allowed FIFO (a stop sign at every corner) or fixed duration green and red signals. What the simulation would show is that traffic immediately piles up at the bottleneck intersections. However, we know that we can dramatically improve traffic patterns with more sophisticated rules that vary the duration of the signal, based on the number of cars queued up (batch sizing), feed cars onto the freeways at an even rate (release and utilization rules), and divert cars to less crowded avenues (alternate routes).

In other words, a simulation package must support the type of intelligent rules we have discussed, or it will give misleading results and poor schedule performance.

Pros and cons. The pros of this approach are similar to those of the common sense approach: It is relatively easy to explain *how* a schedule was constructed (what factors were taken into account) and then adjust the rules to take additional factors into account. The team can meet and *explicitly* list their joint priorities and decide how these will be implemented in practice. The approach is potentially far superior to the local dispatching rules as it handles optimization principles and considers global conditions (equipment status and WIP levels) downstream. Therefore, it tends to generate "good" solutions that can be easily understood.

The cons of this approach are that it is still not optimal and its exact performance characteristics under any operating conditions cannot be predicted. While we hope that the solution is "superior," the optimal solution may be considerably better, giving a competitive advantage. Also, rule sets may give unpredictable results—the application of a variety of rules at a variety of operations under a variety of actual conditions cannot be completely characterized as easily as a very simple rule set (as FIFO or SPT). Therefore, the facility will take some time to understand exactly how a rule set really performs. (Note that there is a similar problem in optimization: Depending on the objective function chosen, the solution may be quite different). Characterizing the sensitivity of the optimization model to the objective function, constraint set, and parameters is also extremely complex for large problems.

Overall, however, it is a major improvement over the first generation of common sense/simulation systems for constrained and/or uncertain facilities, blending the concepts of optimization with those of simplicity.

TABLE 3.17 Summary: Intelligent Rule-Based Scheduling Versus Best Practices

Best-Practice Area	Evaluation
1. Resource feasibility in plan and execution (equipment/labor/parts/storage)	Can ensure resource feasibility in planning and execution through release rules and dispatching if the users choose rules that do so.
2. Leadtime feasibility	This requires operationally scheduling all required tasks or operations prior to accepting/committing to an order delivery date.
3. Considers leadtime variance	This requires using a leadtime fractile that gives the desired level of customer service (instead of the mean). Otherwise, not considered.
4. Considers product quality variance	This requires using a product yield factile that gives the desired level of customer service (instead of the mean). Otherwise, not considered.
5. Trades off capacity utilization, cost, customer service, speed	Done through the construction of rules. No optimization.
6. Uses up-to-date accurate information	Typically used with an MES to provide real time data.
7. Is followed	If the actual dispatching versus scheduled is tracked.
8. Notifies user/predicts when the schedule can no longer be met	If used with simulation.
9. Easily understood and modified	A strength of the approach.

To summarize the intelligent rule-based approach to scheduling versus our best-practices model (Table 3.17):

1. This approach allows the user to develop and execute resource-feasible plans through the release and dispatching rules. However, users could also build release and dispatch rules that would not consider key resources or keep a steady flow of work to the bottlenecks.

2. Leadtime feasibility is more of an organizational issue as discussed before. To obtain feasible schedules, we must know of and schedule all key tasks prior to committing to a due date/availability. If we simply accept orders and set due dates based on a standard leadtime, we'll have infeasible schedules.

3. This approach can consider leadtime variance. We need to use the desired fractile (Fig. 3.4) of the leadtime distribution for planning purposes. Otherwise, it won't be considered.

4. This approach can consider product quality variance if we use the appropriate fractile of the yield distribution (Fig. 3.5) for planning purposes. Otherwise, it won't be considered.

5. All trade-offs of cost, utilization, speed, and customer service are accomplished through the rule sets chosen. We can prioritize our trade-offs. We cannot guarantee maximum or minimum performance on any criterion (eg, no job later than one day; at least 5000 units produced) or determine if any solution exists that would meet our criterion and constraints.

6. These rule-based systems are typically used with an MES. Otherwise, the data does not exist to support the desired rules. They have a symbiotic relationship. The MES supports the use of these rules, and the scheduling package/approach creates incremental return from using MES.

7. To ensure that schedules are followed, we can provide a real time dispatch list that explains why the jobs are ordered that way and monitors whether the schedule is followed.

8. When these scheduling rules are tied to a simulator, we can predict performance and alert users to schedule problems.

9. The rules are easy to understand. What is more difficult is to understand their performance in a stochastic world of unpredicted events. We must simulate the rules to see how they perform under varying conditions.

Or, overall, most of our best-practice criteria can be met with judicious selection of rules. Alternatively, with poor selection, most will be violated.

Doesn't MRP/ERP already do this? A common question I'm asked is, "Doesn't MRP (or its renamed successor, ERP) already do all this? After all, that's why we bought it in the first place."

Let's first look at what MRP does and then grade it against our best-practices model like any other scheduling approach.

MRP takes orders or forecasts demand and subtracts out actual and projected finished goods inventory (from production already under way) to get a net demand or forecast. Then you master schedule that demand across plants to get your worldwide response to net demand. Then that master schedule in each plant is exploded (through the multilevel bill of materials) to get a requirement for additional production of each product and its components.

But MRP makes a very critical assumption in this process: that the availability of parts and rough-cut capacity will ensure that orders will be produced to standard leadtime and standard quality. The standard leadtime and yield are givens and are *independent* of the plan.

ERP is a macroview of the enterprise that focuses on order flow. It is a *material* view of manufacturing. We launch, track, and cost work orders.

Scheduling and MES are microviews of the plant that schedule jobs to finite resources of equipment, labor, and parts in detailed time buckets.

ERP systems really assume scheduling will be done during execution. If they were designed to do scheduling, they would also track equipment status at a minimum. In practice, many ERP vendors suggest or partner with scheduling package vendors to provide more powerful scheduling capability. Unfortunately, *neither* system has the *data* needed for scheduling—for assuring customer service in the supply chain. When we measure ERP against our best-practices model (Table 3.18), we find:

1. In terms of ensuring resource feasibility in plan and execution, ERP does rough-cut capacity planning, but typically on one resource. It does not ensure resource feasibility during execution. Job priorities are typically set by due date.

2. Leadtime feasibility is not normally guaranteed. Orders are taken by an order entry system against an available to promise or available capacity based on standard leadtimes. Actual scheduling of the order is done at a later time.

3. Leadtime variability is not normally considered. Planning is done using a standard leadtime.

4. Product quality variability is also not normally considered. Planning is done using a standard yield or rework rate.

5. ERP systems do not typically trade off cost, utilization, customer service, and speed in their planning or scheduling. Some indicate bottlenecks that must be manually rescheduled. Most dispatch by due date order.

TABLE 3.18 Summary: ERP Approach Versus Best-Practices Guidelines

Best-Practice Area	Evaluation
1. Resource feasibility in plan and execution (equipment/labor/parts/storage)	Ensures rough-cut feasibility plan typically for one resource. Doesn't ensure execution feasibility.
2. Leadtime feasibility	Typically sets a commit date based on standard lead-times and available inventory or capacity to promise. Typically doesn't have a feasible schedule prior to commit.
3. Considers leadtime variance	Uses standard lead times; ignores variability.
4. Considers product quality variance	Uses standard yields; ignores variability.
5. Trades off capacity utilization, cost, customer service, speed	Usually schedules by due date to rough-cut capacity or the scheduler handles bottleneck rescheduling manually.
6. Uses up-to-date, accurate information	Uses up-to-date accurate information on job location, status; no visibility on equipment tool status.
7. Is followed	Doesn't track if the schedule is followed.
8. Notifies user/predicts when the schedule can no longer be met	Schedule runs normally; give exception.
9. Is easily understood	Easy to understand, hard to use.

6. ERP systems normally have accurate, up-to-date information on parts and work in process, location, quantity, and status. They typically have no real time visibility on equipment and tool status or productivity.

7. They normally do not track whether the dispatch schedule is followed on the floor.

8. The scheduling capability can be rerun at any time, and exceptions (to meeting schedules or capacity overloads) are usually noted.

9. The logic is easily understood by users but hard to use in practice given a large number of products and operations.

While this summary may seem surprisingly negative, ERP systems were never *designed* to be execution or scheduling systems. They were designed and architected as *planning* systems. In this role, they need complementary support that many ERP vendors now offer or support through a separate scheduling system. What they *still* lack is the data required to support scheduling!

Implementing Change—Their Recommendation

The final assignment of the course for the team was to develop and then present their approach to achieving 99% customer service.

Strangely enough, the group (except the scheduler and planner) had never thought of scheduling as a big deal—it seemed simply a matter of working on the hot jobs first and then the others. There was usually enough work in the factory to make their volume target number each month and as a result, there was little feedback that there *was* any scheduling problem. Now they were amazed at all the things they had on their list as necessary to achieve a 99% level of customer service. They really wondered if management would accept the proposal when they saw the costs involved. Customer service did *not* seem to be free—not when there was still so much uncertainty in their facility due to yield losses, equipment breakdowns, and rework.

Their list of recommendations were as follows:

1. To assure delivery to schedule, use a *planning* leadtime that was surpassed only 20% of the time (instead of using the average leadtime). Key customers' jobs would be given higher dispatch priority so that their 99% leadtime would fall within the overall 80% fractile leadtime.

2. To assure delivery to order quantity, use a *planning* yield that was exceeded 95% of the time (instead of using the average yield) for JIT or key customers on made-to-order products.

3. To make these recommendations affordable, make a major effort to improve *quality*—of the process, of equipment uptime, of raw material on time deliveries. They would discuss this with the quality group. This was a *key* objective for scheduling success. They realized that they could not afford to schedule *around* the quality problems that were the *root* cause of many of the missed shipments. Unfortunately, quality issues were outside their scope. The team would have to rely on the quality group's success to achieve their customer service stretch goal. It bothered them that such a key concern was outside their control. It seemed the most critical factor to their ultimate success, to decrease the variance of the yield and leadtimes.

4. Buffer the bottlenecks with scientifically calculated levels of work in process to keep their utilization high enough to meet the production plan.

5. For the highest runners, layout dedicated flow lines, run using an asynchronous Kanban approach.

6. For the remaining products, use a functional layout and the intelligent, rule-based scheduling approach. That approach seemed to give them the most flexibility in meeting the team's objectives as a whole. In fact, they had developed a first pass at the rules as a team—trading off their various priorities.

7. They wanted to explore using a commercial MES package to complement their MRP package that could help them implement these approaches with real time data.

Their Rules for Scheduling: Scheduling as a Team Sport

The dispatching rules developed represented the first time that they had agreed explicitly on how to balance their own individual scheduling needs (Table 3.19). Given their customer service charter, they settled on the following rule set: Any job for a key customer that had less than a day of dynamic slack remaining was given highest priority. Second came jobs that would feed bottlenecks that were running low on work. This would help meet overall schedule performance by keeping the bottlenecks fed. Third were jobs that could use the existing setup (if it was a major setup) and were running close to their schedule. Fourth were engineering runs. Fifth were all other jobs, in FIFO order.

In addition, they would not process a job if its next operation was overloaded with work or was in the "down" state—for equipment problems, quality problems, etc.—or if that job was on hold for any reason.

Everyone was reasonably pleased with the rules. They all basically agreed with this priority ordering—first came key customers, then capacity utilization, then using major setups, and then engineering experiments. It seemed fair as long as the rules would be *followed* and not overridden by "loudest voice wins."

For release rules, they used the ones in Table 3.20.

First they made sure that the key customers' jobs were always released into the factory on or before their release date. In addition, they added jobs in due date order up to one day's capacity at the bottlenecks.

TABLE 3.19 Intelligent Scheduling Rules

Job Priority Level	If Job Characteristics Match to	Priority Category
1	If the job's customer is key and its dynamic slack is less than 1 day	Key customer
2	If next operation the job goes to is a bottleneck where WIP is less than or equal to minimum level	Feed starving bottleneck
3	If current setup type at this operation is major and the job's dynamic slack is less than 2 days and its setup matches the current setup	Major setup match
4	If the job is for engineering	Engineering
5	FIFO	Normal
	Don't Process the Job if	
1	WIP at next operation the job goes to is greater than or equal to maximum level	Next WIP too high
2	Next operation the job goes to has status plus down	Next open—down
3	The job is on hold	Job on hold

Then they packetized by board width at wave solder and ordered the packets by increasing width to take advantage of the setup cycle order at wave solder. Finally, within packets, they ordered jobs by due date.

They would continue to use their existing rough-cut capacity planning technique for evaluating demand over the next 6 to 12 weeks. They believed that many, if not all, of their problems had come from not releasing work in a steady flow to the bottlenecks. These new rules would address that issue.

Fine Tuning the Rules—Rule Order and Parameter Settings

While they agreed that, *on paper*, these seemed to be the best set of release and scheduling rules, they also knew that the proof would be in their operating performance. Therefore, they also agreed in advance to measure the performance of the rules at the key operations and *adjust* them accordingly if they were not performing as desired.

TABLE 3.20 Release Rules for Each Line (High Volume Process/Mixed Process) Bucketize—Daily Release

Release:

- All jobs of the key customers due out within the leadtime and not yet released
- All jobs in due date order up to one times the daily capacity of the bottlenecks

Packetize:

- Packetize by board width at wave solder

Order packets:

- From lowest to highest width

Order within packets:

- By due date

Release:

- Up to capacity in packet order

They decided to use the measures shown in Table 3.21. They used the concept of parameterizing the rules and moving their order. If, for example, the utilization at the bottlenecks were too low, they would increase the buffer WIP targets or consider switching scheduling priority rules 1 and 2. If the key customers' customer service was too low, they would increase the minimum dynamic slack allowed before expediting to two days. If there were still too many major setups, they would increase the priority of "early" jobs to allow them to move up in queue and share the setup.

If engineering runs took too long, they would move up that rule to position 3, trading with rule 4.

In this way they could play with the *rules* and not try to develop *individual* schedules for each job. The instructor had explained that many schedulers take a schedule that was not "acceptable" and then try to correct it job by job.

In practice, this was very difficult to do, given a large number of jobs and operations. When an individual job was shifted, it affected the schedule of every other job as well, producing unpredictable overall results. The scheduler was better off adjusting the overall *logic* than trying to find the right schedule job by job. That was akin to trying to solve a three-dimensional New York Times Sunday crossword puzzle.

TABLE 3.21 Measures for Evaluating the Scheduling Rules

1. Planned vs. actual utilization at bottlenecks (if too low, increase the buffer WIP; consider switching scheduling priority rules 1 and 2).
2. Customer service for key customers (if too low, increase the minimum dynamic slack used in prioritizing jobs to 2 days).
3. Number of major setups (if too high, increase dynamic slack to +3 in prioritizing jobs).
4. Engineering experiment leadtime (if too long, move the rule priority from 4 to 3 [switch with 3]).

Predictive Rather Than Reactive Measures

By now the scheduling team had also found the desired approach that would predict scheduling problems rather than detect them after the fact—simulation of the schedule through the facility. If they had an MES with an intelligent scheduler tied to the factory status, they could project the results of the rules on the current factory status. This would let them foresee problems in utilization, customer service, engineering experiment deliveries, bottlenecks, or WIP buildups. If fine tuning the rules did not solve the problem, they could look at more costly alternatives of overtime, subcontracting additional shifts, alternate routes, reallocating equipment, or sending work to other facilities. Again, these could be simulated and evaluated.

It struck the team how much variation or uncertainty affected their predicted scheduled results. If uncertainty was the rule, their schedules would be nearly meaningless in another hour as the whole situation changed again. It was clear that scheduling performance was dependent on quality. Also, it struck them how archaic their factory floor tools were. With an MES, they could instantly view their progress to schedule, be notified of WIP buildups, equipment breakdowns, or late jobs, and reschedule the line. They relied on day-old batch reports, manual totals, and questionable data.

Finally, the last issue that bothered the team was leadtime. Their concern was voiced by a team member. "You know if we could get the leadtimes down, this whole scheduling problem would be much easier."

"These long leadtimes give us the big variances. If every job only took a day or two to be processed, then if we were a little late or early, it wouldn't be a big deal. But with our leadtimes of four to seven weeks, a 20% swing in leadtime performance for any job can mean a week or two schedule delay. From what I see, most of the leadtime is spent waiting in queue. If we didn't have so much work in process, we could do the jobs in a few days!

Then we wouldn't need so much scheduling effort and rescheduling and expediting. The job would be done before we even thought about rescheduling it. All we'd have to do is keep the bottlenecks busy."

Scheduling performance improvement had turned first into quality improvement and now leadtime reduction. The team couldn't argue with this logic. In fact, now they weren't sure why they needed four-to-seven-week leadtimes. "We always have used them," a planner pointed out. "They seem to work, more or less. We've simply kept them in the MRP system all these years."

"But if we go to short leadtimes, then we'll need to get our suppliers to supply us in that same short leadtime; otherwise, we won't have the parts needed."

"Wouldn't that be a joke! We're pushed to be world class and we push it right through to our suppliers!"

"But we could still forecast out six to eight weeks for planning purposes, like ordering raw materials, but only start orders 2 days before they're due. Then we wouldn't have to change our suppliers leadtimes."

"Why should we?" the planner exclaimed. "Do you know how hard it is to forecast out that far? If we only had to forecast out a week or two, we could produce *exactly* what people really want and not what we *hope* they'll want. I'd vote for that short leadtime. Then we wouldn't have so much forecast and leadtime errors."

"Wait a minute," a supervisor retorted. "I need that leadtime to make sure I can keep everyone occupied. Without enough work in process, my guys and gals (he stammered, looking at the operator) will run out of work."

"Not if we fix the quality problems. That's what's killing you. We need to fix the cause and not the symptoms. WIP is just a crutch because we can't schedule with any certainty.

"It's up to us how much we start into the line. We can easily bring down the WIP levels to anywhere we want."

The group was gaining something—self respect. They could all feel a new sensation. They were contributing in a basic and profound way. And they were learning and growing as team players, as problem solvers, as thinkers, and as contributors; and it felt good. It wasn't necessarily easy, but it did feel significantly better than their previous parochial focus. Each was bringing specific expertise and gaining respect from the others. For the first time they felt that they were improving something, not just running in place to keep the status quo. It was the start of their believing in continuous improvement as a philosophy.

Their Scope Defined

While they still hadn't decided what to do about one of their outstanding issues (should they worry about the indirect departments' schedule performance, such as maintenance and engineering?), they had now set their own scope. They understood what caused customer service problems and what they had to do to *really* eliminate them. They would ask the company to drive the quality and leadtime reduction issues as far as they could. Then the scheduling approaches they had learned could give near 100% customer service without outrageous added cost and leadtime.

Since their team encompassed *all* functions, they felt that they could present their approach without fear of having ignored some critical issues or special interest group. Now they just had to work the liaison with the quality team. The drive for perfection in scheduling was in great shape!

Joe was frankly surprised by the turn of events. He had viewed the customer service assignment as an exercise in increasing their finished good inventories until they could meet all orders for key customers from stock. Scheduling was something he (and seemingly everyone else) had little training in or knowledge of. The discussions on handling uncertainty had been the most interesting. Now he had a much better appreciation of what the Japanese had done, and also where it fell short.

All in all, it was turning out to be a more interesting assignment than he had anticipated. He was actually looking forward to presenting to the executive team their progress. All in all, it had been a very good few weeks. Now he had to see if all this actually worked! Somehow he though it might; not because of the theory, but because, for the first time, the whole team was behind one approach. Maybe, just maybe, they would pull this off.

4

Achieving Stretch Goals Through Best Practices in Cost Management

The Top Management Tool: The Budget—an Opaque Window onto the Factory

Bob was frankly still seething after the management meeting. As head of finance, he took personal responsibility for PCB Co.'s fiscal prudence. Having that questioned was difficult. Having that seemingly proven as no prudence at all cut to the quick. He was angry and frustrated, but ultimately he was a top professional. He had unleashed some venom during their staff meeting. Now it was time to see where his tools had failed him.

Ultimately, during his soul-searching, he had reached that conclusion. His failing had been in the selection and continued use of his financial tools. For years now, he had relied, as had Diversified, on standard cost budgets. Each year, he had gained in expertise until he was now a certified master of standard cost budgeting. He could pin down a variance in minutes. He personally knew every aspect of every budget for the last ten years. He knew where the "soft spots" were, where they could eke out a positive variance when they needed one, where someone had signed up for the impossible, and where they were at most risk historically. As proof of his mastery every year they came in slightly under budget, and he sent Mark Ritchards off to corporate safe and sound and, occasionally, even a hero when another division had fallen on hard times.

In truth, he felt he had failed Mark and the team this time. Every year "his" budget, the one he slaved over and nursed in periodic times of fiscal ill health, had passed muster. This year it had detonated like a live hand grenade. His tool standard costs that he depended on had become undependable and maybe even archaic. He didn't like it, and he didn't have to pretend he did. But, what he had to do was fix it or find a new tool they could depend on. That was his responsibility, and he was going to deliver on it.

He started at the beginning, carefully trying to understand why their budgeting process was no longer valid, and in doing so, what to replace it with. He certainly realized that it was no help in one key area—reducing their two fastest-growing cost categories—indirect labor and depreciation. For the last five years, both categories had been steadily rising. Each year, their equipment grew more sophisticated, more capable, and more expensive. As line widths shrank, capital costs grew. New insertion equipment, new testers, new pick and place machines—all had price tags that seemed to double every two or three years, and with capital budgets came depreciation and higher operating costs. What could he do to get their capital costs under control? His standard costing system gave no help there. The second area of concern was indirect labor. Every time he turned around, there was a new requisition on his desk for another head count. All were extremely urgent, well justified, and pure overhead. They needed more people for environmental and OSHA reporting. They needed more customer service reps to handle their JIT customers. They needed more expediters or schedulers to ensure JIT deliveries. They needed more maintenance techs for the new equipment. There seemed no end to the need for additional head count that each department was requesting. Again, the standard costing system gave him no help with whether these hires were justified. It only reported on what they would cost the company. Maybe some new ideas in budgeting or planning approaches were in order.

Bob began his review with the putative budget proposal for next year. It was divided by department and within department by type of expense or line item. Each month he would be reviewing it, each department explaining any and all variances from it. He took the budgeting task extremely seriously. He was an accountant. His world was a set of financial numbers rather divorced from physical product, facilities, or people. It was a set of plan versus actuals with the dreaded variances that had to be explained to his satisfaction.

Budgeting itself was a laborious task. It began with a forecast from marketing of what was going to be sold by product line by quarter. Then any product price raises planned were factored in as well as possible product discounts depending on the competitive situation. This led to a revenue projection and a forecast volume by product line.

Each department was then given this forecast—units and dollars—and, in turn, responded with their plan for the upcoming year, including new hiring, key promotions, any new capital purchases (with justification), key new programs, expected average raise percentages (disseminated by the personnel department), and expected price increases in purchased services and raw materials.

In general, each of the department managers raised their budget by at least the percentage increase in revenues so that the rollup was always over the guideline for profitability and left no margin for exceptions. By the second round of individual meetings, he had "beaten back" each department budget to a livable level (his view of the procedure), and he could bring it up to management for a first unified review. He knew that as long as he grew expenses slightly slower than he grew revenues, Diversified management would, after much poking and prodding (especially on capital items) eventually let their budget go ahead, confident that their "contribution" (profits) would allow some increase in operating profits. If they were in doubt, they would simply delay his capital expenses and then his hiring plan.

This was an annual process that he had been through for years and had, as do most CFOs, become quite expert in winding his way through the shoals and narrows of sharpened pencils.

This year he had expected the same scenario. He had increased margins by .2% to 12.4% by cutting out a new tester for two quarters. He knew that capital was regarded with greater suspicion than head count. If expenses ran faster than revenue increases, they could freeze hiring, eliminate consultants and even, selectively, rationalize head count (the finance department's words for a layoff). However, once a capital item was purchased, it was an ongoing burden (literally and figuratively) on expenses, as it had to be depreciated each year without fail and be properly maintained.

He had felt comfortable with his budget being approved from a "budgeting process"—he had stayed within all guidelines on wage increases, promotions, travel, use of outside consultants, and all the other myriad of guidelines corporate finance put out to guide the budget to a seemingly safe harbor and berth. But the more he looked at the budget, the more he realized that it had become somewhat of a mystery to him. He saw the head count, he saw the equipment, and he saw the materials budgeted, but why they *needed* it if there was a more efficient way to manufacture (if they were truly being productive or just predictable, or linear) was not all clear. It was just assumed (and seemed true) that expenses needed to rise at a similar rate to unit output and/or revenue. That was the way they had run things for years, and it was accurate enough. They did spend the extra budget, and in fact, everyone clamored for even more budget. But if you asked him if he *truly* understood the budget—if he really believed that they had to spend all that money that there was no more efficient way to operate—he didn't know. But looking at the budget, he knew that it met his needs—to be able to justify the annual operating plan to his management. In a sense, it provided him as much a cloak to shield him from their management as it seemed to shield him from what budget he *really* required.

But during this introspective and retrospective review, he was becoming increasingly uncomfortable with this budgeting process. He didn't have a clue how to improve it! He was hearing good things about the quality team and the improvements they were finding. It just seemed to him that budgeting for resources this way hid the *waste* in the operation—that it tacitly approved doing things the same way they had been done every year. He wondered if there weren't a better way to approach budgeting that gave a clear view into the current efficiency of his operations so that he could budget for value-added activities and not for continued waste.

Reflections on the Budget

During this process, Bill pulled out his budgets for the past eight years to see what had changed. As he reviewed eight years of his "budget educational process," several things struck him.

First of all, direct or touch labor had decreased as a percentage of his total expenses from just over 30% to now under 12%. Most of the high-volume insertion had been automated by the large Universal machines (he remembered the battle with corporate over those purchases, along with the increased maintenance, electricity, floor space, and spare parts expenditures they had required). The new designs used fewer parts and more ASICS. The visual inspection and automated testers replaced heavily labor-intensive visual inspection and several loading and unloading test operations. It struck him that they certainly were putting *a lot* of effort into tracking direct labor efficiency for 12% of costs.

Material had stayed constant at around 25 to 30% of costs. The number of components had shrunk, but they had dramatically increased in average purchase price with more complex components and lower power consumption devices. However, indirect labor, equipment, and facilities (the category overhead) had increased from 40% to 58%—more than *half* of the cost he budgeted. Ironically, he had no measures of productivity or efficiency for the single biggest cost component!

Second, in reality, he had not been surprised by Diversified's pressure on them to justify not shutting down the plant and moving the operations offshore. Each year he fought an increasing instinctive knee-jerk reaction that they *should* close down the plant and move manufacturing offshore as so many of their competitors had already done—that a U.S.-based electronics operation *couldn't* be competitive with overseas costs. To be honest, this was something he wondered about himself, which he would never have admitted to anyone else. But he knew that the cost of shutting down the plant as well as the relative inexperience of his management in overseas operations had kept this at an intellectual level and not imminent action until corporate had forced their hand. Personally, though, he wished he had a response to their comments that labor costs in Thailand were one-tenth of theirs. How could they be competitive if this were true?

Finally, he was struck by the increasing number of overhead departments they had created to *improve* their production results. In the last eight years, they had added a quality group, a scheduling group (to assist production planning), a departmental MIS group (to give local support), a test engineering group to develop test programs, and so on. They had managed to proliferate more organizations in the name of "efficiency," each who needed office space, management time, meetings to coordinate with the other groups, personal computers, and secretaries, and yet their cost structure only grew more top heavy.

There it was—eight years of creeping overhead seeping into the status quo. It wasn't something he was proud of. It had just happened that way, year by year, operating plan by operating plan, without much thought. Was there some way to control this bloat? Was there another way to approach budgeting and costing? He wasn't sure. All he knew was budgeting, as they had done it for years, with absorption standard costing to earn the budget dollars by production.

Systems for Tracking Versus Systems for Improvement

The more he thought about (or more accurately, agonized over) budgeting and costing over the next few days, the more he saw it as a tool of the *status quo*. By itself, it normally assumed that the business practices of the past would be continued with proportionately more of each input required as output or revenue went up. The bottom line was that it seemed to violate the new principles of measurement. It gave him no specific view into improvement possibilities.

Budgets had been used for years as a road map. They told you where you had to head in hiring, capital, facilities, and so on, and then let you see where you were during the trip. Unfortunately, it gave no indication of whether that route was an inefficient way to get there (to that output and quality level). Therefore, he needed to find a new way to look at evaluating these budgets—a system that reflected the need and potential for constant improvement. What the lecturer had discussed was the need for new measurement systems. The current ones told you *after* the fact how poorly or well you had performed to your budget (your internal view of standards). The new ones they needed prevented waste, detected problems as early as possible, gave tools for continuous improvement, and told you how well you were performing against *perfection*. Now he needed a way to carry these concepts into budgeting. He certainly couldn't call an accountant for help—they would simply audit the figures to make sure they were completely accurate (even if they were not competitive!). What he needed was a new model for budgeting.

He called several friends until someone had a recommendation: a totally new approach to budgeting. While he waited for the material to be sent to him, he reviewed how useful the current budgeting scheme was with his department financial managers.

Supervisors and the Budget

His managers were hard at work on reviewing last month's budget versus actuals when he walked through their offices. He watched as they read through detailed printouts giving their actual versus earned versus

TABLE 4.1 Direct Labor Cost Variances

Department: Manual Insertion		
	Labor efficiency	
Earned hours		= 1440
Indirect clock hours		= 260
Direct clock hours		= 1480
Performance	= Earned/direct	= 1440/1480
Productivity	= Earned/total	= 1440/1740
Labor variance	= 40 hr. @ $18.00	= $720.00

planned (budgeted) expenses for direct labor, direct material, tooling and allocations for maintenance, computer usage, and all the other indirect line items. He watched one manager go through the direct labor section slowly (Table 4.1).

Since there was a performance variance of some 40 hours (over $700), he now had to explain the problem. To do this, he had to go through the detailed worksheets of the insertion crew which had been key punched and rolled up on the computer. He pulled these out of a file cabinet and began looking operator by operator for any large variances. After more than 30 minutes, he had found some likely input errors. Now he reviewed the actual computer detail to reconcile the two! After another 45 minutes, he had found six input errors (one negative, the rest positive for him) and several questions he had for the lead operator at the work center.

Next he went through the "budgeted" product mix and volume versus the actual product mix and volume. For budgeting purposes, overhead was allocated based on direct labor hours. In the yearly budget, they had planned on producing more of the higher direct labor products. This meant that the standards for the year assumed a mix that had allocated more overhead in this month than was actually earned. In fact, they had run more of the low direct labor products, earning less overhead than had been spent or budgeted.

In reality, they had probably spent what they should have for the *actual* mix. But now that supervisor had an unfavorable variance. The CFO could remember his days as a finance manager, asking production to run the products through at the end of the month that he had personally calculated would keep his variances low, even if that wasn't the mix planning he

TABLE 4.2 Rework, Scrap, and Inventory Variance Reports

Quality Accounting	Units	$
Planned rework	168	$ 336
Actual rework	174	$ 348
Variance	6	–$ 12
Planned scrap	74	$ 370
Actual scrap	90	$ 450
Variance	16	$ 90
Planned inventory	960	$1920
Actual inventory	890	$1780
Variance	70	+$ 140

asked for. You got a "nice going" for meeting or exceeding budget contribution. He had spent almost none of his time trying to improve the standards. Once the budget was set, you shot to meet it or beat it by a little—not to kill it. Otherwise, in the next year's budgeting cycle you'd pay the price: you'd be seen as a sandbagger and your budget cut accordingly.

Bob knew the drill.

The manager then filled out error correction reports, which would update the run for a revised monthly accounting. Next he went on to the rework, scrap, and inventory reports (Table 4.2). They had run slightly higher rework and scrap but lower inventory due to a shortage of a key component needed to start jobs. They would miss the output schedule this month on that product. However, overall, the total cost variance was positive, as the lower inventory valuation offset the higher scrap and rework costs.

Now again, he watched the supervisor review the actual scrap and rework counts reported against job tickets to see what was scrapped/reworked and why. Unfortunately, there was no notation of the cause on the job ticket. For that, the supervisor needed to pull out a separate quality report.

This process went on for most of the day as each total was checked against the detailed entry forms for input errors, then for explanations to allow a written variance report. Finally, the manager took a stab at comparing the actual product mix to the budgeted mix to explain away any further variance.

The sheer waste of this process that he had watched (and done himself) for years all of a sudden struck him. They spent all their time filling out paper justifying slight deviations off a budget that they couldn't really justify. They spent hours filling out forms that were only useful to a corporate financial system miles away. Somehow they had become slaves to a financial system that was of little help in running the factory competitively. They spent all this time outputting paperwork to justify themselves instead of using it for improvement. And where had it gotten them? But what was the alternative? How could they combine the need for financial data—for measuring progress to budget—with the real need for data for improvement, and, therefore, better actual product costs.

He was looking forward to seeing the material that was being sent. As he left, the manager was now moving on to review maintenance, setup, tooling, materials, and other indirect cost categories (that were allocated) variances. It really struck him how much time went into justifying the status quo—*even if it wasn't competitive!* The budget became a huge magnet drawing them back each month to the same conclusions. There had to be a way to break its hold on them and see the factory in a more *operational* light—the way product was made or equipment used or materials consumed—some way that gave *real* insight into the way to *improve* operations and not simply chart their use of the budgeted funds. Even the review and reconciliation process seemed a waste. He wondered if he was going crazy, but what he did as CFO seemed less and less relevant at this moment.

There had to be a better way to collect, analyze, and report on the financial data. He saw a day a month just wasted on the reconciliation process with no positive outcome. There had to be an approach that was real time without these long lags.

What he wanted was to move away from batch after-the-fact financial tracking system to a real time *improvement* system. What he wanted to do was add value as a CFO.

Accounting Systems—In the Past for the Past

When he came in the next day the material had arrived. It went through the options for accounting or budgeting, starting with the standard cost accounting system that they used today. In that approach, used for the last forty or more years in the company, standards had been set for each product's use of direct labor, by labor class or type, and direct materials in-

TABLE 4.3 Indirect Departmental Budgets

Maintenance department budget	$500,600
Production scheduling	180,000
Facilities	1,200,000
Manufacturing engineering	780,000
Quality/reliability	210,000
Tooling	480,000

cluding waste and recovery (or rework) percentages. This information was in the product specification, repeated in the MRP II system in the bill of material and summarized in the standard cost system. Any other indirect costs were reported and budgeted by department such as maintenance, production scheduling, facilities, purchasing, manufacturing (or support) engineering, and so on. These departmental budgets included their own labor, materials, capital, and other "owned" costs. In addition, there would be an overall administration cost—of plant management and possibly finance and data processing (Table 4.3).

Once these budgets were approved, standard costs were assigned to each product by combining the direct labor and direct material costs with a standard cost for all the "shared" indirect costs not attributed directly to individual products. For example, in Table 4.4 the product standards are derived.

Marketing has projected sales levels for each of our products (or product families sharing nearly identical material and labor content). Using our engineering standards, we can calculate the total planned direct labor hours (or dollars) and direct material dollars. From the budgets for each indirect activity, we can "allocate" how much indirect cost each direct production activity "absorbs," by allocating it based on either direct labor or material costs.

For example, the total maintenance budget is $500,600. If we allocate it based on labor hours (a total of 600,000 for the year), then the "burden" would be $.83 per direct labor hour of a product. Therefore, product A will be burdened by $4.98 of maintenance per unit produced. In turn, we can allocate all the other indirect department costs to either direct labor hours or direct material to derive a standard cost per unit of each product. We also could have allocated or distributed these costs based on direct machine hours (perhaps a better factor for allocating maintenance).

TABLE 4.4 Calculating Product Cost Standards

		Projection: Number of Units	Direct Labor	Direct Materials
Product	A	300,000	6 hours	$10
	B	240,000	7 hours	$12
	C	160,000	5 hours	$19
	D	90,000	8 hours	$10
			Budgeted	Hours/Dollars
			Total direct labor:	600,000/$9,000,000
			Total direct material $:	$12,000,000
Indirect cost allocation process: Maintenance:		$500,600 divided by 600,000 direct labor hours = $.83/direct labor hour		
Production scheduling:		$180,000 divided by $12,000,000 direct material $ = $.015/ direct material $		

In general, for companies making standard products (as opposed to pure job shops where every job is custom), standard absorption costing is the most common approach to translating budgets into product costs and later, comparing "earned" dollars against budget/actual costs.

The Fallacy of Standard Costing for Decision Making

As Bob was beginning to understand, his standard cost approach had no view into the likely sources of waste in this budgeting/cost allocation process he kept reading. If we were extremely profligate in our use of maintenance or scheduling resources, all we have done is allocate that waste across all our products. So, to begin with, the standard costing approach offers no real insight into where the real waste in our budget is. It is simply an allocation of the status quo from one lump sum into per unit costs of product so we can compare actual cost to budget based on a volume of production activity.

Unfortunately, this allocation is not necessarily even "equitable." It is based on the assumption that direct labor hours, direct material consumption, or direct equipment hours are an "accurate" allocation scheme or surrogate for usage of these indirect resources. In the years prior to 1960, direct labor and material did often account for more than 70% of production costs. However, over the last 30 years, they are representing a smaller and smaller fraction: direct labor is often 7 to 15% of total cost and direct materials 20 to 30%. Or, in most high-tech companies, overhead costs now represent the majority of costs. Therefore, we may be allocating a pool that is three to four times the size of the *allocation* basis. This means that slight variations in actual direct labor usage by products earn/cost 3 to 4 times that variation!

In a sense, we are now really blind as to where the majority of our costs are *really* going by product. The allocation scheme should, in principle, be reversed—allocating the rather small direct labor cost based on *true* overhead cost.

In fact, there is considerable evidence that direct labor allocation schemes lead to inaccurate cost allocation among products. Consider the more and more common replacement of labor-intensive operations with highly automated equipment for machining, robotic assembly, visual inspection, packaging, painting, material transport, and so on.

The consequence of these replacement decisions is a dramatic increase in *indirect* costs—of maintenance, capital depreciation, electricity, spare parts, and so on. But under our allocation scheme, the products we have automated now have lower labor content and so pick up a relatively low percentage of the indirect costs that really are *directly* attributed to their automation. On the other hand, the older labor-intensive products pick up the majority of the automation costs that they never use, implying that they suddenly have a lower margin! The allocation scheme leads to inequitable cost allocation, and therefore, possibly to unwise product decisions.

Another common example of this inequitable allocation problem is the situation of the new proliferating product family versus the "old" stable high runner in a mature and cost-competitive market. In our case, we may have a board that we run in very high volumes almost continuously. In fact, we keep many machines informally dedicated to this product and have regular material replenishment policies to keep the line stocked. We have also just initiated a new family of flexible boards, each now being produced in fairly low volumes and requiring substantial supporting engineering efforts (as the flexible boards are more difficult to reliably surface mount components without adhesion failures).

As we introduce the new family of boards, our indirect budgets and costs increase substantially in most departments: purchasing, which now has a whole new set of parts to order; scheduling, which has many more runs to schedule; setup, which has many more changeovers to schedule; manufacturing engineering, which has substantially higher support required to work out the process problems; quality, which is working on new reliability tests; and so on.

Imagine the surprise of our high-volume product line to now be told that with these higher overhead costs now allocated to them that they are now barely profitable and will have to find substantial cost-reduction potential if they are to be continued as a viable product line. It is inequitable allocations such as these that lead companies to literally divisionalize so that all costs allocated to a product line are also truly under their control. Otherwise, as in this last case, the budgets to be allocated go up but are allocated (in total dollars) mostly to the high-running product which uses little to none of the added cost.

So, as we can see, standard allocation or absorption costing doesn't lead to an accurate operational view of costs—in fact, it distorts them! It is therefore neither a good tool for accurate product costing nor "waste removal." Yet it is a primary tool whereby companies have made major decisions on product line profitability and buy versus make. Therefore, a new product costing approach, activity-based costing, was developed.

Activity-Based Costing (ABC)—Another Panacea

In response to the inaccuracy of standard cost allocations based on direct labor, material, or machine hours, activity costing has been raised as a more accurate product costing technique. In activity-based costing, each major indirect cost is still put in a pool. The key difference is the care put into the allocation scheme, which tries to allocate cost based on a most accurate surrogate for *actual* usage of that indirect resource, and the choice of pools, to separate them out by more homogeneous customers for that resource.

For example, we might divide our single maintenance pool into two pools—one covering the new automated cells and one covering the older, more manual lines. Similarly, if we thought that there really were different manufacturing engineering groups—one dedicated to the new flexible lines and one to the older products—we could break out the single cost of manufacturing engineering into two pools (or as many as were required to accurately represent fairly independent groups within a department). For example, schedulers can be assigned or allocated by product family, maintenance techs by class of machine or technology, and manufacturing engineers by product family or process technology (surface mount versus through hole or flexible board products versus regular boards); facilities may be assigned by area of the plant; and so on to more accurately allocate costs.

Once we have chosen the pools that truly represent "homogeneous" product/line/cost/resource utilization, we can then decide on the allocation scheme. For example, we may allocate production scheduling cost by *number* of runs or orders scheduled and not unit volume, as the cost may be related to how often a product is scheduled and not how many units at a time are scheduled.

Purchasing costs may be related to the number of parts in a product and not its dollar value, as each part must be ordered, inspected, paid for, and inventoried.

Some costs, such as inventory holding cost, may be allocated based on both *direct* material costs and leadtime. The high-volume products on dedicated lines may flow through the factory far faster than the small jobs that share facilities.

The approach of activity-based costing (or budgeting) leads to a far more *accurate* picture of standard product costs and *what* activities are used in each product. However, from an improvement picture standpoint, activity-based costing is simply a more equitable way to allocate the waste! In other words, it is simply a better allocation method for the *current* cost structure (status quo) than any real view into the *waste* in the factory improvement. When allocating the maintenance department budget to automated and more manual operations, we still cannot easily judge what the maintenance department budget *should be*—what it could be in a world-class company! We still have not met our budgeting goal of assisting in getting costs under control.

The problem still remains that in both of these approaches we have been solving the *wrong* problem. We have been deciding what the "real" current costs of our products are in our current facilities using our current methods, systems, and procedures rather than what they *could* or *should* be. This is an example of type III error—the right answer to the *wrong* problem.

What we really need to do is *first* decide what the target costs of our products in our "improved" facilities using improved methods, procedures, and systems could or should be. *Then* we can use these more accurate allocation schemes to measure what our realistic competitive product costs could be and use these costs to make key decisions on *how* to use our budget—our limited resources.

Every manager is faced with an ever-growing list of cures for all that ails them. ABC has been added to a long list of acronyms that includes MRP, MRPII, ERP, CIM, TQM, SMED, JIT, OPT, and TPM, as well as quality circles, operator empowerment, flattened organizations, decentralization, and so on. The question is which are fads and which are really innovations (if you know all these acronyms, there probably is no hope for you). The easiest way to differentiate between the fads and true revolutions is which keeps the status quo (only painted a different color) and which significantly improves the organization's competitiveness.

ABC in these terms is more fad than revolution. It is the next logical step on the path of standard costing. It is not a program that reduces costs or would change your basic competitive position.

Bob continued his reading with great interest.

He now saw the problem with his financial system—it simply took the current reality and sliced it up different ways. With activity-based costing, he could be more confident in his product costing accuracy. It just wouldn't help him competitively! All it did was make his allocation of waste more accurate. What he needed more than ever was a budget "improvement system" to let him know what his product costs *might* be able to do (if he could wring out the waste) and where he should focus his cost improvement efforts.

What he needed was a *manufacturing*-oriented accounting system—a continuous improvement-oriented accounting system. But he could see that first it would have to incorporate a *theory of manufacturing* and not a theory of accounting. It needed a new view of the plant. He read on, hoping for a better answer.

A Model for Improvement: Value-Added Costing

To determine how to eliminate waste from the factory, we need to examine the source of our costs *prior* to allocating them (equitably or inequitably). In other words, if this were a camping trip, *before* we start putting items in each person's knapsack distributing the weight load fairly, we need to eliminate the items we don't really need! Once we have distributed all the items, the focus shifts to each knapsack (our product costs) where it's much harder to see the waste in our departmental operations.

Instead, let's look at the resources that make up our total budget and see where real inefficiencies currently exist. Our approach is fairly simple. The total current budget for the facility and therefore for our production, is the sum of all the resources used in their facility times their cost/unit. The target *possible* budget (*improvement goal*) is the sum of all the required resources times their cost/unit times their *value-added efficiency* factor, the percentage of time that resource adds value (versus waste) (Fig. 4.1).

To determine possible product cost, we could allocate this possible or improvement budget, using activity-based costing by product where we allocate the resources to the products that consume them. But first we will look directly at the resource/department and how *efficiently* that department itself is run/staffed/planned. Then, after we have set improvement targets, we'll look at the current and *possible* product costs.

What the equations in Figure 4.1 say is that the real immediate and controllable waste in our budget comes from the *non-productive* (*or wasted*) *utilization of each resource in the facility*. If we could look (either on a sampled or continuous basis) at how our *resources* are expended and categorize the activities as productive or waste, we could identify the *real* potential for cutting costs in our budget. In other words, we will focus on the productive *utilization* of each and every resource in total and then look to eliminating that waste in our continuous improvement programs. This means that our focus is on the resources and *not* the products.

Second, we can then look at our cost per unit of each resource versus our competitors' cost or the cost at alternate locations (such as an overseas plant) by separating this issue out from our *efficiency* in utilizing what resources we already have; we separate out the *underlying* causes of any cost differentials we suspect between efficiency and unit cost/resource.

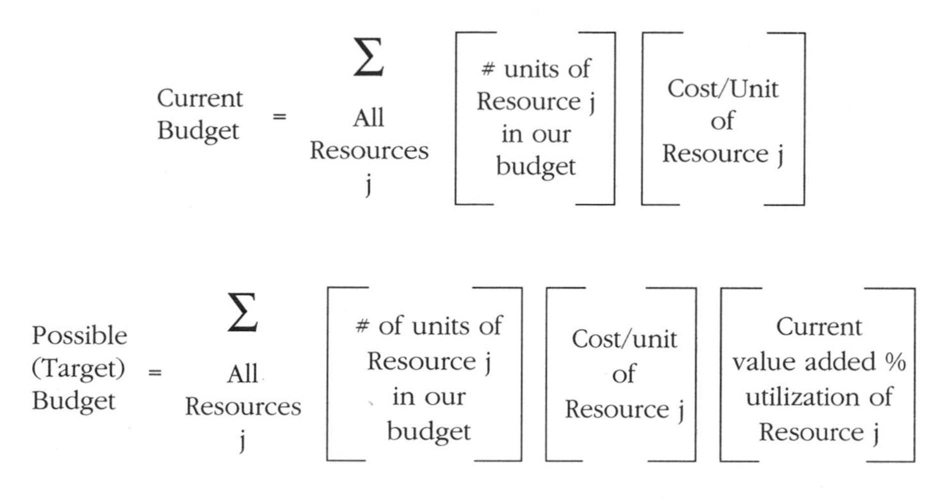

Figure 4.1 Total Current and Possible Budget

The final analysis would look at alternative production technologies for making the product—substitution of alternative machinery, tooling, processes, labor skills, etc., that could give cost savings at higher (or equivalent) levels of quality, reliability, flexibility, etc. In this approach, we would have to look at the resource requirement for products under each alternative production technology. This can be done outside our "improvement budgeting" approach.

In summary, we are going to outbreak our budgeting analysis (and cost analysis) for improvement into three components: how *efficiently* we use all our current resources (the five elements of manufacturing) in manufacturing, how competitively we obtain these resources, and how we can improve the *effectiveness* of the manufacturing process by changing the process or product specification (as reflected in ratio and type of the resources required for manufacturing). (See Table 4.5)

For example, we may look at substituting operator activities for maintenance, setup, or repair activities or substituting automated visual inspection equipment for manual inspection. Bob thought back to the consultant's presentation initially on value-added utilization. This was right in line with that presentation, looking at their value-added utilization. This was just applying the concept in setting budget targets. He continued reading. The next section covered resource efficiency in more detail, discussing equipment utilization.

TABLE 4.5 Best Practices in Budgeting

Concept 1: Efficiency	Tool
Elimination of all non-value-added activities/ utilization of each resource	Value-added utilization chart
Concept 2: Competitive Costs	
Minimizing costs for each resource purchased	Benchmarking
Concept 3: Effectiveness	
Selecting the most productive/efficient/effective methods for each value-added activity	Benchmarking

The value-added efficiency of equipment. In order to measure the value-added efficiency of equipment (and tooling), we need to monitor how every hour is actually utilized. This will give us an exact breakdown of hours by category of usage versus most budget scenarios, in which we assume a certain production level translates to 100% utilization at standard utilization. Such a breakdown is given in Table 4.6 for one of the bottleneck workcenters in our factory. For a machine that is scheduled six days a week, three shifts a day, and considered "110% utilized" (at standard), we can see that, in fact, we only get 32% production efficiency! In other words, two-thirds of the time we have no value from the machine from a customer's or *production* viewpoint. If we add in engineering and test runs (as additional customers), we are still productive only 40% of the time (assuming that those 10.1 hours are completely efficiently used)!

It is common for this value-added productivity number, even at the bottleneck, to run *under* 50%. How is this possible when we usually *see* work piled up in front of it and seemingly constant activity? The answer is that many of our "efficiency wasters" visually look productive (or at least active), such as producing scrap, doing rework, being set up, calibrated, maintained, or doing test runs, or engineering experiments. Even a repair being performed looks like activity. The machine is only seen to be "idle" for 16% of the time when it is not scheduled to be utilized (on Sunday). So visually, there is activity 84% of the time. It's just not productive activity—only "budgeted" activity.

Bob continued reading, beginning to see budgets in a new way. His old approach, trying to shave off a few percentage points of cost each year,

TABLE 4.6 The Value-Added Utilization of the Bottleneck Work Center

	Utilization		Time in Category	
Perfection		Category/State	Actual Hours	Percentage of 168 hrs
x	1	Production	53.8	32%
	2	Rework	5.0	3
	3	Scrap at this operation	2.4	1
	4	Scrap at a later operation	6.8	4
?	5	Test runs/engineering	10.1	6
?	6	Setup	13.4	8
?	7	Preventive maintenance/ calibration	16.7	10
	8	Broken, awaiting repair	8.4	5
	9	Waiting to be repaired (no parts)	5.0	3
	10	Idle (no work)	7.0	4
	11	Idle (no operator available)	6.4	4
	12	Idle (no operator scheduled)	24	14
	13	Being repaired	5.0	3
			168	100%

Current "perfect" utilization (x)	32%
Current questionable perfect utilization (?)	28%
"Waste"—*total*	40%
"Waste"—commissioned (unscheduled)	14%
Waste—omission	26%

was "status quo budgeting." What this approach suggested was how to set "stretch budgets."

The goal of a world-class company is to raise that value-added productivity number as close to 75 to 80% as possible, assuming that we need the slack time for flexibility, experiments, and maintenance—the only two other value-added components of our list are done as efficiently as possible! We may deliberately want even more excess equipment capacity to allow maximum flexibility. However, our goal is to eliminate all *non*-productive activities on our list (Table 4.6); all remaining categories should exist by design and not by mistake.

From looking at this list, we can see that at this work center, we have potentially budgeted close to 60% for nonvalue-added activities. We need to eliminate this waste from our operation before we consider moving the factory operations (and this level of inefficiency) to another location where labor may be cheaper. The point is that there is considerable *operational* waste under our control that we can eliminate before it is necessary to change the structural resource cost disadvantage that may also be present in our current location.

If we look at the major causes of equipment inefficiencies, we can break them down into: quality problems, setup procedures, maintenance procedures, repair procedures (including dispatching, stocking of parts, and operator maintenance), and production scheduling. We will look at developing an improvement program for each one in turn.

Improving Machine Efficiency Through Improved Product Quality

Near-perfect product quality is a critical goal in a budgeting program. In many companies, product quality problems are not found until final testing. This means that *every* operation's added value was *wasted*, leading obviously to inefficiency of the equipment (even if the work at that equipment was perfect). Therefore, the first step in any equipment efficiency program is automatically achieved by the success of a quality program—the elimination of scrap and rework. This illustrates an example of *indirect* causes of equipment inefficiency: the equipment may be working perfectly and, in fact, producing efficiently, but because of problems *elsewhere* (in scheduling work to that operation, in dispatching a repairman to that piece of equipment, in carrying out maintenance on schedule, in processing the product properly at subsequent operations, in using the correct materials, or tooling at that operation), that equipment's efficiency is wasted. This is just another reminder of how manufacturing is a *team* sport requiring the coordination and synchronization of all five elements of manufacturing for complete efficiency and effectiveness. It is not sufficient that any *one* department be successful—the whole factory must be successful to achieve a competitive position.

So the first improvement program involves elimination of scrap and rework (categories 2, 3, and 4 in Table 4.6).

Part of any quality initiative is a program in design of experiments to verify our hypotheses for process improvements. The natural variation in the process parameters is a rich source of data for analyzing the relationship of product quality to the process settings. However, this analysis must be augmented by systematic experimentation to prove these observed relationships statistically (within a chosen level of certainty). Many companies use single variable experiments with ad-hoc sample size. This approach is often neither accurate nor efficient. If the sample size is too small to conclusively prove or disprove the hypothesis, no value has been gained. If the sample size is too large and/or the experiment could have tested a variety of hypotheses, it was inefficient.

A common practice in many companies is to take a normal run quantity and divide it in half, with each half used to test a different processing condition at a *single* operation. Sometimes intuition is best supplemented by a scientific underpinning. It is possible that we could easily prove or disprove the hypothesis with half of this sample size, as well as test several other hypotheses at other operations—leading to a more efficient use of equipment (as well as other resources). Scientific design of experiments improves our efficiency of category 5 in Table 4.6.

Improving Machine Efficiency Through Improved Setup Procedures

Most industries view setup costs and time as a "price" of production: they are a natural result of the equipment design and our operating procedures. The only control we have over setups, it is assumed, is through scheduling—careful sequencing of the products run on the equipment to maximize use of existing setups or minimize changeovers. This is a natural goal of any scheduling approach. However, we can go well beyond this limited improvement to attack the setup effort itself through an approach initiated by Shigeo Shingo called SMED (single minute exchange of die).

SMED is a systematic methodology to reduce setup times by dividing the setup effort into external and internal work—work that can be done away from the equipment itself in preparation for the setup and work that must take place on the equipment itself. In the simplest example, we can make sure that all the tooling required, work instructions, and personnel are at the machine prior to beginning the setup. This eliminates lost time waiting for the necessary resources. In more fundamental work, we can work with equipment vendors to simplify the setup or changeover effort. This minimizes our lost utilization from category 6.

Improving Machine Efficiency Through Improved Maintenance Procedures

The area of maintenance is just beginning to receive the attention it deserves. In the past, maintenance was usually performed at preset intervals (as specified by the equipment manufacturer) by the maintenance crew. Operators were uninvolved in this process. They either moved to other equipment or remained idle while maintenance occurred unless it was done off-shift.

Today there are several new approaches to maintenance to improve both quality and efficiency. These are: statistical analysis of the optimal maintenance interval, triggering of maintenance based on product statistical quality control, or process control alarm (for setting the maintenance interval or ad-hoc maintenance); routine maintenance performed by the operator (freeing up the maintenance crew to work on equipment improvement); separate maintenance shifts for the entire cells, line, or facilities so as not to disrupt production intermittently; and computer-assisted maintenance programs with graphics and video specifications or computer-aided instruction and guidance.

When the manufacturer sets recommended maintenance intervals for an equipment model, they are doing so on an "average" basis for all units at all points in their life. Actual requirements may vary by individual unit and its age/usage. Instead we can use statistical analysis on product quality measures and/or internal process monitors to detect when a machine (or tool) needs recalibration, possibly before or after its normal schedule. In Figure 4.2 we can see that by monitoring the temperature deviation, we can ensure that the process stays within our specification bounds (needed for product quality) and also decide when to recalibrate the equipment and reset the thermocouple.

Similarly, for equipment with catastrophic part failures, we can monitor frequency of breakdowns to try to set a calibration, maintenance, or part replacement strategy. In Figure 4.3 we can see that 95% of breakdowns for part failure occur after 14 hours since the previous failure. This would lead to two programs—one that might schedule maintenance at a 14-hour interval and a second, more importantly, to improve the equipment reliability. This improvement program would be no different than any product quality program where we look for any systematic causes of early equipment failure.

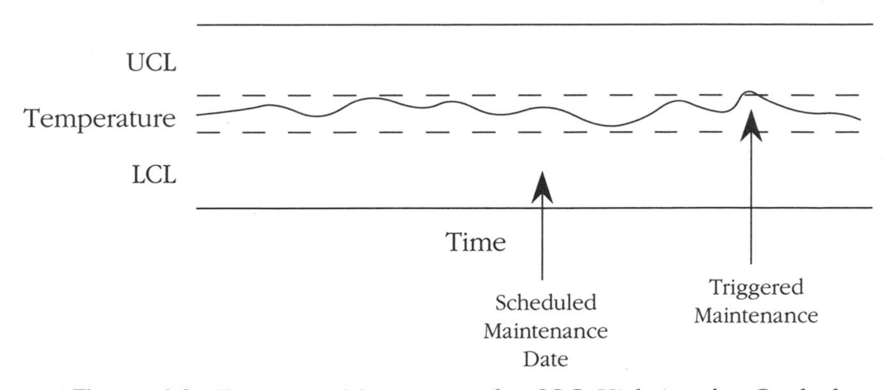

Figure 4.2 Triggering Maintenance by SQC Violation for Gradual Degradation of Performance

Such an analysis may suggest a correlation with the last repairperson (the repair may not have been done correctly), vendor of the parts replaced, the number of units processed on the equipment (perhaps a better indicator for setting maintenance intervals than hours since last maintenance), the operator running the equipment at the time of breakdown, and so on. These analyses are used to eliminate the systematic causes of equipment failure and therefore set optimal maintenance schedules or triggers.

Improving Machine Efficiency Through Prevention of Breakdowns

In Table 4.6 we see that breakdowns currently account for a loss of over 18 hours (categories 8, 9, 13) or over 10% of our total capacity and over 12% of our manned or scheduled capacity. We have broken this total into three types of "waste": broken, awaiting repair; broken, waiting for parts; and the repair time itself. We have done this because each category has a different improvement program.

However, in world-class factories, the real goal is not to improve the repair system but to *eliminate* the need for a repair system and unscheduled downtime. Our goal is to diagnose equipment problems prior to actual failure—catastrophic (as from part failure) or gradual (as from wear or leakage). We would normally look at the most common causes of equipment breakdown and then look to either preventing them (as was discussed in the previous section on scheduling intervals for maintenance) or detecting

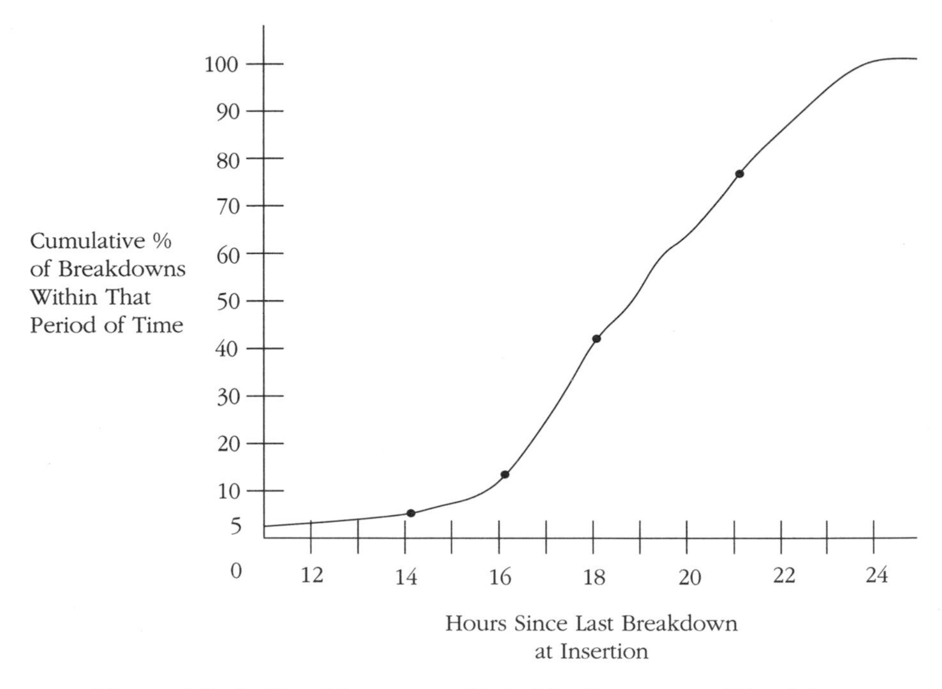

Figure 4.3 Setting Maintenance Period by Frequency of Breakdown for Catastrophic Failure

them prior to failure. For example, in Table 4.7 we see that the most common cause of downtime on the insertion equipment is set up by jams of the components during insertion.

We would attempt to prevent this problem by additional training of the setup crew and/or more exact setup instructions and/or more careful testing of the setup itself. Similarly, we would systematically attempt to eliminate the root causes of each breakdown category. Eliminating all breakdowns is nearly impossible without sophisticated in situ monitoring and measurement equipment that can detect changes in internal equipment (or tool) operations prior to failure. Equipment improvement is also an area that most manufacturers need to work on cooperatively with their equipment suppliers. By carefully monitoring mean time between failure, most common part failure/replacement, most common downtime cause, process variation, and other key equipment characterization parameters, companies can provide their equipment vendors with key data needed for improvement programs.

TABLE 4.7 Common Causes of Unscheduled Equipment Downtime in a Week at Insertion

Cause	Number of Occurrences	Total Downtime
Jammed component	12	128 minutes
Improper setup of equipment		
Jammed component	9	108 minutes
Bent wire on component		
Jammed component	4	32 minutes
Misalignment of insertion to hole by x-y drift over time		
Missing component	1	16 minutes
No component on tape		

Improving the Repair Process

Once equipment has failed, there are four components of the time to repair: the time until a failure is detected and maintenance (or whomever is responsible for repair—possibly the operator himself) is notified, the time until a repairperson is assigned and available to repair the equipment, the time spent waiting for a needed part not in inventory, and the actual repair time itself. We have broken down repair or downtime into these four components because each one has a different improvement program.

While detection and notification of breakdowns to the maintenance department may seem automatic, we have seen many factories with long lags between breakdown and repair. Some lags occur because the breakdown is on the second or third shift, when few (or no) repairmen may be available. Somehow, the information may not filter back to maintenance for some time as the first shift assumes they already know. Other lags occur because of the physical distance from the factory floor to the maintenance department—paperwork on the breakdown may take hours to reach maintenance. Others occur because of the "emotional" distance from the factory floor to maintenance. The operators may not rush to call maintenance, as they don't feel any urgency to get the equipment back up, or they may fear accusations of misuse of the equipment.

It is for these reasons that most Japanese companies and many American companies have made major changes in their repair system. First of all, they put green, yellow, and red lights above each piece of equipment in easy visual line of sight. Equipment that is working correctly has a green light shining. Equipment running but trending into warning limits (of product quality or internal process parameters) has a yellow light, and equipment that has broken down or gone outside of quality specifications has a red light. This allows constant visual monitoring (and detection) of equipment problems. At a minimum, the repair system must ensure an immediate call, electronic mail message, or other notification of a breakdown to maintenance. Also, these companies typically place maintenance and repair groups on the factory floor so that there is no (or minimal) physical distance from equipment operations to equipment repair personnel. They are part of the overall manufacturing team and not there only when problems arise.

Finally, the trend is for the operator himself to be responsible for common or routine maintenance and repair activities. This obviously cuts the lag time until repair to near zero. It also helps eliminate the gap between cause and effect that is so critical to continuous improvement programs. In all such programs, we need to learn the underlying cause of waste or uncertainty so that we can eliminate it. From a production viewpoint, the operator has the primary view of cause and effect in breakdowns and so is a natural candidate to improve them. The operator knows when the setup was changed, when the tooling was changed, if they're using a new work instruction, and if the raw material "seemed" different. However, this added responsibility for operators requires an active training program (as most continuous improvement programs do) over a sustained period of time.

An alternative is a team solving approach, using both the operator and maintenance (and possibly engineering) to troubleshoot the cause of downtime and repair the equipment.

The second source of waste is the lag from notification to assignment and availability of a technician. Maintenance departments have a schedule of maintenance and repairs to perform. When a repair is requested, technicians may already have a long list of previously assigned tasks and, in fact, be in the factory at work. However, all equipment is not created equal. Equipment at the major bottleneck limits production. These pieces of equipment must be given a higher priority for repair, just as a key order may need (until the factory runs on a true JIT basis) higher priority. Yet many factories run on the "loudest voice" concept—whichever supervisor has the

TABLE 4.8 Priority Rules for Ordering Repair Jobs—Example

Ranking
1. Bottleneck equipment broken
2. Scheduled maintenance that if delayed halts production at a bottleneck
3. Equipment broken with more than X units awaiting processing in queue
4. Equipment broken with no alternative operations
5. Rest: FIFO

most influence or highest level of frustration gets his equipment repaired first. Instead, we suggest an agreed-upon rule for setting repair priority. An example is given in Table 4.8. That way, repairs are prioritized as received and scheduled in the order most critical to overall factory performance.

In the example shown, we'll first repair the key bottleneck equipment, then perform maintenances on key bottleneck equipment that can't be delayed, then repair equipment delaying a significant number of jobs, and then repair equipment with no alternative operations or equipment. All other jobs—maintenance and repair—will be done on a FIFO basis.

The Japanese approach is to look to the operators to perform routine maintenance and carry out simple repairs. In addition, they continually train operators and technicians on more and more equipment types to achieve a high degree of flexibility. This alleviates the not uncommon situation of problems with one type of equipment that only one (or a few) technician(s) can repair.

The next possible source of delay is caused by missing parts needed for repair. Obviously, it is difficult to stock all (or even most) parts required for all pieces of equipment. An improvement program here has two elements. First is a careful monitoring of the most frequently replaced parts and a predicted time-to-failure distribution chart to use in both preventive maintenance (to replace prior to predicted failure) and to set stocking levels of these parts. Such an analysis can be found in most reliability and/or inventory textbooks.

Second is a prearranged procedure for rapidly procuring parts that are not stocked—whether it be from the equipment vendor, an independent service group, your machine shop (to fabricate a new one), or a central service group in your company. The key is that there is already a procedure and expectation of how long it will take to procure a replacement part, especially for older equipment that may no longer have large inventories stocked in the field by the equipment vendor.

If your equipment utilization report shows continual equipment downtime waiting for out-of-stock parts, this cost of lost utilization must be included in the operational analysis of replacement part-stocking policies.

The final element of repair downtime is the actual repair time. As we've seen in production itself, where queue time often (or usually) exceeds production time, the actual repair time may be the smallest component of the four elements. However, as we eliminate the first three, it is the one that remains. Many companies are now looking at computer-aided diagnosis and repair systems to both decrease the elapsed repair time and improve the quality of repair.

When Texas Instruments looked at decreasing their equipment downtime in their semiconductor operations, they developed such a system using an artificial intelligence tool. The results were interesting—they cut the mean time to repair by one-third but also improved the mean time between failures by two-thirds! In other words, the repair technicians were not repairing the equipment properly in the first place. On complex equipment, vendors are increasingly looking at computerized aids to diagnose and repair for this same reason: It is hard to train technicians in all the intricacies of the equipment—especially on new models with limited production experience. This computer-aided repair capability will increase as computing costs continually decrease and equipment complexity increases.

Repair time can also be greatly decreased through the proper design of equipment (equipment that has been designed not only for functionality but for servicing) for easy access to and replacement of parts or subassemblies. Again, this type of feedback to an equipment supplier is important in bringing down repair costs.

In summary, our goal is to eliminate equipment downtime by continuous improvement from analysis of downtime causes. However, once a machine fails, we need to respond to it with alacrity and the correct priorities, to ensure it is up with the minimum lost time. Those who do not learn from history are doomed to repeat it. These approaches minimize lost utilization from categories 8, 9, and 13.

Improving Machine Efficiency Through Improved Scheduling

Again, from our analysis of equipment utilization, we see that we may lose utilization due to problems with scheduling work to the equipment and/or scheduling operators to the equipment. Obviously, except at abnormal bottlenecks, we did not need to run the equipment 100% of the available time. Our capacity analysis would show the exact utilization required at any work center. However, a goal of world-class factories is to keep work moving through the factory smoothly (with minimum throughput time or cycle time). This means that our scheduling system should keep work from piling up at one operation while another starts to run out of work for a sustained period. Similarly, it should schedule operators to stations where there is work. We should not simultaneously find operators idle at several operations while others are inundated with work.

The key elements of our scheduling techniques to accomplish this goal are release and dispatching rules that keep the bottlenecks properly loaded continually. Basically, our scheduling rules should allow a modified or intelligent Kanban system that pulls work smoothly through the factory while also keeping bottlenecks fully utilized.

Note that, in general, higher operator cross-training or flexibility will lead to higher utilization as we can shift resources to varying workloads as needed. We can monitor categories 10 and 11 to see when our scheduling rules need modifying, with higher priority put on feeding the bottleneck operations or shifting operators to them.

7 × 24 Operations—The Final Step

Traditionally, many factories ran one shift. As demand increased, they moved to two shift operations. Today, in any capital-intensive industry, companies must consider running seven days per week by 24 hours per day (7 × 24) or around-the-clock operations. There are several reasons for this direction.

First, facilities, equipment, and tooling are fixed assets. If a competitor can successfully run 7 × 24 operations they can achieve the same level of output or achieve a higher unit output from the same facility investment. In that case, their cost per unit is lower than yours.

Second, 7 × 24 operations cut the response time or reaction time to a customer order. There is no "artificial" extra two days of leadtime for orders that sit over a weekend. Nonstop operations eliminate all lags in response. This means a competitive advantage in speed as well.

Third, 7 × 24 operations do not suffer the shut-down/start-up syndrome costs as operations need to "recapture" the process. On Monday morning, we don't lose the setup and calibration time, as well as the initial units, trying to get the process back in operation.

Achieving 7 × 24 operations is difficult but clearly gives a potential competitive advantage in cost and speed. In fact, semiconductor companies now think of 365 × 24 or continuous operation.

Bob's Thinking—Value-Added

Bob finished reading these first sections with a new clear vision. His current budgeting approach was purely based on small steps, baby steps—incremental improvements in cost, speed, or quality. He couldn't change his current budgeting or cost systems. But he could augment them with value-added analyses that would point out "the art of the possible," the real potential room for improvement. The material that the consultant had presented was starting to make sense.

What he needed was a picture of their current value-added utilization of all key cost resources. Then he could project a roadmap of potential budget improvement based on solid analysis. They could compare their value-added utilization to best in class. He no longer was frustrated or angry. He was actually excited. He had a new tool, a tool that would give them a way to set stretch budgets—a new tool for a CFO.

Now he just had to figure out how to get the data he needed. How could he collect value-added utilization? They relied on standard costs. This was going to be a powerful improvement tool for budgeting if they could make it work. Not if, he thought, but when they make it work.

He thought through some of the material the consultant had presented—the analysis of their value added facility utilization (Table 4.9).

Their percentage of "productive" or value-added space (the space required for production equipment and the tool crib) was less than 20%. Conceptually, they could even eliminate the tool crib and place the tools on the floor where they were needed. The rest of the facility was wasted space. If

TABLE 4.9 Actual Facility Utilization

Floor Space Use	Sq. Ft.	Percentage
Production equipment	62,000	13
In transit	62,000	13
Rework area	31,000	6.5
Inventory storage—raw materials	83,000	17
Inventory storage—work in process	41,000	8.5
Inventory storage—finished goods	60,000	12
Tool crib	25,000	5
Idle	120,000	25
Total	484,000	100

they were able to move to a true JIT environment, where raw materials were delivered to the floor as needed, finished goods produced to order (shipped upon completion), and minimun WIP stored between operations, they could eliminate the need for nearly 40% of the facility. This was space used to store "idle" material that earned no return when not in use. He could now see that idle material was no different than idle equipment or operators. It was working capital money, tied up in investment without return.

He saw another nearly 40% of space that had no utility—the space connecting the widely separated equipment (called in-transit) and the space throughout the facility that sat idle. The space created longer hallways, borders around the production area, and storage areas that seemed to fill up with the accumulated waste of years of occupancy (old equipment, tools, parts, and records) that no one used any longer.

He could envision a totally redesigned facility that fit in a quarter of the current one. He had always viewed their facility as a given, a fixed cost on their books depreciated yearly. But now he saw it differently, as a competitor might see it—as a place for creating a competitive disadvantage. If their competitors could achieve the same output with a fourth of the floor space, a fourth of the facility investment, and a fourth of the operating costs of electricity, heating, cooling, cleaning, and security—what a competitive advantage.

And then if he added to that cost differential, the speed differential (of saving the extra in-transit time between operations during production, the extra transport of tools and raw materials, the distance and time be-

TABLE 4.10 Indirect Labor Productivity

Activity	Time Spent (Hours)	Percentage
Work activity	14.5	36
Rework of work activity	3	8
Getting data needed to support work activity	7.5	19
Getting approval of work activity	4	10
Explaining results of work activity	4	10
Transfer of activity to another person	2	5
Follow-up on success of work activity	2	5
Meetings—general	2	5
Training	1	2
Total	40	100

tween cause and effect in operational problems), the advantages grew even greater. And yet, no budgeting exercise had even hinted at these potential savings.

Next he thought about the figures on indirect labor productivity (Table 4.10). Again, this was a black hole in all their budgeting exercises. For direct production, he had a model of raw material and direct labor requirements per unit produced. Indirect labor was a mystery. All he knew was that they seemed to have an ever-increasing requirement for it. It was one of the fastest growing components of their budget, if not the fastest, and he had no productivity measure for indirect labor. What the figures showed needed more investigation, but at first blush it looked like nearly 20% of productivity was lost getting the data needed, lost in the blizzard of paperwork used to run their factories. Their factory ran on paper or islands of databases. They had paper-based travelers or run cards, paper-based equipment log books, paper-based laboratory results, and paper-based spec systems. They had stand-alone facility monitoring systems. Inventory was partly on the MRP or ERP system and partly tracked on the floor. No wonder no one had an accurate, real time view of their operations; no wonder they spent so much time getting the data they needed to schedule jobs, solve quality problems, repair equipment, reconcile their cost rollups, find the latest engineering change notice, or roll out a new product or process. No wonder it took so much time and effort to communicate (nearly 30%) to release the shift schedules, ensure that everyone knew that a specification had changed, get signoff on a proposed change, implement a new maintenance schedule, transfer data between shifts, explain a variance, or find out if a change to the specification was effective.

They were running their factory without real tools. He couldn't imagine trying to coordinate their entire business with a system that integrated their various functions (customer service, finance, manufacturing, distribution) on a single plan. He depended on his ERP system. But the factory had no equivalent system where they could see an unified plan and its progress. He had just heard of such systems—manufacturing execution systems—but knew very little about them. Maybe it was time to investigate MES as a real tool for cost improvement. Perhaps they could even measure the value-added utilization of equipment and materials.

He wasn't angry any more. In fact, he was excited. He felt empowered with a new tool to help the company—value-added budgeting. Now he had to work with the other improvement teams to see that they were focusing in on the non-value-added areas. Now he finally understood how to approach meeting these stretch goals. Now he felt that he could play a leadership role as CFO.

He forgot about calling up his brother-in-law. He was a jerk anyway. It was time to fix the underlying problem, not leverage it over.

5

Achieving Stretch Goals Through Best Practices in Quality—Continued: Technical Capability and Empowerment

The quality group was the first stretch goal task force established, and a real transformation (or metamorphosis) had occurred. After a slow start where members questioned management's real commitment to fundamental change, it had accepted their mission as real and was soaring like a butterfly. As it began to question, even deplore each and every loss of product, each machine breakdown, each component failure, every operator error, it had taken on police-like vigilance. No loss, no rework, no breakdown escaped their attention to be pareto charted and assigned an owner and a priority.

It soon moved from its own circle to start to train more groups, by work center, to carry on the attention to detail. By now the data on how quality varied by element of manufacturing used was being collected at least for initial analysis by raw material and ECN version instead of by product operation, as they had previously collected loss and rework data.

By now there were also some negative feelings arising from the operators, equipment and maintenance techs, engineers, and others, held up to close scrutiny, their quality "on trial" by the appointed task force. The problem reached Bill's notice when it escalated to the point of a fist fight on the plant floor. An operator, fearful for his job, had told the quality reps studying his work to "back off." When they didn't, the argument went from verbal to physical. Bill had even heard rumors of unionizing—of the hourlies organizing to battle back from the "quality assault" on their jobs. Many of them had been doing the same job for 10 to 20 years and couldn't make head or tails out of these sudden changes to their job responsibilities. From an advisory group, the task force had turned into an all-out action force. What was going wrong?

At the same time this situation reached Bill's attention, the quality team was meeting to discuss the deteriorating reaction to their quest. The group overall was irritated by the narrow-minded reaction to their program, seemingly forgetting their own initial negative attitude. One group was arguing for more authority, even to the point of having input to operator, tech, or engineer personnel reviews. One member went so far as to argue for the authority to suspend a worker. "If we can shut down the line for poor quality, why can't we remove the cause," he militantly (and logically) declared, "*whatever* the department." Clearly, a little power was a heady experience. Since Bill had not put bounds on their authority, its limits were unclear. If quality programs took new thinking, maybe radical thoughts *were* in order.

But another group suggested the alternative, that threatening the workers would only harm the quality program. "The goal should be teamwork, not confrontation. Now you want to make us into another 'supervisor' pushing our weight around." The supervisor on the team bristled at this remark.

"Wait a minute," someone said. "Let's look at this problem using the same approach we use to analyze quality problems and stop getting personal. No one is moderating the meeting. No one is taking notes or action items."

There was a slightly sullen silence that lasted a while. "Who's going to moderate?" asked an engineer. Everyone looked around for a neutral party, someone who hadn't yet taken a position. Strangely enough, the most militant spokesperson raised his hand. "I'll do it. You're right. Let's work out this issue like we worked out all the issues so far—as a group. We'll find a solution we like or we'll get some help or we'll do both."

Seeing the most vociferous partisan among them disqualify himself from the discussion changed the whole atmosphere. The points/problems started to emerge, along with likely causes and potential solutions.

"Problem—we don't have buy-in from the rest of the company. It's only us and Bill who are on board."

"Problem—we don't know the extent of our authority."

"Problem—we haven't worked with personnel on how to introduce this program to the rest of the plant."

"That's not a problem—that's a cause of the problem on buy-in."

The discussion went on that way as they sorted through it logically, as they would a quality problem. What was going wrong, they soon discovered, was that they had grossly underestimated (or ignored completely) the transfer of their program to the people who did the day-to-day work, as well as the rest of the managers. There was no carefully thought through implementation plan involving training, motivation, perhaps changing measurement and reward structures, perhaps reorganizing work groups into teams, adding quality monitors for products and processes at each operation, or for indirect activities. In their urgency to find technical solutions and get organized themselves, they had not yet thought through the bigger issue of transferring their "quality R&D" into production.

They all felt comfortable with this diagnosis. What they wanted now was some help. They felt that they were ouside of their area of expertise and saw the need to go back to Bill, personnel, maybe an organizational consultant, or some of the companies they'd visited for advice. They agreed that one subgroup would develop a plan of attack for implementing their quality program across the organization and report back to the team.

There were two more items on their agenda. The second item was customer and vendor partnerships. This was an item one of the members had put on the agenda. This was a common practice. Any member could add issues or items for feedback, advice, action, decision, or study. An engineer raised the issue.

"We've been setting up the quality monitors to ensure that we produce a product that meets our engineering specifications. We clearly want to do that. But I have a question. I've been reading quite a lot about other quality programs, and most of them talk about the supply chain from vendor or supplier through to the customer. Many of them actually invite customers and vendors to participate on their quality teams and programs. Should we, or maybe I should say, shouldn't we involve both of them in our quality program? Shouldn't the customer help define what *they* see as quality? Shouldn't we start to work with *our* vendors on our expectations of their quality? That's my question."

There was a momentary silence while each of them thought through the question. This was a more and more frequent situation. Members on the team, feeling responsible and empowered, were reading articles and books, speaking to friends at other companies, attending quality conferences, joining outside quality organizations, or in other words, continuously improving their own knowledge and capabilities. Changing the status quo meant changing *their* status quo. Actually, most of them liked the change. They felt more powerful and more useful. The work was less routine. Only a few were disturbed by the constant need to question current practices, notions, traditions, relationships, and values.

Here, one of the old or reflex reactions popped up first.

"Talk to a customer? That's marketing's role. They'd never want or let us talk to a customer. In fact, I don't even know who our customers are. It's hard enough just working on our manufacturing quality. What if the customers don't like it? Then what do we do? It's not our problem."

The concept of going to a customer felt strange to most of them. It had taken months to learn to work effectively as a team, and they were all from the same company! How they could integrate a customer onto the team, expose their secrets, problems, and internal conflicts? What customer would want to spend the time to help them and, therefore, maybe help a competitor who bought the same boards?

They could force a vendor to join the team, but would they want a vendor? Vendors were supposed to meet specs at the cheapest price. You always had at least two vendors for key parts or components so that you could play one off against the other on price and delivery terms. And then if one couldn't supply an order, you could get it from the other. Anyway, purchasing would be furious to hear of working with a vendor internally. They were the sole contact point for vendors to ensure that no information slipped out to them on where they stood relative to their competition on price, quality, leadtimes, or flexibility.

The group was clearly uneasy. Every time they reached a certain point and were comfortable, someone raised a new idea, and they had to grapple with more change. This continuous improvement was certainly unsettling.

It seemed that there was no respite, no periods where they could just sit back, complacent (fat, dumb, and happy as they called it), and relax.

"Another study group?" someone ventured. The group nodded. They selected three of the team, headed by the engineer who had raised the question to go off, research it, and present a recommendation. They had long since given up on trying to get everyone to fully understand every issue. Being part of a team faced with an endless set of tasks (or a never-ending set of tasks) meant continually forming project or task subteams with the necessary skills to present back and sell the larger team as a whole on their findings. This required considerable trust in the team process, but it was considerably more productive and respectful of all the talent on the team. While they were uneasy about adding customers and vendors to their quality scope, it did make sense. Many others had actually heard or read similar remarks on their visits or during outside activities.

"Last item—'best of breed'—whose is that and what is it?"

One of the operators raised her hand.

"You know, I've been looking at my station's quality now for the last four months," she began. "Every day I print the traces on the new surface mount cards. I've developed several quality devices. We used to use an overlay screen, then we added an inspection station, and now we even have an automatic visual inspection station on trial."

Everyone nodded. This was familiar to them all. In the past they would have simply purchased the automatic inspection station. Now they actually tried it out to see whether it actually improved quality, how much time and effort it took to operate, and whether it had high uptime or quality.

"I think we're going about this whole quality program backwards, all wrong."

"What!" This was such a strange statement that one of the team members couldn't help the exclamation. "But we've cut quality problems by nearly 80%. Yields are dramatically higher. We've almost met our goal of no defects six or seven times now. How can you possibly say we've gone at this all wrong!"

The statement brought a real surge of hostility from the group. Others chimed in. "We've had a dramatic impact. We've probably done more to improve quality than all the previous attempts put together."

Finally, the moderator raised his hand. "Why don't we let her explain before we judge it? Let's hear what she has to say."

She began. "Look at all of your responses. You're angry, you're upset, you're not really even listening to me because I've questioned the way we're doing our job. Maybe now it's clearer to all of you why the operators and techs are so upset. We're doing the same thing to them that you think I just did to you. We're telling them that the way they've been doing their jobs all this time has been wrong! So let me explain what I began to realize while working on trace quality. The key in putting down traces is to get them exactly in the right place, exactly the right width and length, and without any flaws. We check for all those points with our inspection and then what do we do? We keep looking for any that fail inspection, and then we put all of our effort into trying to figure out why they're not good."

Everyone nodded at this, slightly suspicious that they were agreeing.

"Well, I began to notice that some of the traces that passed were better than others. In fact, a few were nearly perfect, *exactly* at the specifications. But what was I supposed to do with those—the most perfect traces I had ever produced? Simply throw them in the box with all the other boards that were acceptable! In other words, I was supposed to throw away all the knowledge of how to make *perfect* boards!"

The room was struggling with this, trying to follow her point.

"Of course you should pass them. They're all acceptable. What's the difference if some are slightly better than others?" one tech volunteered.

She nodded. "That's what I struggled with also. They all were acceptable, but some were 'more acceptable' than others. So I tried to think about it from a statistical viewpoint. And I eventually came up with this chart" (Fig. 5.1). It shows the distribution of my traces from perfection on the right end of the distribution to scrap on the left side. I simply took how far the traces were above or below the absolute target and plotted it.

"Now look at what we're doing. We're putting all of our attention on the 5% that are outside of acceptable limits, those to the left of the 1-mil deviation point. But we're not putting any effort into the 5% that are near perfect! As I see it, if we could use the same effort to figure out how either one happens, we should put it into figuring out how to make perfect ones every time!"

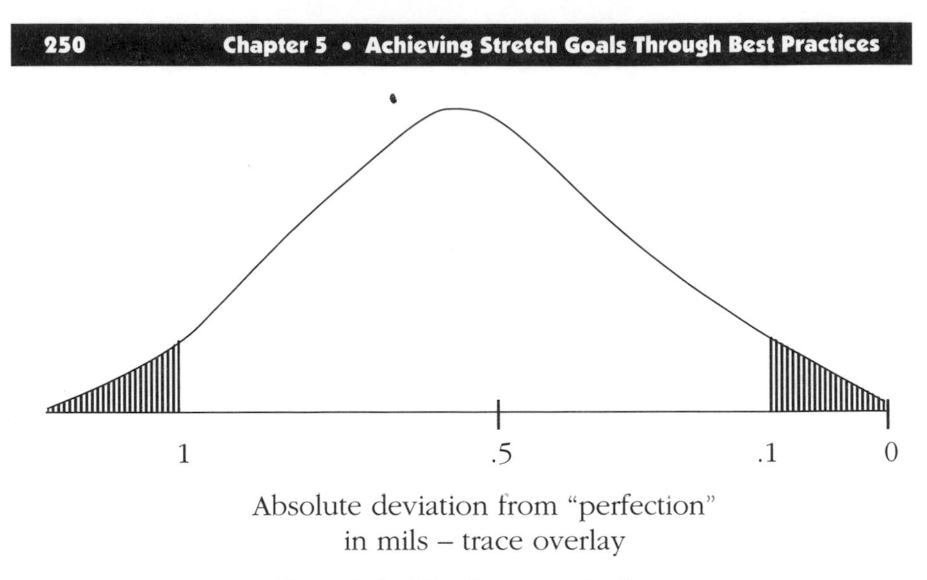

Figure 5.1 Distribution of Quality

This time there was amazed silence. People were thinking this over, looking for the flaw in her reasoning, comparing it to what they were currently doing, and in general, feeling puzzled, foolish, or slightly excited.

"What an interesting idea," someone murmured.

"You see," she continued, "the more I thought about it, the more I couldn't see why we were putting in all this effort on the wrong side of the distribution. We could use the exact same analysis techniques but on the perfect jobs. Then look what happens." She drew a second figure on top of the first one (Fig. 5.2).

"If we can figure out what causes "perfect" traces—what conditions, what settings of the five elements, then we can make near perfect ones every time—we automatically drop off the bottom of the distribution. So we get both results for the same price!

"Better yet, we don't have to go around policing people and jumping on them for making poor quality. We can urge them to look for their best quality and then figure out how to replicate that every time. We can have a contest to find the best equipment settings, find the best raw material characteristics, find the best facility conditions, and honor the person with the lowest variance products. Then we can train everyone else in how they did it. That's how we can get *everyone* thinking about continuous improve-

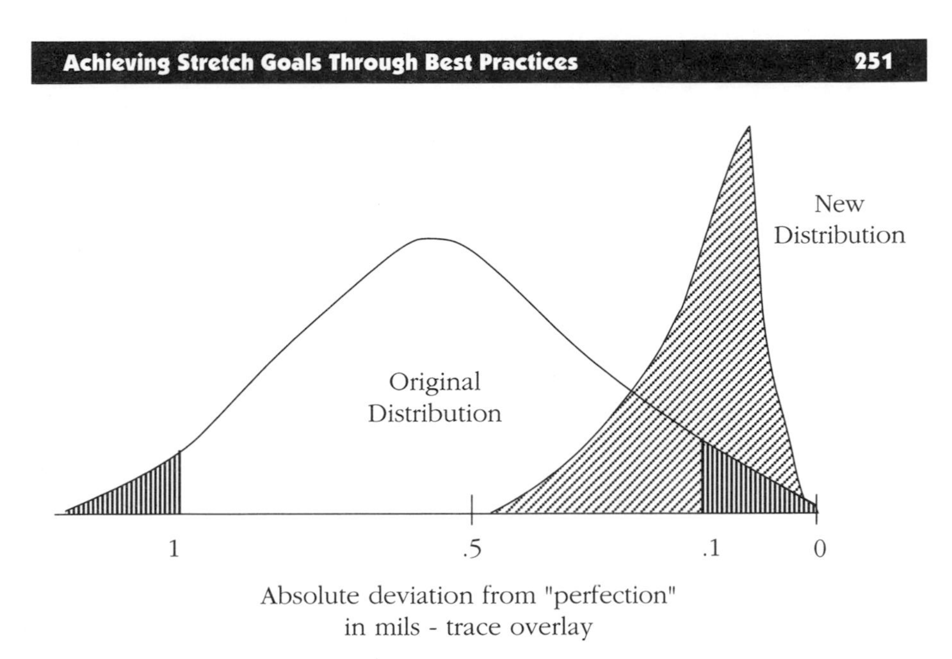

Figure 5.2 New Distribution of Quality Using Analysis of Perfection Techniques Versus Original Distribution of Quality

ment—honor them for their achievements instead of punishing them for their mistakes."

"Why didn't we think of that before?" an engineer asked. "That makes so much sense. It seems so intuitively right. Why did we waste our time focusing on the wrong concept. I'm incredulous that we didn't see it. It's so obvious."

The whole group was struck by that, but no one else had ever raised the concept before. Now they couldn't imagine using any other approach!

"I thought about that as well," the operator said. "I think it's because we never tried to reach stretch goals before. We didn't try to improve anything dramatically. Our goal was to incrementally improve the *status quo.* You know, our biggest fear was that we'd lose the process and not be able to make anything. It was only when I thought about stepwise improvements, or changes to our basic paradigms, that I started to think about learning from the best and not the worst. I think that focusing on stretch goals is going to require changing many of the things we do now and challenge many more of our concepts. If we were this far off base on quality, we must also be way off base on many other things—things you can't do by keeping the status quo."

One of the engineers jumped on the now-forming bandwagon. "This is going to be great! We always shrink the traces for our next-generation processes and products and assume that we'll have to buy new equipment. We don't think of improving the processes between generations of equipment. If we need tighter specs than the equipment can consistently and reliably produce, we simply put down a yield or downgrade factor. Now we could continually refine products. We could even tell manufacturing what to focus on that would give us the biggest competitive advantage. This is great! We need to get on this immediately."

The "Theory" of Technical Capability

What the group was discovering was the use of quality improvement techniques for continuous improvement of technical or process capability. Technical capability is defined as the ability to reproduce or meet exact specifications on product or process characteristics. Companies with greater technical capability can hit tighter or more precise specifications consistently. In a "status quo" approach, we look to eliminate anything outside acceptable specified limits. What remains is all "equally" acceptable—or the mediocre and the excellent are all the same in the land of the noncompetitive.

In a continuous improvement approach to technical capability, we continually look to improve technical capability by searching for "best-of-breed" examples of perfection and trying to discover the conditions under which they can be replicated endlessly. This again is what every world-class athlete strives for: replicating perfection every time. By applying this approach using our product-to-process relationship, we can search for the exact settings of the five elements that give rise to these perfect traces and/or how they differ from the settings for the ordinary or "mediocre" products produced.

In Figure 5.3, we see that the "perfect" traces seem to share a very tight range of facility humidity, equipment temperature, and one particular vendor's solder paste. These may or may not be the only (or correct) process determinants, but they are shared by the perfect trace boards versus the other ones produced.

The key principle is one of "best-of-breed" benchmarking: (1) finding the best performers in every key category of performance; (2) identifying the key causes, conditions, methods, or reasons for this outstanding perfor-

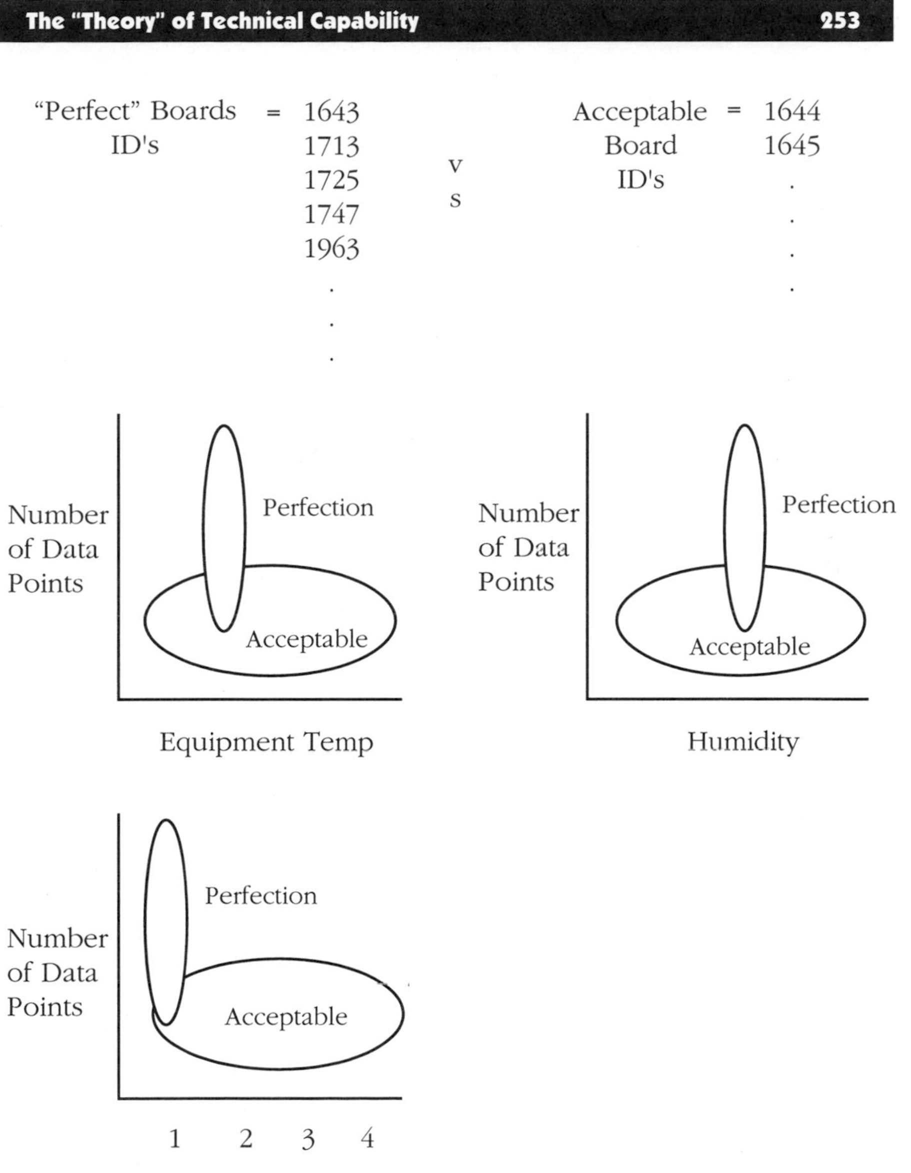

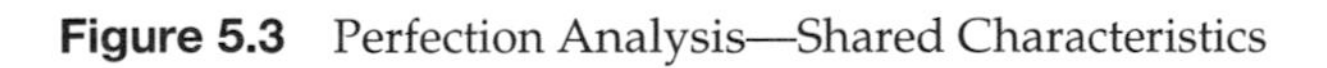
Figure 5.3 Perfection Analysis—Shared Characteristics

mance; and (3) extending those conditions, etc., as standard operating procedure for everyone.

In our PCB factory, we could use this principle to find:

- The most productive piece of equipment—measured by output per scheduled hour
- The most reliable piece of equipment—measured by hours of unscheduled downtime per week
- The most available equipment—measured by hours of available (not broken, being maintained, calibrated, etc.) time
- The most productive engineer—measured by the greatest improvement in yield or rework per product
- The most "productive" supervisor—measured by either the output per available equipment or operator hour
- The most reliable vendor—measured by the percentage of on-time deliveries and/or percentage of units accepted to specifications
- The most reliable process/operator/specification—measured (for operations of similar manufacturing complexity) as the number of ECNs, the lowest levels of scrap and rework, or the lowest levels of operator error
- The most effective maintenance tech—measured by the percentage of unscheduled downtime and/or maintenances done on schedule

The point is to look at the examples of "perfection" for any key competitive measure and bring *all the lower performers up to the highest performer.* This is a powerful approach to attaining stretch goals.

The Organizational Power of Technical Capability

When starting a stretch goal or benchmarking program, it is common to visit other companies' plants, both within and outside your industry. When you do return with benchmark data or anecdotal data (they had four hundred machines and only three were out of service), the immediate reaction from the nonparticipants (and perhaps from some participants) is that those plants are different from ours. People will question whether their products, customers, equipment, salaries, organization, plant layout, working conditions, training, skill classes and levels, culture, and every other factor they can think of to negate the importance, relevance, and *transferability* of what the task force has seen. In fact, some of these points may be true.

However, it is much harder to analytically, abstractly, and dispassionately dismiss clearly superior results at some work center, floor, or plant within your own company where the conditions should be replicable. In fact, in many ways it is hard to identify with a worker 3,000 to 6,000 miles away in a plant you've never seen that supposedly can "beat your socks off." It is much easier to identify with a worker, piece of equipment, job, or tool right in front of you. For this reason, while "best-of-breed" benchmarking should not be limited to your own factory, it is a good place to start with a goal of bringing everyone up to a common level of excellence you already have achieved.

In addition, this is an amazingly positive approach to quality. Instead of hunting for the "guilty" and laying "blame," you are searching for the "excellent" to expand its domain. Everyone is motivated by positive feedback. The practice of searching for the best leads naturally to continuous improvement. We want to find the ones that were perfect; we don't want to dwell on the ones that were ruined. We're curious how to repeat our successes; we're fearful of repeating our failures. We want to be honored, singled out (even if anonymously) for our excellence, and have our methods copied as best practice.

Implementing Technical Capability

As the group embraced the concept of technical capability, they started to look at the range of quality across machines, operators, shifts, vendors, jobs, tooling, and other elements of manufacturing. They had run SPC charts by equipment but never focused on other forms and measures of performance. They were amazed at the variety of performance. It varied by as much as a factor of 10! Just out of curiosity, some of the operators on the team also measured the equipment productivity and quality at insertion and inspection. These samples are shown in Tables 5.1 and 5.2.

They even carried the concept to looking at the time to repair by individual maintenance techs. Everywhere they looked, they found enormous variances among people and equipment, in quality, productivity, speed, and other factors. The only area where there was little variance was in direct productivity per operator. Most of the operators had similar output in

TABLE 5.1 Range of Product Quality and Productivity at Insertion—Weeks 16/17

Automated Insertion Machine	Percentage of Boards with Insertion Errors	Average Output/Hour (Number of Boards)
1	2	38
2	1	44
3	5	36
4	3	40
5	10	22
6	6	38
	Range Factor: 10:1	2:1

boards per hour, but their quality varied considerably. As they thought about it, they realized that the operators were graded on output, not quality, and "driven" to meeting that standard. It struck them given the "natural" variance they'd found everywhere else, a *lack* of variance here was probably *not* a good sign. In fact, there probably were operators who could be more productive but held back to be "one of the group." They remembered from their statistics training that a lack of variance was as suspicious or as big a red flag as too much variance. Somehow the operator measurement scheme had stifled a "best of breed" occurring naturally! Certainly on the automated side (at insertion), there was a significant level of productivity variance. But overall, everywhere they looked, the range from best of breed to worst performer would have meant *significant* corporate improvement *if they never left their own four walls* but brought everyone up to their internal bests.

TABLE 5.2 Range of Operator Quality and Productivity at Inspection

Operator	Percentage of Inspection Errors Made	Average Output/Hour (Number of Boards)
Joe	1	55
Bill	4	55
Tom	.5	56
Ed	2	54
Jill	.5	55
Sally	1	55
Tim	6	58
	Range Factor: 12:1	

At their next meeting, they discussed the enormity of the task in front of them. There was an air of disheartenment in the group.

"As far as I can tell, we, as a group, could work day and night for a year and not make a dent in all the areas we've uncovered. It looks like we'll have to look at nearly every production operation, every group of equipment and operators used in these, every maintenance group who maintains them, every supplier group, and so on. We've clearly bitten off too much. We simply can't implement this technical capability improvement program. It's out of our capability. It's too big."

The engineer who reluctantly said this could see that most of the other members of the team were in agreement with him. They had stumbled onto something big but just were too small a group to pull it off.

"All we could do," he continued, "is really prioritize and perhaps work the most critical areas from a cost or quality standpoint. We could focus on the insertion equipment and maybe the automatic testers since they are so expensive and limit our output. I don't know where to look on the quality side. We could do a pareto chart on overall poor quality causes and see which operation is most critical. Even with limiting our scope to those two or three work centers, we'll be incredibly busy, comparing performance across every machine, operator, vendor, and so on. I can see why companies focus on only the "bad" product—it's much easier than finding the best product, equipment, operators, and so on. We're not set up to do that; we don't have the data accessible to us. All we collect is output, rework, and yield by operation, and product for the MRP II system."

Again, he could see the heads nodding. They just weren't set up to do this and yet it clearly offered the possibility of dramatic improvement. But for now, they'd just have to keep going the way they were—focusing on what was manageable by the team. Some ideas sounded great but just weren't practical, not in the real world. At this moment, they were doing all that they could.

A Shift in Paradigm Required

For the next several meetings, the group discussed some of the other topics they'd been wrestling over: whether to invite customers and suppliers onto their team, and starting a companywide introduction to quality improvement to get wider buy-in to proceed. But something had gone out of the group. There was a terrible feeling of compromise, of not doing the right thing. They knew what they should be doing and they had backed away from it. Somehow, they had lost a lot of the pride and enthusiasm that had driven them—of feeling like pioneers, new thinkers, hard-charging change agents. Now they felt just like everyone else again. They felt like the people they'd criticized in these same meetings—probably knowing what should be done but too caught up in the day-to-day events or status quo to really make any significant changes. They were behaving like people who always had twelve reasons why something couldn't be done and none why something *had* to be done.

Their discussion about meeting with customers and suppliers was very tentative. Some people on the team felt it was premature to meet with customers until they had first done everything they could internally to improve their quality.

"We'll just embarrass ourselves," the scheduler who represented that contingent argued. "They'd see how primitive our quality program is, and then marketing would kill us for screwing up the account. In fact, marketing would never let us see an account until we had presented to them first thirty times and made sure that we'd all wear suits."

"Have you ever met anyone from marketing?" an operator asked. "I never have. Wouldn't they want our customers to see how much we're improving?"

"Marketing would never want to discuss what customers already *assume* you do. They only let design engineering meet with customers."

"I *still* think we should start including customers and suppliers in our quality team. We don't have to actually have them on the day-to-day team. We could start with some 'key customer' reviews where we discuss what we're doing about quality, but more importantly, find out how they look at our quality, where improving it has the most advantage to them, how our quality compares to our competitors, and how they're improving quality. I don't see how that could hurt us. On the contrary, maybe they'd be impressed that we wanted to listen to them."

This was input from the supervisor who was leading the "pro" group. "I'm always impressed when someone wants to hear what I want or need."

"How will we know what they want or need if we don't ask? I thought the point of all this was to live in the *real* world with our competitors, customers, and suppliers. It feels to me like we're retreating again into our own "safe" world. Isn't it time to "benchmark" our quality program against our customers' needs?"

"Okay," the scheduler finally offered. "If you can get marketing to set it up, then let's give it a try. But I personally doubt they'll do it. And I'm not sure what you're going to discuss with your suppliers. I think we should simply tell them what we want—perfect quality parts, at the current price, and when we want them—and see who'll bid on the RFP (request for proposal). That's what our customers are doing."

The supervisor said she'd follow up with marketing and purchasing to set up meetings with a key customer and a supplier, as well as set up an agenda for the group to review.

After that argument they decided to postpone the next item on the agenda—how to take their program companywide. The group didn't have it today. In fact, ever since they had realized their own limitations, the group had psychologically floundered. It was palpable. Their purpose was compromised. They were slightly embarrassed. That, in fact, was why so many of them didn't want to meet with customers or suppliers: they were feeling defensive. If the customers raised the issue of why they weren't pursuing technical capability, it would be very embarrassing to admit that they had given up on that approach as too ambitious a goal. Somehow they couldn't reconcile their desire to be world-class thinkers with their ability to pay for it. The enthusiasm was going out of the group. In truth, it was at a critical crossroad.

Including the Customer: Closing the Customer Chain—Opening Two-Way Communication

Much to the team's surprise, marketing was not a barrier but a strong advocate of the idea. "In fact, many of our customers are telling us that their other vendors are already working more closely with them. They want more access to our engineering and manufacturing groups. In fact, we were going to approach you with the request, but we figured you wouldn't want to do it. We thought you'd see it as our problem to get customer requirements."

Jointly, they selected a target customer to pilot the concept and set up a proposed agenda. They went through several criteria for selection. They wanted a customer with whom they'd had a reasonably good relationship, who was a significant factor in revenues, who already did customer reviews with other suppliers (so they could learn from their experience), who was geographically close (so the exchange wouldn't take an extra two days of travel), and who had existing internal re-engineering or improvement programs. What they were looking for was a potential mentor, and they were candid with their short list before selecting the initial guinea pig. Again, to their surprise, all the target list companies were willing to help if PCB Co. was "serious" about listening.

The agenda they proposed to their customer, Hot Box Electronics Co. (HBEC) was long on listening and short on speaking (Table 5.3). They wanted to get as much feedback as possible on their performance as a vendor, how to run a quality program (especially involving key suppliers and customers), and any new directions HBEC might be taking. They were hoping to also get some idea of what their competitors were up to and maybe a few benchmarks. They never expected what actually occurred.

The meeting was at the customer site. Both PCB Co. marketing and design engineering were well represented, in addition to the quality team (all in suits and dresses). The PCB Co. sales rep for HBEC actually ran the meeting and introduced the players. The HBEC group was eclectic—purchasing, manufacturing, engineering, finance, and MIS were all represented along with the vice-president of HBEC manufacturing. The team felt like they were "playing" grown-up. It was strange to be at a customer site representing their company, ready to get feedback directly from an end user.

TABLE 5.3 Proposed Hot Box Electronics Co.–PCB Co. Meeting Agenda

Review of PCB Co.—revenues, markets, products
Review of PCB Co.'s new quality program
Review of Hot Box Electronics Co.—revenues, markets, products
Review of HBEC PCB vendor requirements
Review of HBEC internal manufacturing processes and strategic direction
HBEC review of PCB Co. as a vendor—key areas for improvement
Review of HBEC quality improvement program
Review of HBEC key supplier program
Potential joint areas for collaboration/follow-up

Many of them had never even thought about the end uses or users of their products—it seemed irrelevant. Their view ended at their departments' walls. Now they were representing their company. They felt both pride and a little worried about what they were going to hear. And with reason.

After they had all shaken hands, gotten coffee, and found seats, the PCB Co. sales rep presented the company overview. Seeing their company presented on color 35mm slides, they felt proud and impressed. PCB Co. looked good on film. There were slides of sales growth, product lines, plants around the United States and distributors in Europe and the Far East, assets, and, of course, people—their number one asset. No one on the team had ever seen this presentation. In fact, no one had ever seen the company presented to anyone. Maybe they were doing better than they realized. It sure sounded good.

For their part of the presentation, one of the PCB Co. supervisors had prepared an overview on their quality project. They used black-and-white overheads instead of color 35mm slides. They presented the steps they had taken—from organizing a multidisciplinary team to outside plant visits to their ongoing quality analyses, and now to their first customer visit. Put that way, it seemed very Mickey Mouse compared to the marketing presentation! The supervisor prepared to sit down but was stopped by a barrage of questions from the HBEC attendees.

"Do you have a product designer on your team?"

"What commitment has plant management made to implement your findings?"

"Have you looked at work teams for the plant floor?"

"Have you looked at your performance review system to build in quality as a key job measure?"

"Has top management participated in the program?"

"Have you benchmarked yourselves against competitors or best-of-breed suppliers?"

"Have you trained your workforce in quality tools?"

"Do you have quality measures at every operation?"

"Does everyone in your company believe quality is key to survival for your company?"

"Do you complete globally or are most of your sales in the United States?"

"When are you going to be a global manufacturer?"

"Have you worked with your suppliers on their quality?"

The torrent of questions piled up higher and higher like a flood unleashed. The team now knew what Custer must have felt like. They were outnumbered in questions, tasks undone, and good ideas coming from every direction. The sales rep from PCB Co. was getting more and more worried, as was the marketing rep. The look on their faces wavered between dread and disbelief. Each was thinking about who they could blame for this.

One person on the PCB Co. quality team was writing as fast as she could. After each question, she'd hold up her hand until she'd finished writing it down, and then let the next question begin. In response to most of the questions, the team said that they hadn't gotten to that yet or that they were still thinking about how to do that. In actuality, they felt like rank amateurs.

Finally, the vice president of manufacturing at HBEC took pity and called for a break. When the two groups, one shell shocked, one now interested, returned, he began their side of the agenda.

Organizational Change—The Organization at Rest Has Great Inertia

The most common question I'm asked is "How did you get your organization to change and how can I get my organization to change?" It is important to realize that, for most organizations, change is an *unnatural* act. Personally and professionally, most of us search for stability and therefore the status quo. We go through a planning phase once or twice a year and then simply try to meet the plan. The concept of *status quo* permeates most organizations. It is embodied in our budgeting approach—developing standards—for operational costs, part costs, leadtimes, yields, and rework rates that remain constant for the year. We "learn the rules" or the culture of our organization and follow them on a steady path to success. In fact, the organization itself depends on this massive "self definition," a set of rules and procedures, relationships, and mores to function as new people are absorbed. In many ways, it is an enormous tanker steaming through roiling waves. And that is exactly why it is so hard to change!

These organizations were suitable during periods of slow or even nonexistent changes. If your industry was heavily regulated, there were few changes, and those that did occur had early warning visibility. In other industries, competitors were familiar with each other's practices (often their executives circulated from company to company through normal turnover) and the rate of change was not dramatic.

There obviously were exceptions—new industries that sprang up with new rules and *tempo* or rates of change.

The semiconductor industry, for example, as well as all the industries it created or made possible—personal computers, workstations, personal computer software, disk drives, and computer-aided engineering—soon challenged existing industries (mainframes and mainframe software). But, in general, the rate of change was relatively slow in our major manufacturing industries (automotive, electronics, chemicals, and aircraft/defense). As a result, "status quo" organizations, methods, procedures, and measures of performance moved a company slowly and predictably along a stable and predictable growth path.

The 1970s and the advent of major foreign competition changed that situation forever. Companies now faced competitors with whom they were not familiar. This meant surprises—plans that had to be adapted to new market opportunities, new potential partnerships, new competitors, new pricing, and new customer expectations. In particular, manufacturers faced a new form of competitor: the continuously improving "lean" manufacturer. This company, who had adopted and practiced and excelled at continuous improvement techniques, improved *at a faster rate* than the "status quo" company who thought of improvement, each planning period, and status quo in between.

The "status quo" company operates traditionally. Operators and technicians are given specific instructions and carry them out identically every day. Correcting errors, quality problems, or breakdowns is the responsibility/problem/purview of the engineer, supervisor, or manager.

Operators are responsible for production volume, engineers for quality, and maintenance for equipment uptime. Changing this massive tanker to a nimble speedboat is no easy task. The tanker has *inertia;* it keeps moving in its same direction for days even after the motors are turned off.

It is not surprising that your company has not reacted well to the quality teams' initiatives. They are *changing the status quo.* They are questioning accepted principles. They are fixing "blame" without everyone's full understanding of the need for change. Many company psychologists think of change as a three-step process—unfreezing, rethinking, and refreezing.

In unfreezing, the organization has to recognize the need for change to begin a real self-evaluation—to question its current methods, organization, culture, and even goals and objectives. To do this, it needs new information and, unfortunately, some major stimulus to recognize the need for change.

In looking at organizational change, it is instructive to look at personal change. An organization, after all, is simply a collection of individuals organized around common goals or objectives (who all get their paychecks at the same address). Individual change is extremely difficult to promote. Normally, many people are unable to change their habits until a personal crisis or disaster strikes—a heart attack, cancer, a divorce looming, or a broken relationship. The status quo is a powerful objective for most of us. Even biologically, we talk of homeostasis—the goal to reach and remain in equilibrium. In an analogy, most people are like clay. They are easily molded when young and get more and more rigid and hard as they age, until something "breaks" the mold.

If you probe for how many of us actually base our personal lives on the philosophy of continuous improvement, the chore of changing the entire organization may be evident. Few of us yearn and take action to continually improve—to grow more knowledgeable, stronger, and more capable or to improve our relationships, our appearance, even our cholesterol, weight, and blood pressure.

Instead, we think about change but rarely do more than update our clothes. We change the paint but not the structure of the house. Once we are finished with our formal education, few continually upgrade their professional or personal skills. We let our *organizations* move us along, train us, promote us, and reward us gradually, slowly.

Therefore, to change the status quo, we need a powerful incentive. Unfortunately, for many, it is literally the threat of plant closure, loss of job, or loss of advancement opportunities that finally creates the opportunity for unfreezing—questioning the current state of affairs. It is hard to arouse urgency in an organization where no immediate threat can easily be seen. In this case, unfreezing requires more time: to discuss the longer-term outlook, to see the parallels with other industries and companies, to see the potential for improvement through others' experiences, and to visualize a new order for that industry.

Rethinking involves developing a new set of capabilities, organization, procedures, systems, policies, relationships, culture, and so on to meet the new external forces reshaping the industry—new competition, external regulation, customer requirements/demands, access to capital, raw material shortages or price trends, technology, and market opportunities.

With the organization ready to think in new ways, groups can prioritize the key areas requiring change and focus on them. However, there will be a strong tendency in the unfreezing period to deny the need for change. It will often be a period of excuses and defensiveness. These will be many statements such as:

- This is a temporary effect due to a slowdown in the economy.
- Let's not panic. We've seen this before and been just fine.
- Those other examples (companies visited/used as examples) are significantly different from us and therefore don't apply.
- We're really not doing that badly.
- This is just a fad that will pass like all the others did. Just keep your head down, nod, and don't do anything.
- Manufacturing has never been that important to us. We've always competed on products, design, distribution, or our customer relationships.

It is unlikely to convince everyone that this is a time when real change is required. It is important, however, to target key management and opinion leaders on all levels to have a strong infrastructure ready for change. That is why most task forces sent out to benchmark other companies have representatives from multiple departments and levels of management.

In rethinking, teams review the current way the organization functions, set goals or objectives for its "new" performance in key competitive areas such as cost, quality, etc., and then translate them into measures applying to each department or work center.

In refreezing, these changes are adopted as SOP (standard operating procedures). Early successes are highlighted to encourage other groups. If the unfreezing, rethinking process were started in one area, it has now moved to additional departments, organizations, work centers, or facilities. Often some rivalries are set up to encourage competition and excitement. Very importantly, progress toward goals is clearly measured, communicated throughout the company, and, if possible, reviewed in terms of end market performance—gains in marketshare, profitability, customer satisfaction, and company expansion. For this process to succeed, progress must be seen or the point of the program is vague. Change must lead to improvement, not to maintaining the status quo.

HBEC's Message to PCB Co.

"Let me be frank. I came today because I was curious about what you were going to say. If this had been strictly a marketing glossy presentation, I would have already left. You guys are where we were about three years ago. We waited a bit too late for my taste. We were starting to lose market share like crazy to the offshore clones and add-on companies. You must have been more protected than we were.

"Frankly, I have some bad news to tell you. We're cutting our vendor list from eight suppliers to three for boards. And you're not on that list. Your quality stinks. But you are close by and we have had a long relationship. I'm blunt because I don't have time to waste. We've improved our quality by a factor of 1,000; you can too. But if you want to be on our supplier list, you're going to have to get moving. We're going to expect JIT/TQC suppliers—no incoming inspection required, deliveries within 4 hours. That's why I'd like to see you succeed. You're almost next door. We could really help each other. But not with your current quality. We don't want to return boards; we want parts per million defect rates.

"We're going away from through hole to surface-mounted boards. They're smaller, higher speed, lighter, and equally reliable. We're going to flexible manufacturing where we can turn around any customer order in 24 hours, any of thousands of possible configurations. We're going to be competitive with any workstation vendor. You can join us or leave us.

"Do you have samples of your competitors' products hanging up in your factories? We do. I want everyone to see and feel the enemy. This is the real world, and I want it intruding at every moment. Time is passing real fast. If you've never spoken with your customers, you're in for some real surprises. If you'd never have come down today, you'd never have figured out why we changed suppliers. You'd just have thought we found a cheaper vendor. What we're looking for are suppliers who make us more *competitive.* If that means cheaper, it's cheaper. If that means faster or more flexible, it means faster and more flexible. It's not enough just to exist anymore. You have to offer your customers a competitive advantage. Good luck. I really mean that."

And the vice president of manufacturing walked out of the room. If he had been trying to have an effect on PCB Co., he was successful. The sales and marketing reps were white as sheets. The quality team was stunned.

The rest of the meeting was very subdued. The quality team listened to the four-hour version of where HBEC was going, an extended version of the four-minute version they'd just heard. For the first time, they felt a real fear—for their jobs. By agreement, they listened, asked questions when appropriate, and said almost nothing.

It was a very somber group that returned to PCB Co.

A Decision to Be Reached: New Paradigms Now (N.P. Competes)

Back at PCB Co., there were two series of meetings going on in parallel. The sales and marketing reps had escalated the results of the meeting up through the chain of command. They were meeting now with the plant manager. The quality team had reconvened the next day. They assumed that plant management must be meeting, and as a group, felt powerless until they got more direction. They felt a little like the boy who opened Pandora's box. But deep down, every one of them had heard the same message. You couldn't avoid hearing it. Either they changed the way the company did business, or they'd lose their customers.

The operator who had been taking notes at the HBEC meeting looked very resolute. She was the same one who had raised the issue of technical capability and argued for it. She stood up and said, "I have a suggestion and want you all to consider it. Hear me out before you respond. All of us have been sitting here overwhelmed by the responsibility for changing our quality. We can't do it. But the reason we can't do it is how we're going about it. I thought over and over again about technical capability. It is the answer. We already produce a product that's good enough—just not often enough. But we agreed that we as a team don't have the resources to implement technical capability. That's true. But the fallacy is in thinking that we're the ones who should implement it. Everyone needs to! We need to get everyone practicing it, or there never will be enough resource to get it done. We can't possibly get it done by ourselves. We need to empower the whole company! That's exactly what management started. They expanded the focus on quality by forming our team. Now we need to expand it to the rest of the company.

"Now I've already thought through what some of you are going to say—that some of the operators can barely read, that many don't care, that we'll lose productivity if operators aren't focused on production. That's all true, but it doesn't change the fact that we need a lot more help. Maybe everyone can't help—not until we teach them or train them or maybe even hire people with the skill levels we need. But many of us can."

And she handed out an article to the team.

The article covered the concept of employee empowerment. It related example after example of major companies who were re-engineering the traditional breakdown of work into the "thinkers" and "doers"—or those who directed, managed, or improved and those who carried out narrowly bounded tasks. When the main task of production was *replication,* doing the same thing every day, then that breakdown of work made sense. But when the main task moved to improvement, then that breakdown wasted the largest potential capability of the company: the energy, experience, intelligence, leadership, and problem-solving capability of its *entire* diverse workforce.

In each case mentioned, the company had re-engineered its operations or responsibilities to tap into this talent pool. Timken, at its new ball-bearing plant,[1] gave complete responsibility for each shift to a seven-person team. The team was independent and chartered to make all decisions. There was *no* team leader. The team was expected to reach consensus, delegate among themselves, and, if necessary, ask assistance from the four remaining staff people—the plant manager, a human resources person, a finance

[1] *Industry Week,* August 1991.

person, and a plant engineer. GE had made empowerment a major thrust for the '90s, unleashing the problem-solving and improvement talents of its workforce.[2] Teams were empowered to find solutions to key problems, and then their managers were given *one minute* to either approve the solution, deny the solution, or ask for more data. The point is to empower people and accentuate that if speed is the competitive weapon, delay out of habit or out of distrust saps competitiveness. The key is to put together a team that you trust to solve the problem, train it so that it has the tools to be successful, and then listen to it once it's done with its analysis. The "one-minute" review is really sending the message, "Put the quality into the team's process, not into Q/A on the team's final decisions."

If there is no real time to review the decisions, then management needs to ensure that they *will* be good decisions and try to improve their decision-making process, not the one-by-one decisions.

The article covered example after example of companies that had made this major shift of responsibility and authority to teams or individuals: at Tektronix, Hewlett-Packard, Motorola, IBM, Ford, and Digital.

There is no better source of improvement direction. Toyota received 3,065 suggestions a *day* for improvements and implemented 94% of them. An American company that size would expect 1,000 suggestions a *year*.[3]

At the best American companies the average number of suggestions may be three per employee per year. For all companies, it's 0.14 suggestions per employee. At Pioneer Electronics it was 60 ideas per employee per year; at Canon, 70; at Mitsubishi, 100![4]

What is the reason for the differences? Those companies *asked* for the suggestions and then acted upon them. They were set up to *utilize* their employees' ideas on improvement. Their employees *wanted* to help. They did not view management as the enemy. The enemy, if there was one, was waste that made the company less competitive.

Each member of the team read the article silently and absorbed the material. At the end, one of the engineers spoke: "It won't work for exactly

[2] *Fortune*, August 1991.

[3] Davis AG. Moving toward high quality and productivity. *Manufacturing Systems*, March 1990, p 54.

[4] Schnitt P. Employees withhold ideas. *SF Chronicle*.

the reasons you've already mentioned. We don't have management's support to do it. Our operators are not capable of doing it. We can't even get agreement here whether we should do it. To me it's a great idea we're not ready for."

There were a lot of nodding heads agreeing with the last point.

"You know what you say to something you should do but don't want to? That it's a great idea—for someone else. It either is a great idea or it isn't. If it is, then the question is not whether or not to do it but how to do it," the operator responded.

Some people nodded at that point also.

"Look, why don't we explore what it would take to pull this off? Why don't we see what each of these companies did to be successful? Then we can decide if we can do it here also." This came from a previously silent supervisor.

"What did all these companies have in common in their programs?"

The group worked on this point for the remainder of the session and, in fact, two more. They called some of the companies listed for updates of their progress and actual accomplishments. They had become comfortable with research outside their four walls. In some sense, they were more comfortable with external discussion efforts as there was nothing threatening about talking to outsiders. It was changing the internal organization that was so hard to do.

What emerged were four or five requirements for successful programs. The first was a real change in management's philosophy. Management had to agree to relinquish or distribute their authority to the teams who were best suited for these improvement programs. Without that mandate, the teams became bogged down in examining what was possible and what would never be approved. Those teams either worked on less critical tasks, on projects that they knew would not question any current practices, or worked so tentatively, continually asking for approval and clarification from management, that nothing was accomplished. In all three case, the teams were eventually disbanded out of their frustration that they were wasting their time.

The second requirement was an urgency for change that was clearly communicated to the company as a whole. Change was not easy. Without a significant reason, too many employees fought change, passively or aggressively. Without a significant commitment and urgency for change, the inertia in any system robs most new programs of initial energy. This urgency has to be clear at all levels in the company, not just at the management ranks or in the departments that deal with customers or competitors. This was a major difficulty in many companies, especially unionized shops, where management initiatives were viewed with suspicion or outright distrust. Educating the participants as to the changes in the customer and competitive environments that necessitated change was a major project alone. It also depended on trying to make each employee see how he or she directly affected customer satisfaction and competitive advantage.

In most companies the employees were far removed from the competitive and customer needs. The team agreed that this was a major issue, even if they got their management mandate.

Third, the teams needed extensive training: in project management and team work, possibly in the subject area for improvement, and probably in problem-solving and analysis skills. In fact, from their research, there were few people in their company or other companies who had received significant levels of training once they had left school. In their own company there were many operators who spoke English as a second language, who read below the high school level, and were functionally illiterate in basic mathematics. These were all necessary skills for improvement programs.

They were beginning to understand why Japan had been able to empower workers naturally. Their high school graduates, in general, were all capable of 12th grade reading and mathematics. Japanese students spent about one-third more time in school each year than American students. Over a 12-year period, this meant four more years of education—the equivalent of a college education.[5] Ninety-five percent of Japanese students performed better on math and science exams than the top 5% of American students. In the United States,[6] an estimated 25 million workers were functionally illiterate. However, only 8% of American firms provided remedial classes in reading, writing, or math. In addition, the leading foreign firms spent up to 6% of payroll training their workers, heavily investing in

[5] Gragg JE Jr. The commercialization of R&D in the Semiconductor Industry: A comparison of U.S. and Japanese Practices. *SRC Newsletter,* March 1991, p 5.

[6] Rayner BCP. Warnings flags are raised: Japan is top capital spender. *Electronic Business,* September 17, 1990, p 27.

direct operators. In the United States, the number was closer to 1.4%, with a narrower focus on about 12% of the workforce, usually the management levels.[7]

The end result was that not enough of employees were capable of participating on continuous improvement teams without substantial additional investment in training.

Fourth, most improvement programs had needed major organizational changes—in measurement and reward systems to reward teams and behavior that benefited the company as a whole, and in organizational responsibilities—to projects or teams rather than to traditional functional departments.

Finally, people had to see this as an ongoing, long-term, dramatic change to the way business is conducted. They couldn't view it as a one-time, short-term fix or fad or aberration that would disappear with new management.

Looking at where PCB Co. was as a company, they looked like 0 for 5. It was inescapable that they had no management mandate, no broad company urgency, no company training program, no possibility of changes to the current organizational responsibilities, and no belief that their project team was a permanent role model. In fact, this last point was most striking. They felt more like an experiment than a permanent function, more precarious than precious. They rarely met with Bill. They had never met with the management team. Now that they had totaled up the score, they all had to agree that they weren't ready yet. Most were secretly relieved. But a few were more than anything sad, feeling that their company was doomed to mediocrity if not eventual extinction. They didn't see how they could compete with the "learning, improving" companies they'd read about, continually getting more competitive with time.

At the same time that they were going through this process, another set of emergency meetings were taking place within Mark Ritchards' staff. They were discussing what was unaffectionately called "the quality debacle." The vice president of marketing was furious. "I can't believe that you let them visit a customer. Manufacturing should *never* see customers—they should stay in the plant and manufacture. If they were supposed to see cus-

[7] Training shapes the workforce of the future. *NCMS Focus,* August 1991, p 8.

tomers, they'd be in marketing." This incident coming on top of their corporate emergency had exacerbated normal tensions between departments to the breaking level. Their responses under pressure were even more defensive than normal.

The vice president of manufacturing would normally have responded, but he was keeping a low profile so that no one would point out that it was his poor quality and delivery performance that had caused the problem. If someone did, however, and he expected that by the end of this meeting someone would, he had already developed his defense: that engineering gave him poor designs (you could see that in the number of ECNs required), that purchasing didn't get him the parts he needed (you could see that from the orders on hold awaiting parts lists), and that sales accepted orders due before the normal leadtime (you could see this in the number of orders started that were already late at their first operation). He felt somewhat secure that he could pass the blame around enough to escape more than a collective guilt. And if the plant manager had let him buy that new insertion equipment, perhaps their quality would be acceptable. So he sat there, concentrating on his various lines of defense.

The plant manager sat there, hoping that corporate would not hear of this latest debacle that would only intensify the pressure to shut them down. He cut off the "search for the guilty" and with a new forcefulness asked the group, "What are we going to do about it?"

The group was unprepared for this question, and a silence ensued while each person switched from thinking about who to blame to how to fix it without taking any blame. Mark's question had surprised them. Perhaps they really were serious about improvement.

Marketing had been about to say, "Keep manufacturing out of marketing's responsibility"—the customer—but somehow thought better of it after Mark's question. He was surprised that they had already shifted away from affixing blame to fixing the problem. This was not like their usual meetings that ran on for hour after hour of blaming one another for whatever was wrong before grudgingly deciding how to fix it.

After another few minutes, finance responded cautiously.

"We need to be sure that we're not overreacting. This could simply be a ploy to renegotiate our terms. After all, HBEC's purchasing didn't tell us this—manufacturing did. And we know that manufacturing isn't authorized to speak for the company. I suggest that we call the purchasing agent and see if he mentions anything. If he doesn't, maybe it was just manufacturing sore over a late order or something. Did we just have a recent problem with them?

"On the other hand, if they are serious, I suggest that we consider some face-to-face negotiations. Maybe if we gave them a special deal, we could eliminate the problem. We could shave 35 cents a board off their price if we substituted a slower memory chip. If that doesn't work, marketing better come up with a major new customer ASAP, or I'm going to have to draw up contingency plans for a layoff. I could do that with about two to three weeks of work."

"Maybe Bob is right," said the vice president of manufacturing. HBEC certainly couldn't change suppliers that quickly. Maybe he had just received a bad shipment and blew up. I could give him a call and talk with him."

"Oh, no! Haven't we learned from our mistakes yet," asked the vice president of marketing. "If anyone is going to call him, I will. And let's not rush into any discussions of price reductions. You only give something away when you're negotiating. Maybe all I need to do is take him out for dinner—say at the new place that just opened—and smooth out what's pissed him off. And no more customer visits! Do we agree?"

The plant manager looked at the head of engineering who'd been silent.

"Bill, any thoughts?"

Bill hesitated for a moment and then said, "No. I agree with what's been said. Hopefully, it was an isolated blowup." He said this without much conviction.

Mark Ritchards was shaking his head. "I can't agree with you. I don't think it was an isolated blowup, and I don't think it's going to blow over. HBEC's needs have changed. They can't succeed with the quality and service we currently provide. We have to fix the problem or lose them as a customer. It's not a price issue.

"Rod, can we get quality and customer service up to the levels they're asking for? Or maybe I should ask Bill and Joe, or all three of you."

Rod looked at Bill. It was clear that he, himself, was not sure of the answer. To his credit, he forgot about blaming any other departments.

"I'm sure that we can, Mark. But I'm not sure how long it will take us or how much it will cost. We may need to change vendors ourselves. We may need to find suppliers with shorter leadtimes and higher quality. We may need to replace some equipment that isn't consistently reliable. We probably need to retrain some operators, technicians, and supervisors. We may have to do more final testing. Maybe add a "fast turn" line. I'm sure it can be done. But I don't know what it will cost and how long it will take. And right now, we probably need to cut our costs, not spend more. We're between a rock and a hard place on this one."

Bill and Joe were quiet, sensitive to stepping on Rod's toes, organizationally speaking.

"Bill, " Mark prompted.

"Well, Mark, as Rod said, it may take some time. But . . ." He paused and seemed to take several seconds to deliberate. Mark saw a momentary uncertainty.

"Bill, if you have any ideas, now's the time to suggest them," he prodded.

"The quality team does have some ideas, Mark. But they are radical. I'd like them to present to this group but only if it can be done in a supportive environment. We may be used to our "give-and-take" but they're not. But they're definitely onto something."

"Joe?" Mark asked.

"I think that the customer service team also has some answers. They've done a good job."

"Let's do it, and soon," was Mark's response. "And Joe, I'd like to visit HBEC. Set up a meeting with their CEO and VP of manufacturing as soon as you can. Is there a chance we can replace that much business in their time frame?"

Joe looked pessimistic.

"I doubt it, Mark, though I'll look into it. It's a tight market right now.

"The Far East suppliers are taking a lot of the new business. I need much more flexibility in pricing." He looked directly at Bob.

Bob looked pained.

"Our margins are already under ten percent. We don't have a lot of flexibility left, especially with these new corporate goals."

Mark looked around the room at his faltering team. This second blow had taken more wind out of their already nearly becalmed sails. He hoped that the teams had come up with some significant solutions. Or he undoubtedly would be looking for a new job. HBEC would be the final blow to his career.

"Bill, Joe, let's set up these meetings with your teams as soon as they're ready. And let's see if HBEC will meet with us."

The quality team was not surprised to hear that the management team wanted to meet with them. They had been expecting harsh feedback about the results of their visit to HBEC. Some in the group expected reprimands. Others feared that they were facing career-limiting reviews—informally or formally. While none believed that they had caused the problem, none believed that anyone else would see it that way.

Inside the company, the mood was grim. Rumors swirled daily—layoffs of at least 20%, pay freezes, perhaps outsourcing of some of the low-margin boards, and whispers that the whole plant was going to be moved offshore. The whole company seemed to know about the visit to HBEC. Bad news seemed to travel a hundred times faster than good news and be deemed ten times more credible. The quality team actually heard about their meeting with the plant manager through the rumor mill before their more formal invitation.

The more traditional invitation came directly from the plant manager, who called them as a group to his office, along with Bill.

He started with a calming and yet electrifying statement. "I know that none of you are responsible for what's happened. If anything, you did us a service by discovering the problem when you did—before we were simply discontinued as a vendor. It's a natural reaction to blame the messenger for the news he or she brings. But in some way, maybe it's a blessing that this is happening when we still have time to do something about it. For the last six months I've really felt that the world had changed but we hadn't. That's why I started these teams—to begin our change process. Now it's been brought to a head, sooner than I would have liked, but maybe it's been longer than we had any right to expect.

What I want you to do is continue to be part of the solution. As a group, you've looked more carefully at our quality issues than anyone has before. I'm not sure if what you've discovered can solve all of the problem or part of it. But it's bound to help. But as importantly, I want to prove to the company that there is more talent here than anyone realizes. What I want you to do is show that we can solve this quality problem with our own people, not in spite of them. I want you to present your suggestions and not be intimidated or limited in your thinking. You have to trust me that I'll back the *process,* if not every one of your ideas."

The group listened and had a range of emotions. Some were distrustful, wondering if the plant manager wasn't simply looking for a group to blame if they couldn't meet the new quality and service requirements. Others were scared that they were going to be responsible for the company's success or failure, doubtful that they were up to the task. A few were excited, sure that this was their opportunity to shine in front of management and gain recognition. And a few felt that they had the answer to the company's quality problems and that this was their and the company's chance to implement it.

They agreed to present their recommendations in one week to the management staff. There was one common emotion. All of them, including Mark Ritchards, were nervous.

The group, strangely enough, did not disagree on what to present. If they had one meeting with management, they wanted to press for what they had all come to believe was the real answer: empowering as many groups as possible to pursue technical capability in every area of the company. What they worked up was an analysis based on their data to date. What they wanted to show was what their own company was capable of if they could achieve their *internal* best-of-breed performance. In other words, if they could bring all their resources—equipment operators, raw materials, tooling, and specifications—up to the best performance in each category, what would the result mean for the PCB Co.?

Their presentation a week later had two parts—the first to answer whether their customers' requirements were possible, and the second, how to achieve them (and why they couldn't under current circumstances). The operator who had fought so hard to institute technical capability was the unanimous choice of the team to present their analysis. Much as she had done at their own meeting, she reviewed the concepts of technical capability—how within their company they already *had* world-class performers, but that as a company, that performance wasn't being understood, honed, and then emulated as best practices. Then she dropped a bombshell.

"As best we can tell, if we were able to achieve our *own* best-of-breed performance across the company, we could hit all of our customers' quality requirements and reduce our costs by at least 20%." The management team was astounded, then disbelieving, and then confused as she explained.

"What we did, informally, was ask each supervisor, technician, and operator which was the best machine, operator, and ECN for one of our high runners that we make daily. Then we simply recorded its performance at that operation last week and recorded its best shift performance. As you can see, if we could get the best shift for each of our best resources for this product *all* the time, then we could easily meet our goals." And she presented the findings (Table 5.4).

The management team reacted predictably. "But those are the *very best* results, not only over all the equipment, operators, and tooling but over the whole week. You're cherry picking. We can't replicate these all the time or we would."

TABLE 5.4 Rough Estimate of Product Quality Technical Capability

	Best Quality at Each Operation—Product X (Estimated Defects per 10,000 Boards)		
Operation	Best Quality*	Planned	Planned Yield
100–Insert auto	0	400	96
105–Insert manual	0	200	98
110–Clean	0	0	100
120–Wave solder	0	500	95
130–Deflex	0	0	100
140–Open/short test	200	800	92
150–Functional test	100	600	94

* Best shift for that operation, over all equipment and/or operators doing at least 50 units of product in a shift.

The operator had fought this battle once before and was prepared.

"Why can't we? If we understood why we achieved them once, then we could duplicate them. Or are you saying that our process is really a mystery we can't understand? I believe we just never spent the time to understand it. We accepted where we were as adequate and stopped improving any further. Those operating conditions met our goals, so that's where they stayed. We didn't *expect* any better performance, so we didn't get better performance. But the numbers tell us we could—and do. But only sporadically and for some of our equipment.

"Let me show you the same analysis for equipment uptime." This time she presented a view of how they could increase utilization by almost 10% (Table 5.5). Between the potential for improved quality and higher utilization, there looked to be almost 20% cost improvement on the table. She continued.

"But as far as we can tell, at this point, we can't achieve these improvements."

The management team didn't know if they were coming or going now. On the one hand, they were shown a possible and totally unexpected solution to their looming disaster. And now it was being snatched away as rapidly as it had appeared. The plant manager saw the confusion and motioned for the operator to continue.

"What we've done is *very* roughly benchmark our *internal* technical capability or best of breed. Now to really achieve these numbers, first, we'd have to actually verify our numbers and find the internal capability for most of our major products and equipment. Then we'd have to understand those practices that give rise to that best performance.

TABLE 5.5 Rough Estimate of Equipment Uptime Technical Capability (Hours/Shift)

Equipment Type	Lowest Downtime	Planned Downtime
Insertion	0	1 hour
Wave solder	0	.5 hour
Test	0	.75 hour
Functional test	0	.75 hour

"Finally, we'd have to institute these practices throughout the company. There are only nine of us, and there are at least thirty major products and a hundred secondary ones, over two hundred pieces of equipment, and over six hundred operators spread over three shifts. The job is way beyond our scope. The only way to accomplish this task is to empower many groups, one at each work center or operation or for each product to do exactly what needs to be done. You need to empower more of the company—to make it a self-learning, improving, and changing organization. And that is a monster of a change. We can't do it."

She went on to explain how their group had scored their preparation for success at 0 out of 5.

Mark thanked them for the presentation and personally applauded their work. Each of the other vice presidents joined him. When the team had left, he turned to his own team.

"What are we going to do?" he asked, knowing for the first time what the answer was. Strangely enough, no one disagreed—there was consensus without blame. It was a first but not a last.

The next weeks were extraordinarily busy and started the true transformation of PCB Co. to a competitive, customer-oriented, continuously improving company, and away from an inwardly focused, budget-driven supplier. What PCB Co. did was start their drive to build an improving and empowered workforce. They decided to form several more teams, taking five of the members from the quality team and putting one on each of the five new teams they had formed. Each team was given a management member as a "sponsor" or advocate to help cut across department lines as needed. Each team was given training in project management, problem solving, and quality methods on an ongoing (or just-in-time) basis.

Equally important, the plant manager met with HBEC and invited them to participate in the improvement program—to critique their progress and offer a customer's perspective. Somewhat to their surprise, HBEC had accepted and offered assistance in sharing their own experience in becoming a customer-driven company.

Along with those immediate steps, their human resources department initiated the start of an employee skills assessment, looking at the skills needed in their company. They began evaluating what their hiring requirements should be by job positions and what their training requirements would be to bring existing workers up to needed skill levels.

The plant manager felt strangely invigorated. Rationally, he knew that they faced major problems. It was not clear to him that they would be able to meet either of their six-month deadlines. In fact, he was dubious that they could. On the other hand, he felt that he had accomplished something of basic and critical importance: moving the organization to be self-improving instead of static. The static organization only moved forward based on top-down orders. The improving organization continually adapted to increasing levels of customer needs and competition. He felt that he had unleashed something positive that ultimately would make them a fierce competitor capable of survival and even dominance in these changing times. But whether or not they could move swiftly enough was the question.

6

Achieving Stretch Goals Through Best Practices in Speed (the Race Goes to the Speedy, Not the Meaty: Time—the Next Enemy)

Mark Ritchards' professional life was falling apart. The corporate meeting had been the first blow. Now HBEC's threat to change vendors was more pressure than he knew how to handle. He concentrated in meetings at the company, forcing himself to remain calm and show leadership.

At home, however, he was a wreck. He found himself staring at a baseball game, uncertain what inning it was or the score. Instead he was going through the "should have, could have, would have" game, bottom of the ninth, down by three runs.

What overwhelmed him was how little time they had to succeed. If they had started these programs a year ago, he knew they would have had a chance. But six months was the blink of an eye. It took them six months to do their yearly budgets!

He felt like he was glued to the starting blocks at a race when everyone else was sprinting out of the box. His corporate legs felt heavy as concrete. It was as if he'd spent all his time preparing for a weightlifting competition and all of a sudden he'd found himself, muscle bound and slow as molasses, in the sprints.

What had gone wrong? He wasn't sure.

He tried to think through his concept of time at PCB Co. Time always seemed focused around budget cycles. Every year, at budget time, they crammed in meetings, budgeting exercises, and rebudgeting adjustments until right before the deadline when the budget was sent off to corporate. Once it was accepted, time returned to the normal set of rhythms based around a set of meetings, reviews, reports, off-sites, and other pre-scheduled operating procedures that set their schedule. They met monthly to review plans and budgets; they went offsite every quarter for departmental reviews; they had their annual winners' circle for salespeople, and on and on with regularly scheduled events. Time just . . . well, went on. Each year they set some new targets—to improve profitability, quality, or get a new product out—and then everyone worked at a set pace to meet their plans, reviewing them at their meetings.

What had gone wrong with this paced, measured, regular, predictable set of cycles? He really wasn't sure. Somewhere along the way, the rules of the game had changed. Someone had compressed the time clock. Someone had started to count speed. Someone had blindsided them.

What panicked him was that there was no "concept" of speed in his company, no urgency for speed. In fact, it was quite the opposite. Speed was synonymous with sloppiness, with rash behavior. Everything had to be studied, reviewed by endless committees, tabled, and retabled. Budget variances generated urgency; so did negative flow. Speed? It wasn't even one of their concepts. How was he possibly going to change that? How was he possibly going to meet their deadline for change? He could hear his management team now, each so used to moving at their own rate or pace, each so used to their own autonomy and freedom, each so unhurried. He had to change their culture. They moved at a snail's pace. Or more accurately, they moved at *their* pace—somewhat removed from the pace of their competitors, customers, or market. Momentum would carry you for a while, but then you were left further and further behind accelerating competitors.

What were they going to do? He just wasn't sure. It was a long night as he lay in bed. It went without a trace of speed. He just couldn't figure out why they had no urgency, no focus on speed.

Originally, things had been fast. When they had first started up the board division for the corporation, there was a real urgency to get their products out, a real sense of the need for fast results. They were willing to compromise. They knew that they couldn't have perfection in the first release of every product. They were reasonable. From discussion to design change to prototype was usually days, sometimes even a day or two. Those days were certainly gone. Now every new idea was discussed and sent off for a feasibility study, cost analysis, spec development, and purchasing regulation. Everything kept taking longer and longer to do it "right," by the rules. And the rules and procedures kept getting longer. Sometimes it felt like all their rules and procedures weren't protecting them but strangling them.

And as the rules and procedures got longer with proposals, presentations, studies, review, and approvals, the organization just naturally adjusted to how much longer everything took to do—new products, new vendor qualifications, order deliveries. They just accepted it and worked with it. No one viewed it as evil or waste to be stamped out. It was like the weather—you carried an umbrella instead of questioning why it rained. You planned for twelve-week customer order cycles, six-week engineering change notice review cycles, three-hour setups, five-day vendor delivery leadtimes, two-week planning cycles, and five weeks to close the books.

Somehow time had become a "valueless" resource—its marginal cost was pegged at zero. You could use as much of it as you wanted.

As he tossed and turned, he tried to see when the urgency had been siphoned off, what had killed it. Once, a long time ago, they'd had it. Now it was dormant, dead, or gone. In its place remained an endless set of policies and procedures. Why?

As the hours crawled by, he went back through their evolution, from fast to slow, responsive to bureaucratic, alert to inert.

Somewhere around five in the morning, as he thought again about all his direct reports yapping at him, he began to have an inkling of where their speed had gone. And as the sun started to rise and dawn broke, the truth was emerging in his mind. *He* had killed off the speed in his organization with a set of fiefdoms that dissipated any urgency in their parochial concerns and a limited or nonexistent view of end customers and competitors.

As they had grown, he had taken the team, originally focused on their customers getting out new products, and given each member a bigger and bigger separate organization. Engineering, manufacturing, finance, sales, and marketing—all were now larger than their original division! And each one had, in turn, created layers of managers and begun focusing on their *own* interests (be it engineering or manufacturing—organizing how much of the budget they got, how many people they could hire, how much capital they were allocated, or how many offices they were assigned). Each had taken on a life of its own—self-focused, isolated, except at interface points. And with that growth, the customer had become lost from immediate sight, buried within each department's *own* view of what was good or bad for the customer, of what was important. Each department required more time to review the other departments' suggestions that impacted it, and formulated rules and procedures to govern the interaction. The time urgency vanished as the real end customer and even the intermediate customer was invisible inside the imposing offices or space of their department. Few of them even saw the end customer anymore. What they saw and heard were *each other*—the members of their own department.

Speed had been replaced by greed (self-interest) and distance (emotional, physical, and managerial).

As he drove to work, his determination grew. He had caused the problem; he could fix it. He would assist Rod and join the task force on speed. He'd start on the production floor, at least to see why jobs took so long.

Somehow he knew that this project was going to be the hardest one, that it was going to have the greatest impact on the organization. Everyone related to quality, cost, and customer service. No one wanted to be "rushed." No one was going to like the organizational changes that might be necessary. No one except their customers, that was.

Peggy, the quality team operator, was nervous but resolute upon the call to appear at the general manager's office. She hoped that she wasn't going to face personal retribution for the customer visit gone awry or their suggestion to empower the workers. Maybe management had seen that as threatening or arrogant—to suggest that they didn't know everything that was possible. Anyway, the longer she was on the quality team, the more determined she was to do what was right. She had been given this responsibility, and she was only doing the best job that she could.

She was surprised by the warm greeting from the general manager who looked like he hadn't slept all night. His eyes were red, and in fact, he looked awful. But what surprised her most of all was his request. He was creating a team focused on increasing speed, and he'd like her to join it. He also wanted her to suggest anyone else she thought would look at the problem objectively and creatively. He needed people who would become as passionate about reducing delays as she had become about improving quality.

The first thing they did was take a tour of the plant floor. He realized how infrequently he actually toured the floor—not to walk through to meet one of his managers or hold a staff meeting in one of the conference rooms here, but an in-depth walk-through of the manufacturing line to discuss their manufacturing capabilities and problems. He was very removed from the "physical reality" of the floor. He didn't recognize many pieces of the equipment. He didn't know why there were racks of boards in one area, idle machinery in another, and boxes of components marked "hold."

Basically, it was foreign to him these days. He didn't understand the way manufacturing really worked anymore. He'd always believed he didn't have to—that's why he had a Vice President of manufacturing. But maybe there was something between minimal knowledge of their practices and the ability to run it himself. The problem was, at his current level of understanding, he didn't know how "good" they were. Were they world-class (obviously not), average, or even below average?

He didn't share any of these feelings with Peggy. He just asked her to walk him through the plant and explain where their speed was lost as best as she could from her viewpoint. Before the team was formed, he wanted to see what was involved, which disciplines were needed.

Their tour confirmed the consultant's numbers. As best he could tell, the actual processing time of the jobs was about 8 hours along with a 48 hour burn-in period (which no one seemed sure was still needed or needed for that length of time or for all products). That meant a total of 56 hours or slightly over 2 days of processing time for a process that they scheduled at 6 weeks (or 42 days)—a 20:1 ratio of nonvalue to value-added time.

Where did all that time go? There certainly was no lack of material. There were boxes, racks, and totes of boards everywhere. And yet there were also idle operators and equipment.

The factory was laid out into functional areas (Fig. 6.1). They had a raw material/components warehouse near the loading dock, a staging area

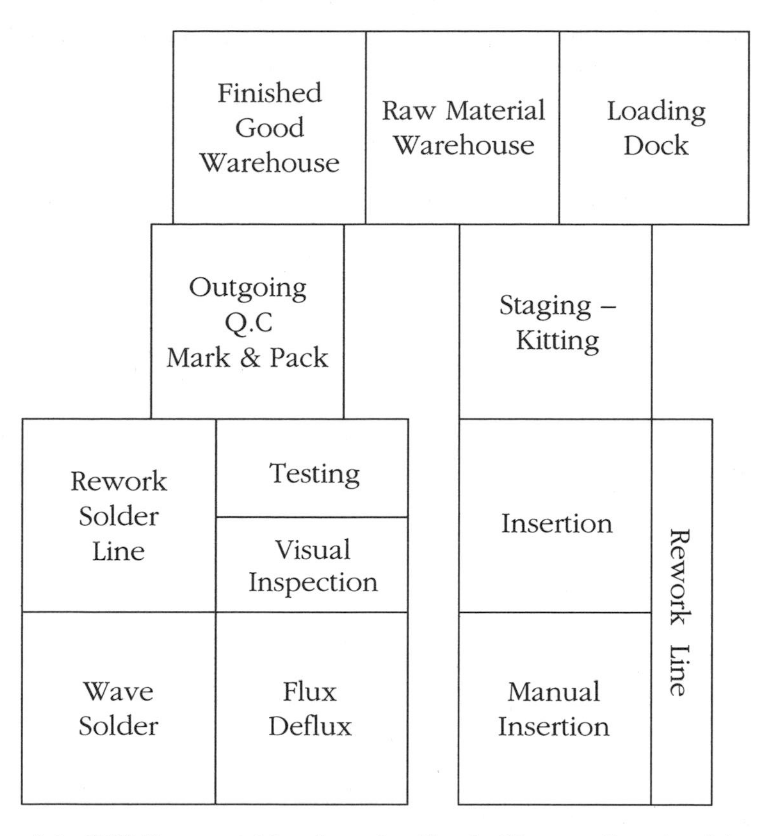

Figure 6.1 PCB Company Manufacturing Physical Layout: Functional Areas

for placing the components onto rolls of tape prior to insertion; the insertion area, staging for automatically inserted boards waiting for manual insertion (and/or missing parts and/or needing rework already). There were also the manual insertion stations and rework stations.

Then there was the flux stations, near the wave solder machines and the deflux or cleaning station. Also nearby was the "bed of nails" test area for checking for opens and shorts.

After that, there was the functional and environmental test room with its burn-in ovens. Another rework line handled both test areas. Finally, there were the mark-and-park areas with final quality control and a finished goods warehouse.

Nowhere to be seen were the maintenance, scheduling, engineering, finance, personnel, or facilities departments. Each had people on the floor, some frequently, some infrequently, but their offices were elsewhere. Somewhere.

Lots of things struck him as he toured the facility—all the materials/work in process queued up, waiting; the separate colored exception labels—missing parts, wrong parts used, customer cancelled order, rush order; the general low-key pace of operations—totes were moved periodically, but much of the time nothing moved from operation to operation. If there was any sense of urgency, it was too subtle to detect. Nervousness, yes, at his walking through the factory. Quizzical glances at Peggy, the operator, wondering what was going on. Some downright hostility and some indifference (he thought he heard a few low murmurs of laughter in areas). But no urgency.

What he didn't see was any of the speed he was use to: at a McDonalds, where the help competed to serve the customers faster than the next counterperson; at his local bank where the managers assisted to speed up customer lines when they grew long; or at the supermarket where new check-out stands opened to keep lines under four customers.

No, there was no "customer" in sight here, visually or emotionally. The only "time" was the time clock, telling the operators when they were on break, lunch, or out of there. Whatever management techniques or measurement schemes they used, they'd certainly managed speed right out of consideration.

They had to rediscover it. They had to dig it up wherever it was buried—under procedures, apathy, mistakes, or acceptance. More than anything else he could see that they had come to accept the status quo, and with it, slow deterioration of their competitive position.

What struck him was how obvious it was watching the factory floor that time was "wasted." You could see it in the slow movement of material, in the piles of work in process, and in the idle equipment. All around him valuable time just piled up and was wasted. And what about the *indirect* functions—supervision, accounting, scheduling, maintenance, and engineering? He had *no* idea of their pace—their speed, their reactive velocity. Visually, there was no easy way to see it! They certainly didn't measure it, put their performance outputs on the walls—two hours to repair a broken inserter, twenty minutes to schedule a new order, and seven days to close the books monthly. He was completely blind to their speed. Everything was under the influence of budgets. They counted activity and WIP to measure cost. Nowhere in their budget did they consider speed.

He had to find a way to highlight speed, to find out where the waste was, and then empower people to change it. Otherwise, ironically, they were rapidly going to go out of business. There was no doubt about that. And he had to do *his* job quickly! He needed a tool that would naturally lead to increased speed.

A thought occurred to him. Why not ask the head of Hot Box? Maybe he could both get some help *and* show their sincerity in wanting to change.

He'd call as soon as he got back to his office.

Hot Box Helps

The general manager was pleasantly surprised at Hot Box's response. The plant manager there was not only willing to help; he was happy to help. "We need you to become a JIT/TQC vendor," he explained. "If you need some help to become one, that's okay. My help will be repaid in both our successes! No one can be a successful company today without successful key vendors and customers. Your boards could give us a competitive advantage. Your engineering is excellent. But that's not enough. If you can fix your manufacturing problem, you will become faster and faster, and soon you'll be giving us a competitive advantage in our products. So it's well worth giving you a hand. Besides, it will give me an idea if you're really going to make it."

The last sentence again froze the pit of Mark's stomach. Sometimes even he was too far removed from their customers' needs, concerns, and most importantly, satisfaction with PCB Co. He relied on sales and marketing to watch over their customers. Recently he'd heard of a new practice companies were initiating, a practice used heavily by consumer goods companies—customer review boards. In consumer marketing, a neutral third party conducted interviews with consumers comparing their reaction to various competitive products. In this variant, the company management invited in groups of customers to critique them, face-to-face. It was a daunting experience, and apparently, just the sort of experience that PCB Co. probably needed. The customers compared you on cost, quality, customer service, speed, technical capability, flexibility, and general attitude to your competitors (if they knew them) and/or to other suppliers they considered relevant.

He had to admit, he'd also lost sight of the customer. He rarely visited them. Periodically, he made purely ceremonial appearances at customer visits to their plant. Ultimately, there was no one to blame for their dilemma but himself. And if he'd gotten them into this situation, it was his job to get them out of it. For that matter, did he have a choice?

The "Czar of Time": Measuring Speed

When Mark went to visit Hot Box, he was introduced to Lou Jacoby, a short man with a firm grip. Lou was, according to the HBEC plant manager, their "czar of time." He was educating everyone on the importance of time-based competition and teaching their people how to become world-class racers. They actually had the HBEC Olympics, where every six months they gave corporate awards for excellence in time.

"Didn't this make people rush and make mistakes?" Mark immediately asked.

Lou smiled. "Well, first of all, speed is measured on when the task is done *correctly*. So if quality is poor, speed will by definition be poor as you'll have to fix it or do it all over again. We think of that as dropping the baton or missing a gate. In fact, all we did initially was get rid of the time wasters that have *nothing at all* to do with the work itself. But we're getting ahead of ourselves. So let me take you through our race to competitiveness.

"Electronics is one of the fastest-paced industries you can imagine. Our products are competitive for perhaps two to three months. Then there's a new 'hot box' on the market, and we're scrambling to catch up to it." ("If we're scrambling, by the way, it means our vendors need to scramble too," he said as an aside.)

"We think that this is going to happen in every industry, by the way. The advantages of speed in terms of competitiveness and profitability are overwhelming as you'll see. It just happened in our industry first because the first company to use the latest semiconductor or disk drive immediately had product performance leadership position. So everyone scrambles to be first. But all customers want differentiated products. The first supplier who can meet that need in your industry will lead it. Our first step in becoming faster was to *measure* speed.

Measuring the Cost of Speed

"What we found was that we didn't have much information about our speed. In manufacturing, we knew when we started a job and when it was completed. We took this data and reported it as average leadtime or throughput time by week or month.

"However, as we thought about it, this approach had three problems. First, it didn't show us the variance or distribution of leadtimes. We only knew the average. Therefore, for planning purposes, we didn't know what leadtime to use if we wanted to guarantee delivery with a certain probability (Fig. 6.2). It also didn't tell us how much improvement was needed in reducing the variance of leadtimes (or variability) that was a basic cause of late deliveries."

Mark nodded at this explanation.

Second, leadtime measured this way was a lagging indicator. We knew our leadtimes *after* jobs were completed. We didn't know if leadtimes were getting better or worse for the jobs in the line or what it was by operation. Therefore, we could only react long after a problem had arisen at an operation.

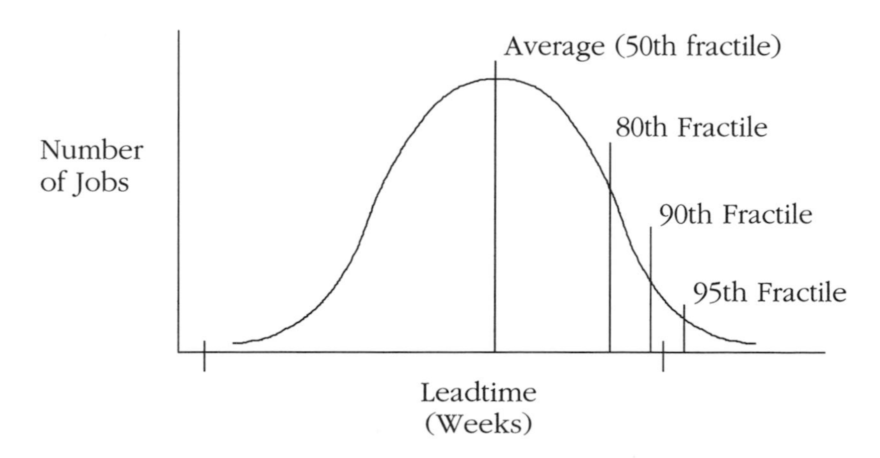

Figure 6.2 Distribution of Manufacturing Leadtime

Finally, we had no idea of what *caused* the leadtimes or their variance. Since we didn't know the root causes, we didn't know how to fix the problem or in what order to address the problems. So our first step was to measure speed at each individual operation or work center by cause (delay) with both averages and variances by job.

That took a new data collection procedure or system, as our current MRP shop floor control system didn't even begin to collect all of that data. We had to do it by hand initially. More recently, we implemented a new plant floor system—an integrated manufacturing execution system (MES) that tracks the jobs as well as equipment, tooling specifications, materials, and facilities.

That gave us the data to begin understanding where our speed or time went (Table 6.1). As you can see, we spent too much time in *non*-value-added activities—in queues, waiting for setup, missing parts or tooling. I suppose there were no surprises, but now we had actual data with the number of occurrences (or frequency) of any problem.

However, we were in positively great shape in analyzing our manufacturing leadtimes compared to understanding our "indirect" speed—the speed of all our indirect activities such as maintenance and repair, sign off on engineering change notices, and so on. We didn't even collect the averages of these activities. All we had were anecdotal instances of how long these activities seemed to take. The first step was to start collecting the major components of time for each key indirect activity (Table 6.2). We had

TABLE 6.1 Measuring Production Speed at an Operation

Average Leadtime by Cause	Hours	Percentage	Number of Occurrences
Production	1.50	20	300
In queue	2.50	33	285
In transit from previous operation	.50	7	300
Setup	1.0	13.5	206
Test/measurement	.50	7	300
On-hold—quality	.40	5	96
On-hold—no tool	.10	1	51
In rework	1.0	13.5	123
	7.5	100	For 300 jobs

TABLE 6.2 Average Indirect Activity Leadtimes by Major Component at Insertion

Equipment Repair		Hours
Time until maintenance is notified of a breakdown		.50
Time until technician is assigned		.50
Time until technician arrives		1.00
Repair time		1.00
Time until accepted by manufacturing		.50
	Average time	3.50
Equipment Setup		**Hours**
Time until setup department is notified		.50
Time until technician arrives		.50
Setup time		.50
	Average time	1.50
Material Review Board		**Hours**
Time until job is assigned to MRB		.5
Time until engineer is assigned		48.0
Time until engineer collects data		48.0
Time until MRB meeting		72.0
Time until job is reviewed		2.0
	Average time	170.5

never really focused on indirect activities. They were handled in the main by exempts who were on straight salary.

As soon as we had the data, initially collected by sampling, we found that we had the same situation as in manufacturing—too much time spent in nonvalue-added activities. We had typically a 2 to 10:1 ratio of value-added time doing the activity to nonvalue-added time—waiting to be notified of the activity, for resources, for approval, for data. Clearly, there was *a lot* of speed to be gained that would translate to higher utilization or higher productivity or faster leadtimes.

At first, the exempts were offended that we questioned their efficiency, but by empowering *them* to look at their speed and improve it, they were satisfied that our goal was competitive advantage and not punitive measurement.

They had never really thought about the *cost* of "slowness." Their activities "took what they took." What also became clear to us was that the problem wasn't lack of measurement alone. To some extent that was a symptom of the *real* problem. The real problem was lack of interest or concern about speed. It wasn't *important* to us. We had never really focused on speed. It wasn't how we competed. It was just a given. It was "part of the process."

"So how did we increase our speed? Well, the next step was to think of speed as being under *our* control and not a *given* or state of nature. The example everyone gives, of course, is having a baby. That takes 9 months no matter how much of a hurry you're in. But that analogy has almost nothing to do with the activities in your factory. So the first step *after* we developed an urgency toward speed was to *stop* treating it as a given."

Lou continued. "The problem in your company, I'd wager, is that you've grown to treat speed as a given, not under your control. You have lost sight of it as a *cost*, or even better, as a competitive advantage. Speed is, to us, a weapon. Everything I can do faster than you, I will win. Think of speed as how fast you convert your resources (or cost) into an advantage (or revenue or lower cost). The faster you do this, the shorter the lag between expense and customer satisfaction, the greater the profit or advantage. Think of dollars lying on the ground. Speed is how fast you can pick them up. If that sounds far-fetched, think of it this way: Every activity you do expends resources. That's your *cost*. Everything eventually generates revenue. The faster you are, the sooner you convert *cost* into revenue or improvement. It really is *that* simple. But most companies have completely lost track of that viewpoint.

"Why? Because we as individuals are far removed from either the cost or the revenue side. We only see a small part of the *cost–revenue chain*, so we don't see how we impact it. And most of our current systems and procedures only make it worse. They measure us against our current or standard leadtimes and the current cost to revenue lag, called the budget. They don't focus on cost "aging"—how fast we turn cost into revenue or how fast we *could* convert it."

"Here's another way to think of it: It's like people "owing" you money. Wouldn't you want to collect it as quickly as possible? That's what order management and manufacturing speed can do. Think of growing your revenues. That takes new products, new channels, and new cus-

TABLE 6.3 Time-Based Savings Analyses

Cost $150	Revenue $250	Profit $100
1. Order to customer: Order entry Order acceptance Raw materials Equipment time Labor time Shipment time Invoice time	Order shipped and collected	By standard accounting $97—by time-based accounting $\text{Cost} = \frac{\$150 \times 12\% \times 2\text{ months}}{12}$ or "Speed" time saving $3.00/unit
2. Major ECN Experiment: Discover problem Collect data Design experiment Run experiment Change specifications Implement change	ECN results in ½% higher yield Revenue Not measured—pure cost	Standard accounting—not done Time/speed savings = $150 × .005 improved yield = $.75/unit/month Ex: 1 month; 100,000 units = $75,000
3. Major machine repair—bottleneck: Machine goes down Reported to maintenance Assigned to technician Gets info (history) Gets tools Diagnosis Repair Gets parts Test Runs test units Backup	1 hour more uptime gets 1 hour output or less overtime 1 hour more uptime reduces job's queue time 1 hour Not measured—pure cost	Standard accounting—not done Time/speed savings = 60 units/hour × $100/unit profit = $6000/hour at bottleneck

tomers. Imagine again that you could improve the cost–revenue chain now. You could grow faster with less investment, with higher return on your money. Once you see speed as the conversion rate of cost into revenue or improvement or of demand into customer satisfaction, then everyone begins to see its importance. Let's be even more specific and convert speed into dollars."

"What I've prepared is a "time-based savings" analysis that we do internally. Our standard financial systems don't really see the cost-revenue lag so we have to do these ourselves. I've listed three examples of different types of "cost-revenue lags": revenue itself, "improvement received," and "waste decreased."

"In the first example, let's look at the way we used to treat a customer order. The total cost to produce a workstation add-on board was $150; the price we sold it at was $250, so the profit was $100. The standard leadtime was two months from order to shipment/invoice (not even receipt of funds, the way we should measure it). If manufacturing came in on "budget" ($150, of cost) and at standard leadtime (two months), then we were "on budget."

But now, let's look at a time-based accounting view of this transaction. From a time-based view, or cost–revenue chain, we've spent the money two months ago (worst case) and now have to tie up *our* money until we can get paid. So the cost of our money at 12% value/year is $1.50/month/unit. Our "*real*" profit is not $100 but actually $97, since we also have to pay for our capital being tied up for two months. Now we see the value of questioning our standard leadtime—$1.50/unit/month.

Now 3% of profit may not seem like a large amount, but right now, our financial systems don't reflect this cost of leadtime. We do have accounting for our inventories but not for our speed. So it's not always easy to understand the value of speed. And with our operating margins usually falling below 10%, an additional one or two operating percent profit (since there will always be some leadtime) would add 10 to 20% to our total operating profit.

In our second example (making a series of design or process improvements) we don't even have a way to account for the expected yield improvement. We simply treat engineering as an overhead function and have a planned yield or learning curve that we build our budget or plan against. But instead, let's look at a *time-based* accounting for yield improvement. Right now it normally takes three months to go through a learning cycle. That's the time from when we start working on a process improvement until we complete our ECN. Now let's look at the value of speed. If we could reduce the three-month yield improvement cycle by a month, we'd see (for example) a half percent improvement in yield a month sooner. This would be worth .005% (the percentage of yield improvement) × $150/unit scrapped × the number of units produced per month (say 100,000). The total value of time is $75,000/month!

What we see from these examples is that we *can* quantify the value of speed to us—in either lower cost or higher revenue or both. In our last example, we examined the time value of speeding up repair activities. Again, at a bottleneck, we can calculate the additional $ of output possible from the factory by an added hour of production—in our case, $6000/hour/bottleneck equipment!

The point is that the leadtimes we take as "given" all have "revenue" associated with them, revenue that we are delaying (and usually don't associate directly with our tasks) until we become a time-based competitor.

Table 6.4 gives a longer list of the activities we take for granted as "overhead" and the potential "revenues" associated with them. In a time-based company, we must consider the "real" cost of time and the "real" value of speed.

"What was the point of all that? To make it clear that you *must* think about speed. That old saying 'time is money' couldn't be more accurate. The reality is that when one of your competitors is faster, you'll be left behind. Imagine in our example if you continued doing four experiments a year, learning quarterly, and I could learn monthly. I'll be improving at 6% a year to your 2%. How long before my costs are significantly lower than yours; my new products launched before yours; or my new equipment, operators, and facilities producing before yours? In today's world, it's not the

TABLE 6.4 "Revenue-Producing" Indirect Activities in the Cost–Revenue Chain

Activity—Example	Associated Revenue
New product introduction Customer order delivery Sale to customer Facility construction	Revenue from customer received sooner
ECN Operator trained to full productivity Equipment qualified	Improvement benefits obtained sooner
Equipment repaired Job taken off hold Facility "out of spec" condition eliminated	Waste eliminated sooner

big who eat the small, it's the fast who eat the slow. They literally eat the food before you can get to it—the customer's demand."

The plant manager's head was spinning. He thought about all the time-based *waste* in his company. He had never considered time as having a large "marginal" value. Things took as long as they did. As long as they were profitable, he never thought about *how* profitable they could have been—how much of a competitive advantage they *could* have had.

They weren't fat, dumb, and happy. They were slow, dumb, and now about to be very unhappy.

Lou saw the dazed expression on his face and gave him a minute to collect his thoughts.

"Interesting, isn't it. The longer we do this, the more we realize we need to think of manufacturing in a whole new way—as a lean, fast, precise, learning, improving, and hungry machine. We have to eliminate all our complacency. This is a race, a contest, a competition. Yearly budget improvement cycles with status quo in between is *not* going to win. We post our times and shoot to improve them just the way a racer does. Everyone runs this race, not just production. Speed is good for everyone!"

Mark heard everything that Lou had said. He believed the advantages speed would bring. That's why he had come personally, to see how they could speed up their improvement programs. But he had one major nagging question, one major doubt.

The real stumbling block for him was whether you could "rush" activities—whether speed would lead to competitive advantage, or in fact, just rushed shoddy decisions made for speed's sake, without careful, reasoned analyses.

He tried not to be defensive, but he was afraid that's exactly how it came out.

"Aren't you rushing decisions? Don't people feel so much pressure to be quick that they start to make *any* decision, simply to "look fast"? Aren't you putting unnecessary pressure on people to work faster than they're used to, or in fact, paid to work? Don't people feel that they're on an endless treadmill with the motor speeding up constantly? I'm not sure I or anyone else would want to work in what sounds like a piece parts sweatshop environment."

His questions were honest. He couldn't believe that speed was all positive with no downside.

Lou had obviously heard these questions before. He didn't wince; he didn't change his pose or expression.

"Does it strike you that these were the same questions you or others raised when you instituted your quality program? Was it really possible that you were eliminating sheer waste? How could you have allowed all that waste all those years? Well, times changed. What wasn't valued years ago is today. Where maybe you were once miles ahead of your competitors, they *had* to run faster, and they did until they caught up. When we started our time-based program, I had the same questions asked over and over again. Good questions. Won't this affect the quality of our decision making? Won't this have some hidden cost we're missing?

"I asked people, what is the value of having a job wait in a queue for two days prior to processing? Does it improve the quality of the processing or is it sheer waste? We're not necessarily trying to make decisions themselves faster; we're simply trying to eliminate all the delays that not only don't add value—they take away value.

"Let's go on to the second topic: where your time goes."

Examples of Where Your Time Goes—The Time Wasters

"What I've done is take three activities, before we eliminated the 'time wasters,' processing a job, improving a process and repairing a piece of equipment, to show you where your time goes (Table 6.5).

"They range from physically intensive (processing a lot or job) to information processing intensive tasks (improving a process). All I've done is categorize *where* the time goes for each task.

"Look at processing a lot. What strikes you is how little of the time is actually spent processing the lot. Of the over six weeks the job *spent* in our factory, over 90% of that time is *waiting* for manufacturing to add value. We wait in queues at the equipment, we wait for tooling, on hold, for parts we're missing, in rework, or in transit. If you didn't know better, you'd think we must be in the storage business since that's what we seem to spend most of our time doing.

TABLE 6.5 Where Time Goes: The Components of Waste

Processing Time Components	Equipment Repair Time Components	Engineering Process Improvement Time Components
From job release:	From equipment down:	From process problem:
In processing	Notify maintenance	Detect problem
In queue	Assign to technician	Notify engineer
In transit	Technician arrives	Assign engineer
On hold	Get equipment history	Engineer arrives
In rework	Get equipment manuals	Engineer finds:
Waiting for parts in storage	Interview operator	Equipment log
Setup time	Diagnose problem	Run cards specifications
Filling out paper-work	Get additional parts and tools	
Getting tools	Test equipment	Engineer sends job to lab for analysis
Getting parts	Run test units	
Measurement	Notify production	Engineer analyzes all the data
Storage as finished goods	Operator returns	Develop hypothesis
		Design of experiment
		Run experiment
		Change specific temporarily
		Change specific permanently
		Sign off on specification
		Communicate to floor
Physical processing	⟶	Information processing

"Now everyone agrees that if we could eliminate most of this waiting, we'd certainly not affect the quality of our manufacturing process. In fact, many people feel it would be much better, as we could get faster feedback on quality, not suffer any damage from excessive handling, and so on. But no one would argue that longer waiting times improve our process. In fact, the entire movement to JIT, group cells, and focused factories was aimed right at eliminating that waste in storage space, inventory holding cost, handing, or lost equipment utilization.

"Now let's look at repairing a piece of equipment. Where does that time go? Nearly one hour is actual repair time. But there are another three hours of something: waiting for a technician to be assigned, waiting for the technician to become available, waiting for him to get the needed information, tools and parts, or waiting for production to accept the equipment as functional. Do any of these add value? Do any of these lead to a "better" repair? Not in the least. If we could have the technician there immediately, provide him with all the necessary information, tools, and parts, and be sure that the equipment was repaired correctly, we'd get the same "value-added" result (the repair) with none of the added cost or lost time, the *non*-value-added activities.

"Now let's look at a purely information task, pure analysis and decision making, fixing a process problem. Look at the data—the percentage of time the engineer *actually* works on solving the problem is less than 15%! The rest of the time is spent waiting for someone to detect the problem and assign it to him, and for him to get *all* the data from the five or ten systems that he needs to reference, write up the new specification, convince others to approve the change, and then get it onto the floor in all the specification books. That's where the *real* time goes.

"Does anything strike you? In almost every case we've looked at, we could cut the time for a task by 80% and *not even touch the time* allotted for the value-added component! We don't need to *rush* the real value-added task. We simply need to get to it sooner and get on with it once it's done. Most of our activities are filled with "time wasters" that we *simply accept* as the status quo. We're wasting time!

In general, we can break down every activity, task, or job into a set of steps. What we need to do is examine where we waste time at every step. We've done that and categorized them into eight types of time wasters.

The time wasters exposed. What we found was that we could categorize the various types of waste into seven root causes and so develop typical approaches to eliminating each one. The time wasters are shown in Table 6.6.

The first time waster is a lag in communication of needed information or action. As examples, these are time delays between when an order arrives and the production floor knows about it; from when equipment fails to when maintenance is informed to when a specific person is assigned to repair it; from when the lab data is available to when the engineer receives it; from when a job is delayed to when customer service and production scheduling is informed; from when a process goes out of statistical control to when the deviation is detected and assigned to a person for correction; or from when there is a new ECN approved until the floor knows about it. All

TABLE 6.6 The Time Wasters

1. Communication lags or "knowledge" distance—lags until a problem/alarm is detected or information is available to a key person (the handoff lag) or a problem assigned to an individual, or a task or a solution communicated to the end customer.
2. Physical distance to resources—to equipment, tooling, the next work center, or logbooks (information), etc.; time spent getting the resources and information required for the job.
3. Knowledge gap—gaps in the skills or knowledge required to do the task by the primary person that requires another person to complete the task (knowledge, job, or responsibility fragmentation).
4. Nonvalue-added steps in the task—storage, waiting, setup, transit, data collection duplication (filling out same data elements on multiple forms).
5. Making too much product or making it too early, leading to idle time and storage.
6. Variances in quality, equipment uptime, operator availability, raw material availability, etc. that cause nonstandard flows, reprocessing, scrap, idle time, or other deviations in the task's completion.
7. (Manual) work which could be redesigned, assisted, or automated for faster and more efficient completion—often found in planning, reporting, scheduling, analysis, communications.
8. Approval lags ("responsibility lags or distance") lag until someone else can approve the decision or outcome, in the department or outside the department.

of these are examples of unnecessary time lags until a problem can be identified, assigned to a person, or solved. They are due to "information speed" delays. These types of communication lags also lead to unnecessary uncertainty (one organization has to anticipate or forecast where there is no real uncertainty). For example, shipping may guess at what products are going to be run at insertion when the insertion supervisor has that schedule. Maintenance may have to guess at the production schedule when it's available. Both of these are examples of communication lags. Interestingly, so is JIT. We are artificially not telling the previous operation information we have on what we are currently running.

The second time waster is physical distance (or lags) to obtaining the resources required for the job, to the five elements of manufacturing: equipment/tooling, material, personnel, information/work instructions, and facilities. Common examples are poor layout of facilities, with work stations unnecessarily distant; tools and raw materials stored in areas far removed from the production floor; work instructions kept in an engineering area in a separate building; maintenance areas similarly removed from the floor; paper records that are hard to access; and so on.

The third time waster is a knowledge or skill gap that requires use of another person(s) to complete the task—waiting for a technician to set up the equipment; waiting for the engineer to diagnose a machine problem; waiting for the maintenance technician to perform a standard preventive maintenance; waiting for a lead operator to select the next job; waiting for the material replacement operator to replenish inventory; and so on. These are examples of job fragmentation that delays the completion of the major task unless perfectly timed or coordinated, as on a paced assembly line where every step has been choreographed to avoid any delay. Unfortunately, specialization (or job fragmentation) usually results in waste—idle time for one or both of the players waiting to be needed or to be available.

The fourth time waster is a nonvalue-added step in the task/process itself—moving from an operation into a storage or waiting area/queue; long in-transit steps between operations, testing steps that should be done using sampling or replaced by process control monitors; data collection that involves duplicate entry onto run cards, operator logs, machine logs, station logs, and reports; long setup and teardown times; large experimental runs where a carefully designed experiment would provide the same results with less time and material.

Today we think of many of these steps as "necessary" evils or just part of the manufacturing process. In a time-based company we continue to question every use of time to see if the "revenue" warrants the cost or if the cost can be decreased or eliminated.

The fifth time waster is overproduction or early production—making more than was ordered or producing sooner than needed. While this may not seem like time waste (the excess will be held in inventory and sold later), excess production affects both direct costs and time. The obvious direct cost is inventory holding. We tie up our working capital or "cost." In addition, we have the cost of storage areas to hold the inventory, handling costs (personnel and equipment) to move excess inventory to and from storage areas, and data processing costs to track the inventory. We also face potential damage, theft, and other losses associated with inventory.

But the effect on time is equally important. We tie up our equipment making product not yet needed, extending production leadtimes. This means longer leadtimes for current orders or forecasts, translating into longer waits for customers and larger forecast errors.

The desirable approach is to look at reducing setup times so that shorter runs are economically viable. While setup times may seem a "given," most companies have been able to reduce them by 25 to 95% through use of single minute exchange of die (SMED) ideas and techniques discussed by Shiego Shingo.

The sixth class of time wasters is "variances" of all types—equipment breakdowns, product rework and scrap, operator lateness, technician error, and vendor quality problems, and so on—that render our plans obsolete. Production is halted, work must be rescheduled, additional production is started to replace what is lost, and steady flow is halted to the succeeding operations. Often the most common approach to dealing with variances has been to build a "just-in-case" plant within a plant to deal with them or buffer their impact. Production schedulers "buffer" their planning lead-times, leading to elevated work-in-process levels and parts inventory levels. Production managers try to keep work center inventory levels high to keep their operators and equipment fully utilized (usually their primary measure of performance). Plant managers arrange for excess capacity on the most suspect tooling and equipment.

All these "just-in-case" activities "steal" time while adding no value. They are only necessary because we are unable to run the operations within tight statistical process control limits. Worse yet, we begin to expect these variances and use them as a *given* to explain why we cannot run a "fast" operation. The key to a fast team is a team where each member has confidence that all the *other* activities not under their control will occur on schedule. Otherwise, there is no point to speedy, on-time performance on their part. Why "rush" to move jobs forward if you know they're only going to wait at the next operation. Why "rush" to provide a tool or raw materials if they will not be needed for hours or days. Manufacturing is a team sport and variances break down the smooth operation of the team, reducing it to quasi-independent units that buffer themselves against unreliable team members.

The seventh time waster is manual procedures or tasks that could be sped up by either automation or redesign. Examples of the first include: manual or visual inspections; placement of parts; report preparation, planning, and scheduling; and determination of physical inventory and location of jobs. Examples of the second include: processing at poorly designed work stations (that require excessive movement to reach tooling, equipment, and parts); work instructions that are overly complex or arcane; and assembly designs where assembly is complicated by poor placement of parts on the board. Each can be redesigned to dramatically cut task time (and usually increase quality as well; complexity usually means a greater chance of error).

The eighth time waster is a fragmentation of authority or responsibility or job function so that the primary person or team has to wait for outside review, approval, or authority to take decisive or final action. Examples include:

An operator puts a job found to be out of specified limits on hold and waits for an engineer to review it

An engineer modifies a specification and must get a list of reviewers to approve it

A manager has a supervisory behavioral problem and must wait for a human resources person to sign off on any action

A maintenance technician completes a repair and must wait for a production supervisor to "accept" it

A production task is divided into many small tasks done by separate individuals or departments (setup, calibration, processing, quality checks, etc.)

Note that many, if not all of these examples, seem to represent a logical series of checks and balances to ensure that mistakes are not made. However, think about the commonly accepted approach to quality—that we want to *build it* into the manufacturing (or in this case decision) process and not simply *test* it in. In many of these examples, we are testing it in—having someone else double check or confirm that the decision or work was done properly. In fact, many times this approval or inspection or review is done more perfunctorily than the original work! The inspection tasks become, in reality, getting on someone's schedule for their signature.

Eliminating the Time Waster: Re-engineering the Process—Best Practices in Speed

Lou then put up a "Time Waste Busters" chart (Table 6.7). "As you can imagine, the cure for most of these is pretty straightforward—at least conceptually. The difficulty is getting the *organization* to change, to give up

TABLE 6.7 Time Waste Busters

Time Waster	Time Waste Buster
1. Communication lags	1. Move the support people to the work; communicate problems instantly.
2. Physical distance	2. Locate the resources at the work; make data easily available.
3. Knowledge lags	3. Move the support people onto the team; train the workers to do these tasks (if possible).
4. Nonvalue-added steps	4. Eliminate them by work flow/scheduling/single data entry.
5. Overproduction/early production	5. Eliminate them by reducing setup times, scheduling orders when needed.
6. Variances	6. Eliminate them by quality tools.
7. Work/task inefficiencies	7. Redesign for efficiency; automate as needed.
8. Approval lags	8. Move the decision makers onto/next to the team—expand responsibility of the work team, if possible.

or take more responsibility, and to match the *task* requirements to the team's resources and authority.

In the case of communication lags, approval lags, and knowledge lags, we're usually fighting the last forty years of specialization, management hierarchies, and departmentalization. It seems that everyone has become differentiated. No one simply wants to be a team member. We want to be in the production, maintenance, engineering, or scheduling departments. We want to be supervisors, managers, or directors. We want to have our own office areas and meetings. These changes have divided up responsibilities so much that a "simple task" (producing an order) now involves twelve departments passing paper, producing status reports, and meeting internally and interdepartmentally to re-coordinate the task they cut up so finely. Scheduling takes the order and notifies production and the store room (kitting); production notifies the tool shop; maintenance schedules their activities and coordinates with production; engineering works on improvement programs and coordinates ECNs. Everyone has managers above them who need to approve their decisions.

We simply said, let's try to get everyone on *one* team that gets the task done with minimal waste. Let's see that the interdepartment and intradepartment communication, approval, and knowledge lags are minimized or disappear. We built some teams around the product—those on a dedicated line—encompassing all the functions needed to make and improve the product. Other teams were built around the "process" or operations, at specialty work centers or shared steps, such as special machining work centers for prototypes that all product groups share.

Those teams can work together in one location with one goal—to make our products cheaper, better, and sooner, the same goal that both the company *and* our customers have for us.

"What a nightmare organizationally! You see, what we were doing was taking the traditionally "weakest" or lowest power group (production) and saying that everyone else was a *service* to it. That wasn't the way the others saw it. When we moved engineering into each product area and out of central engineering, the director of engineering had a fit. The same thing happened with the maintenance, scheduling, and accounting departments.

"Now maybe if speed weren't an issue, we'd never have done this drastic a step. But if you want speed, you want focused teams. If you want speed, you want clear objectives. I'm sure we sacrificed something, but it isn't clear what yet. We left the central organization in charge of "central" functions—career path planning, promotion criteria, hiring and training programs, running best of breed practice forums, and collecting and disseminating these guidelines/practices. But we're even looking at moving those functions to teams drawn from the individual areas instead of a separate department or hierarchy.

"What we've done is flatten the organization for speed, to prevent information from having to run up and down the organization and approvals be made from afar. The challenge has been to train and mentor the work teams to *make* good decisions and learn to manage themselves. The former managers are now team mentors and trainers.

"In the case of physical distance (time waster #2), we simply need to move the resources to the place of work. In general, we want to shrink all distances. Distance has no value, only cost. In our case, we built specially designed work cells, designed by the production team, with all tooling and parts right at the cell. What was hard about doing this? Organizationally, those departments screamed their heads off. They saw it as taking away their responsibility and authority. They owned the tools and parts, and the production groups were either borrowing them (tools) or "buying" them (parts). By keeping them at the work cells, there was no tight "control" or ownership. But, for the life of me, I couldn't see what the location of tools had to do with *teamwork*. The tool shop and stock room had somehow become kingdoms over time. They didn't see themselves as part of a team. They didn't see that the *team*, the company, owned all the resources. They would still be responsible for calibrating and repairing the tools and making sure that there was a steady flow of raw materials to the floor. But somehow they had completely lost track of the fact that *they* were a *supplier* to production; production was their *customer*. They treated production like the enemy, not the customer. There was no concept of team at all! Neither side seemed to trust each other. Production ordered more parts than they needed before they needed them because they were worried that the stock would be out of what they needed. The tool shop felt production took poor care of "their" tools. Everyone had lost sight of their *real* customer and the *real* cost of time. We moved the parts and tooling *suppliers* to production. We grade the entire *team* on overall speed. It's not perfect. They still argue sometimes about who's at fault, but it's much faster and improving!

Nonvalue-added steps (time waster #4) really couldn't be eliminated until we also attacked the variances and organizational design. Most of those steps (except setups) seemed to be in response to crossing borders in "organizational" territory and variances. When we redesigned production into product lines, that organizational waste quickly disappeared. But variances were still killing us. A machine broke down, work piled up here, and dried up there, so the supervisor wanted buffer stock. A vendor order was late—the production area wanted more raw material inventory. We just kept working away at the major variances one by one to kill the reasons for these "just-in-case" time wasters. We're getting there.

"Oh, and once you grade the production line on "time-based revenue"—and they can increase their profitability with speed—*they'll* drive those nonvalue-added steps away—fast. Once the variances go away, then we could schedule using shorter leadtimes and have confidence that schedules would be met.

"Overproduction or early production required getting help on reducing setup times. We worked with our equipment suppliers, engineering, and maintenance people to reduce setup times at the key equipment. We discovered lots of tricks and techniques that no one had ever taken the time to think about.

"Finally, the team itself has kept improving each task's efficiency—to increase its profitability.

"There really were several keys to accomplishing this improvement. First was to focus on time as a *real* waste—to make sure everyone thought about "time-based revenue" for their customers and measured speed as diligently as we did quality. Second was to consider reorganizing to put the resources needed for the task on one team. This way we could avoid conflicting measures of performance and huge coordination times and costs.

"Finally, we empowered the teams—with responsibility and authority, with training and mentoring—to continually improve their performance. We forced them to come to the forefront and contribute to the solution actively. Not *everyone* can do that, but we built *every team* to be able to do that.

Time Waste Charts—A Tool of Empowerment

Most of the really hard work is getting an organization to adapt to a new type of competition—a time-based competition. But we used a very simple tool just to convince people that we *could* dramatically increase speed *without* rushing decisions. We call it the "time chart," and it simply flow charts each process and documents which of the eight forms of time waste exists at each step (Table 6.8).

Let's look at a time-charted indirect activity or process—equipment breakdown and repair.

Then, using the time waste busters, we redesigned each subtask in the time chart to either eliminate it or improve it. The redesign was done in phases since some changes required physical changes to our facility or new technology (Table 6.9).

But most importantly, we added a step 13. Our step 13 was a continuous-improvement loop. We looked at the root cause of every breakdown after it occurred to see if we could have prevented or predicted it. We wanted to not just speed up the "business process" of breakdown and repair; we wanted to eliminate it. Was the breakdown caused by operator error—incorrect use of the equipment, incorrect setup, incorrect tooling used, running it at the wrong speed, running a product that it wasn't qualified to make, and so on? Was it caused by a technician error—incorrect setup, calibration, or repair done previously? Had it missed a scheduled preventive maintenance? Often the repairperson was under great pressure to repair a piece of equipment and either did not perform all the required checks, assumed the first likely cause of breakdown they verified was the only cause,

TABLE 6.8 Time Chart for Equipment Repair

Subtask—Description	Time	Resource Lag–Physical Distance	Communication Lag	Approval Lag	Knowledge Lag	Nonvalue-Added	Over/Early Production	Variances	Task Automation/Redesign
1. Maintenance learns equipment is down			Typically 1 to 8 hours						
2. Repair assigned to technician (schedule)			Typically 1 to 4 hours						
3. Technician learns of assignment									
4. Technician "free" to do repair			As scheduled						
5. Technician gets equipment history/maintenance lot, special tools (if needed)		Typically 1 to 2 hours							
6. Technician interviews operator									
7. (Optional) Technician finds last 10 batch; run cards									
8. Technician runs pre-set diagnostics 8a. Pinpoints problem									

8b. If diagnostics fail to indicate problem, call for senior/specialist or continue trouble-shooting					Typically 1 to 2 hours				
9. Repairs equipment 9a. Parts available		Typically 1 hour							
9b. Parts not available—ordered								Happens 30% of the time, typically 24 hours	
10. Supervisor tests repair, fills out acceptance				Typically 1 hour					
11. Production accepts repair				Typically 30 minutes					
12. Operator notified of machine available			Typically instantly						

TABLE 6.9 Redesign of Equipment Repair

Process	Cause	Redesign
1	Poor communication to maintenance	Phase 1: Instant electronic mail to maintenance department to notify of breakdown Phase 2: Maintenance department moved to production floor Phase 3: Operators do basic repairs/technicians assist training operators and doing complex repairs
2	Can't see all technicians' schedules	Phase 1: Schedule board put on line Phase 2: Automatically given (and prioritized to best technician availability Phase 3: Operator does work when possible
3	Technician doesn't know of new assignments	Phase 1: Technician can be paged in emergency Phase 2: Technicians see their updated schedule after each job Phase 3: Operators do repairs with technician's assistance
4	Technician assignments aren't updated "in real time," so urgent jobs may be done later than they should	Phase 1: See step 3 Phase 2: See step 3 Phase 3: See step 3
5–7	Tools and equipment logs are in maintenance; much of the data is spread over 3 sources—equipment log, batch tickets, operator knowledge	Phase 1: Keep all data at the equipment Phase 2: Move maintenance to production floor Phase 3: Integrated MES; operator does basic repair
8	Technician has insufficient knowledge	Phase 1: Beeper to alert senior specialist of need for help Phase 2: Continual training of technicians and operators on most common problems; continual improvement of equipment to avoid these problems Phase 3: On-line protocol/expert system to assist in diagnosis
9	Not having right parts inventoried	Phase 1: On-line inventory of all parts; agreement with suppliers on 8-hour shipment of parts; analysis of common parts needed Phase 2: Preventative maintenance to replace most commonly failed parts Phase 3: Look at other equipment/processes/suppliers for faulty equipment/predictive maintenance

TABLE 6.9 Redesign of Equipment Repair *(cont.)*

Process	Cause	Redesign
10	Supervisor has always signed it off	Phase 1: Agree on QC test for acceptance if possible, done by technician Phase 2: Acceptance test by technician/operator/supervisor with continuous improvement done at sign-off Phase 3: Operator repair and sign-off
11	Production has always insisted on veto potential	Phase 1: Production agrees on acceptance test and technician does it Phase 2: See step 10 Phase 3: See step 10
12	Production moves operator to other tasks while the repair is done (not a problem)	Phase 1: As is Phase 2: As is Phase 3: Does repair

or didn't do all the post-repair checks to confirm the diagnosis and repair. Other times production asked to forego a required maintenance to finish out a shift of work, and no one remembered to reschedule it. Was the breakdown due to a faulty part? Was the equipment a "lemon" that did not meet it's specified mean time between failure (MTBF) goals or representations by the equipment supplier.

Another thing we did as part of our improvement program was to start to work with equipment suppliers. We started on our most critical pieces of equipment, to select and negotiate with suppliers on the basis of guaranteed minimum equipment availability (total time less scheduled and unscheduled downtime), minimum MTBF stocking of key parts, setup time-reduction programs, technician and operator training programs, and joint equipment improvement programs.

We started to collect critical data on MTBF, scheduled and unscheduled downtime, part failures, common problems, and operator complaints, and shared this data with key equipment suppliers on a biannual basis.

We moved to calculating an overall equipment efficiency index for judging equipment based on output rate times availability. Some equipment may have been slower but had much higher availability.

All of the effort in step 13 was to prevent and eliminate steps 1 through 12. And that's the real key to improving the repair process—to try to eliminate the need for it.

These steps also forced us to look at our enabling tools. In a speedy and continuously improving company, you need new tools and a lot of data. In our case, we added a manufacturing execution system (MES). It electronically linked all of our departments. It kept the specifications for all tasks. It provided on-line protocols and expert systems to guide technicians and operators through each activity. It kept the status of all technicians, equipment, and parts inventories, and it kept the history of each piece of equipment.

I never realized how much paperwork we used in our indirect activities nor how it prevented rapid communication and problem solving. Once we had the equipment records on line, we could calculate all the equipment efficiencies. We could analyze the failure rate and pattern of any part. We could instantly provide the operator or technician with a full history of each piece of equipment.

Since MES also tracked and scheduled production, it could schedule maintenance at times that least disrupted production.

I can't imagine how we ran this place before we bought it. Oh, yes I can—poorly, blindly, reactively."

Mark smiled at the last comment, but inside he winced. Undoubtedly how HBEC used to run was exactly how PCB Co. ran today. Three initial reactions were forming from this visit.

First, he realized that he had been focusing on the production side of their operations—the actual manufacturing process. He had really not given any thought to the indirect activities—maintenance, scheduling, process engineering, facilities finance, planning, and so on. Yet, they were constantly increasing. Each budget seemed to call for more indirect personnel, and he had no idea of their quality, speed, or productivity. He had done nothing to try to measure or improve these.

Second, he had never thought about manufacturing—direct or indirect functions—as requiring its own system (what he now knew was called a manufacturing execution system or MES). He knew that the company needed systems. They had just purchased an extremely expensive and complex MRP or ERP system that was supposed to provide all the information needed. At the time, manufacturing had not been very interested. Now he understood why. It wasn't a tool that was designed specifically to support their needs. It seemed to assume that manufacturing simply needed parts availability to meet their targets. It certainly had never given them the insights into their real value-added performance nor the tools to improve it to which he had since been exposed. He had to have Tom, his IT manager, take a look at MES and expose his staff to what it might do to help them, if they still had a plant to run.

Finally, he had never really thought about trying to speed up or automate the indirect activities. His process engineers were continually looking at how to automate the manufacturing steps. Somehow, they had not been focused on their own activities—the shoemaker's kids going barefoot as it were.

From what he had heard, from both his consultant and now from Lou, was that he could dramatically improve indirect labor productivity, speed, and even quality by trying to "automate" the tasks—provide the data needed instantly on a computer, so that time and productivity weren't lost searching for and through all their paper records for data; provide automated tools tied to that data for scheduling, yield data analysis, reporting, inquiries, and notification of alarm conditions for the same purpose; provide computerized protocol, specification, or expert systems to guide technicians through setup, repairs, calibrations, lab tests, and so on.

A few numbers stuck in his head. His consultant has estimated at least 20% of his indirect productivity was lost searching for data. How had someone put it? Data was the "raw material" for an indirect, for a knowledgeable worker. Without easy access to it, they wasted time, his company's time, getting it. And he had no idea how much more productivity and quality he could gain by giving them data and automated tools. He'd get three benefits at once: added speed, added productivity, and added quality.

Lou kept silent while Mark thought to himself. This was not the first tour Lou had given. He could see when the light suddenly turned on in a visitor's head, and he needed a moment or two to think through what was suddenly illuminated. It was the first time he had seen a real smile on Mark's face during their tour.

Mark's next question for Lou was, "How does this compare to what Japan has done with Kanban and JIT?"

What Japan Did

What Japan did was focus on eliminating all waste in the *direct* manufacturing process. The list (Fig. 6.3) shows the concept of waste in Toyota's JIT/TQC approach.[1] What they did was make direct *manufacturing* as efficient as possible by:

Shrinking distances between operations (waste in conveyance)/distance

Continually eliminating all causes of quality variance (waste of variance and correction)

Empowering the operator to perform as many functions as possible, as efficiently as possible (waste in processing)

Minimizing setup time (waste of overproduction in large runs)

Replenishing inventory as needed (waste of inventory)

Moving jobs as needed (waste of waiting)/queue

Grouping operations into a cell of single function (waste of motion, waste in conveyance)

"Together, I think of these as the efficient, focused assembly line with quality. We link operations in the shortest distance with a focused support team of production, maintenance, scheduling, and engineering running with minimum variances. This is lean, direct manufacturing and results in the shortest *production* leadtimes and highest quality. All obstacles that increase leadtimes are eliminated.

"However, what Japan did not focus on were the *indirect* functions of scheduling, maintenance and repair, costing, management reporting, and other overhead or support functions Their primary focus was on the *product*—the value added to the end customer. The same focus on efficiency in indirect functions was ignored. In fact, the goal was to eliminate the need for many of the overhead functions by simplifying the manufacturing pro-

[1] Going for the Globe by Michael Stickler, pg 33. Production and Inventory Review, Nov. 1989.

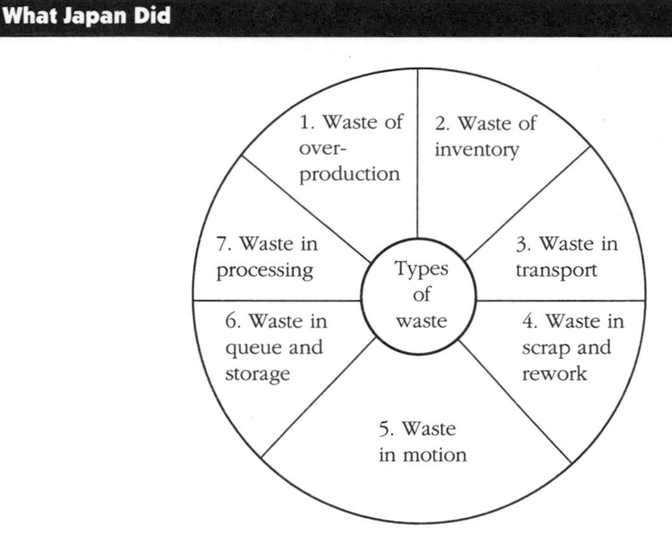

Figure 6.3 JIT and Waste at Toyota

cess. For example, if there are dedicated manufacturing cells by product type, scheduling is not needed. If a separate fixed maintenance shift is set between production shifts, maintenance scheduling is not required.

"The overhead functions remaining were done manually. Most automation was focused on the manufacturing process—to eliminate manual steps/direct labor. Very little automation of tools was provided to the overhead functions.

"As a result, production was dramatically improved, but indirect functions were not. With the increasing product and process proliferation, the complexity of the indirect functions—scheduling, quality improvement, process development, for example—also increase greatly. With more requests from customers and regulatory agencies for proof of compliance in process, quality, safety, and waste disposal, data requirements grow.

"Therefore, the next area for increasing speed, quality, and efficiency will be in the indirect areas. Japanese factories tend to be much more heavily automated than their American and European counterparts (with more robots, vision systems, and flexible manufacturing systems). However, they have much less data or information automation than their American counterparts. This data is the "raw material" needed for most indirect activities, and the computer network acts as the "material transport" to move information, alarms, and communication with minimal lags between team members. Productivity tools for scheduling, equipment maintenance, reporting, and costing form the automation for heavily manual tasks.

"Japanese companies clearly recognize this. In a recent survey,[2] the number one priority in manufacturing companies in Japan over the next two years is the integration of information on the plant floor—the basis for speed in detection, analysis, improvement, and execution. I expect that they will expend the same effort to eliminating waste (and therefore, reducing leadtimes) in indirect activities as they did to eliminating it in direct manufacturing. With that speed comes gains in all competitive metrics.

A Recap: Time Equals Money/Quality Goes Through It All

Mark returned to his office with many emotions set swirling. He felt that he had been given a set of simple but powerful tools and concepts for his company. He felt he had an approach to generating the speed he now believed was vital to successfully competing and ensuring their and their customers' success and survival. (Imagine if his customers' competitors had "speedy" vendors and his customers didn't—would he have thriving customers to buy his products?!)

But he was confused by how all the initiatives he had started fit together. Quality seemed at the heart of it. They had to eliminate variances to get quality, speed, customer speed, cost (he assumed), and technical capability. Technical capability, in manufacturing, was really a focus on the *success* side of quality rather than on the failure side. So he could put the two together under the topic of quality. But how did the others relate? He picked up the report from the costing team and started to review it for a second time. How could he implement all these programs? Improvement-based costing appealed to him as much as speed. If he could be the low-priced supplier, that might be as "competitive" as being the "fastest" supplier.

He flipped through Table 4.6 on utilization of equipment again. Equipment was their largest investment—high value-added utilization was

[2] Waseda Bu, Insead study, 1990 International Manufacturing Futures Survey.

critical to low cost. Then he looked at Table 4.9, utilization of facilities, and Table 4.10, utilization of personnel.

As he looked at these charts, something was familiar. He stared at them for a minute until it hit him. These were the same "waste" categories as in Hot Box's *speed* analysis! How was that possible?

As he sat back and thought about it, it began to make sense. Any "waste" of a resource meant "time" was wasted, and that's exactly what *utilization* was—the use of *time*! Speed was just the average for one example or event. Utilization was the *total* overall events as seen by the resource (Fig. 6.4).

If a job was "slow," then that delay had to show up in the "utilization" of their raw material resources and/or equipment resources and/or

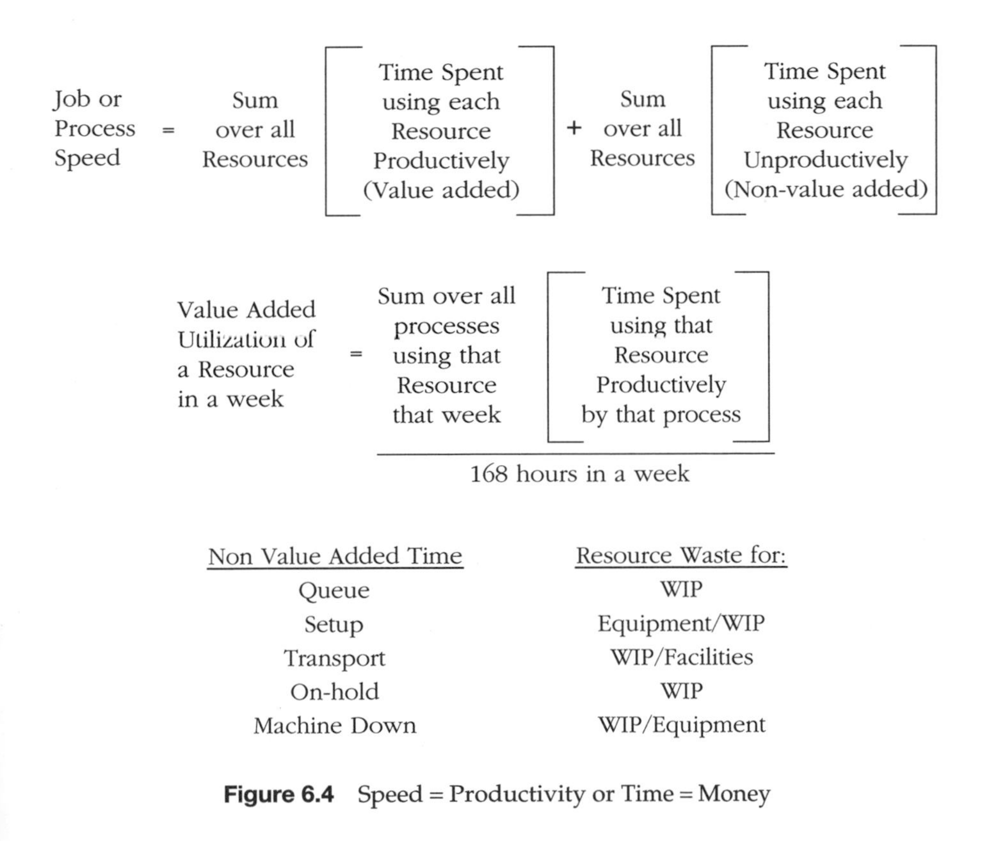

Figure 6.4 Speed = Productivity or Time = Money

operator resources. If there were *no* delays, then the only problem could be with overcapacity. So in other words, time *was* equal to money. If they worked on eliminating every *delay* or time waster, they'd automatically eliminate the cost as well. Every waste of a resource was wasted time as well!

So now he had only *two* key elements to work on: quality and speed. With quality and speed, he'd automatically get lowest *cost* as well. He'd have the *fastest* delivery of a customer order possible. And then he realized that quality variances were just another time waster. In fact, there really was only one program to work on: speed. It truly encompassed all other improvement areas—quality, cost, and customer service. Finally, he had some clarity into their problems and where they needed to go. They had to become a time-based competitor. And he knew the tools for the workforce to use.

Feeling his mind become clearer and more determined than it had been in a long time, he finally saw the future—how an empowered company could chop and slice the waste away. He saw how value-added processes could bring customers the highest possible value and allow competitors the least competitive advantage. He had the answer. Now all he had to do was implement it.

The Unexpected Benefits

Over the next six months, the company made their attack on time a focused priority. Members of the original quality, cost, and customer service teams were used to seed a variety of teams for "key competitive processes"—order entry, manufacturing, distribution (to cover the customer leadtime cycle), equipment maintenance and repair (to cover their main capital cost), engineering process improvement, and new product and process design—as examples. All were empowered to improve speed. Each reported or were assigned to one management "mentor" who could assist across departmental boundaries or approve resources. All received training in project management, teamwork, analysis, and when necessary, presentation and writing skills.

Some teams were more successful than others, perhaps because of their own skills or because there was more immediate opportunity for improvement, but all teams contributed.

However, as the six months went along, the plant manager noticed some unexpected benefits that began occurring. As they cut their customer leadtime and increased their speed, quality in *all* processes increased.

Reduced overhead complexity. Mark, for once, heard fewer and fewer complaints about late orders from customers. He had visited the scheduling department to compliment them on their improved performance. The head of the department, normally the most harried man Mark knew, was smiling. Here was a man who was caught between the proverbial rock and a hard place—between Scylla and Charybdis—between two groups he had no control over, the customer and manufacturing—smiling!

"I wish I had a camera. I can't believe it—you're actually smiling!"

"My life is worth living again. I've given up Rolaids. Ever since we cut our leadtime by 50%, first of all I only have *half* the number of orders in the factory to schedule, track, and manage. That *alone* would make my job half as difficult. But with half the leadtime our forecasts are more accurate and customer orders don't change as often. And with fewer problems in the factory—fewer jobs ruined, fewer machines down for days at a time—we don't spend all our time rescheduling the schedules. We can actually produce a schedule that's *reliable* for a few days. Production has stopped calling it "the schedule de jour" or the hot plate schedule.

"You're making this job worthwhile again. We're starting to believe our own schedules. We can actually spend time to develop a good schedule, one that balances our workforce and facilities. We can suggest to marketing what orders we could take to fill up the facilities; we can evaluate rush orders to see if we can produce them profitably.

"We could do even better if we had a little more help in the form of some scheduling tools. Now that we don't spend all our time reacting, we'd like to do more analyses—the "best" mix for profitability, the right mix of labor force skills, the areas to look at adding equipment—but we don't really have anything except a personal computer, and it's not even hooked up to our own computer system. I'm going to look at a few scheduling packages I've heard about and see what it would take to implement them, if we can figure out how to feed them the data they need.

"And, I've also noticed that as our mix of products varies, we get a very uneven load on the departments. You know, since we all got started on continuous improvement, I started to read a few articles on planning and scheduling and even went to a seminar or two. I realize that I've never really had any formal education in planning or scheduling, just a lot of on-the-job experience. And I was thinking that if the operators were a little more flexible, we could even out the loading more easily. Anyway, it's just an idea. But this stuff is great! Now we can schedule instead of spray and pray."

This was a pattern that emerged as Bill spoke to more of the indirect functions. With less firefighting and "traffic management," each could *plan* an effective schedule that increased utilization even more. Maintenance said the same thing. With fewer "emergencies," they could plan out a maintenance schedule that was complementary to the production schedule and best utilized their people. They could plan training without it being disrupted by rush repair jobs. With fewer unplanned repairs to schedule, managing the scheduling was easier, freeing up part of a clerk's time for analysis of equipment performance for improvement tasks.

But again, the same two themes arose. First, maintenance really could use a database on the equipment, to start looking at best and worst performers, to set correct maintenance schedules and spare parts levels, and schedule maintenance activities—maybe even electronic mail for instant communication and notification. Second, they were thinking about more cross-training. The maintenance department manager noticed now that as the mix of maintenance jobs varied, some of his techs were too busy and others were idle, waiting around. Before, they'd always *seemed* busy, rushing to fix a broken machine, looking for a part, or trying to find a machine manual. Now their value-added productivity was much higher, and he found a periodic imbalance in the skill mix he needed. He had idle techs at the same time others were overscheduled. He was thinking about ways to fix both problems—some maintenance software packages and a serious cross-training program.

The pattern was a pleasant surprise to Mark, but in retrospect, shouldn't have been. What he saw was that by improving quality and shortening leadtimes, many overhead tasks went away *altogether* or were simplified. If the leadtimes were cut significantly, not only were *nonvalue-added* activities (such as storage and queue time with its related handling, movement, and recordkeeping or overproduction) eliminated, many "value-added" activities were reduced, such as the effort involved in scheduling or handling customer inquiries or costing or forecasting/planning due to the reduced number of orders and/or leadtime and/or changes to the plan. In general, shortening leadtime seemed to reduce *complexity*. Cost, as he thought about it, must also be a *function of complexity*. Reducing leadtimes by cutting out nonvalue-added steps simplified associated overhead costs of management, control, scheduling, costing, and reporting. In fact, in their company, since overhead costs exceeded direct costs, any direct costs eliminated more than *double the savings* by the concomitant reduction in associated overhead!

He thought that relationship—cost as a function of complexity—was one he really should raise with everyone in the company. As he thought about it, every new part they simply *had* to use in the latest board they designed (instead of using an existing part they already ordered) meant a new part number, new specs for purchasing, additional orders, new storage areas, new qualification tasks, and new engineering knowledge. All from just adding one new part.

More certainty/teamwork/positive attitudes. The second message Mark kept hearing or, more precisely, what he heard less and less of was interdepartmental squabbling. Previously, production blamed production control for schedules that kept changing, requiring too many setups, overtime, and expediting. Production control blamed sales and customer service for taking orders in less than the agreed-upon leadtime. Engineering blamed production for not following their specifications, which production said were incomplete, hard to read, or ambiguous, as well as completely out of date. And so it used to go.

With much shorter leadtimes and schedules changed less frequently, so *joint planning* made sense. A plan for the week with agreed-upon production, maintenance, setup changeovers, training, and engineering experiments all spelled out actually had a reasonable chance for success. This led to more teamwork as their plans were successful. When no schedule lasted for more than a day, no one could really judge if any joint efforts paid off. Meetings became more of a joint effort rather than a daily exercise in determining who won and who lost.

It seemed that no one had worked well together under the old "launch and expedite, schedule and ignore, yell the loudest" world that engendered too much uncertainty. Now they had time to discuss improvement, not just argue about survival.

More positive customer relationships. Mark had always received customer calls. He'd received calls about late orders, short quantity orders, orders quoted outside that needed delivery dates, defective parts received, prices quoted that seemed out of line, and so on.

Their first publicly announced step toward speed, done two months ago, had been to cut their quoted leadtimes by 30%—a significant statement. Internally, they now believed they could cut it by up to 80% on high runners where they were looking at dedicated lines and 50% on low-volume boards. More than half of their leadtime wasn't even in manufacturing. Furthermore, with some new agreements with their *own* parts suppliers, most parts were available within days. They'd simply had to renegotiate their *own* needs as a JIT/TQC supplier with *their* suppliers.

The response was uniformly well received. PCB Company had always been known as a top engineering company. Now people were seeing them as a JIT/TQC or TQM vendor. Their marketing department had just initiated a marketing campaign explaining the changes they had made to better serve their customers.

They were projecting that within 6 more months, their unit costs would decrease over 30%. Mark had originally been certain that they would need a layoff to *actually* see the bottom-line improvement from their productivity improvements.

But three things had happened. Natural attrition had brought down head count. Turnover had slightly increased from discouraged groups of workers in all departments on several management levels who had not responded well to the empowerment efforts. Some saw it as too much responsibility, some as a threat to their autonomy, some as too much change. They left, over time, to find more comfortable (if not more secure) jobs. Usually, they were not missed. Many were the loudest warriors who couldn't accept an end to the interdepartmental wars.

Second, some costs had been eliminated or reduced, such as overtime, inventory holding costs, and contract labor, which also contributed to the bottom line.

But finally, sales was able to win some incremental business with their shorter leadtimes. That allowed them to grow into their added capacity.

Mark felt that they were still overstaffed but was now confident that they would grow into their staffing. He felt like an overweight runner who had lost weight, now felt faster and more competitive, and was getting compliments on it.

The rate of improvement/learning—type 3 error. As the leadtime went down, Mark began to feel another unexpected benefit, their *rate* of improvement was increasing. As production leadtimes decreased, so did the time until a problem was discovered, fixed, and eventually eliminated. This led to leadtimes further decreasing (without that type of variance), so the rate of improvement increased.

Similarly, as the leadtime for process improvement experiments decreased, their number per year went up, and therefore, the *rate* of yield improvement increased. Decreasing production leadtimes only further decreased experiment leadtimes as well (though the new rules for scheduling put improvement experiments as the top priority jobs).

It seemed to Mark that the whole organization was moving at a different pace. Suggestions or ideas that would have been belittled, ignored, or studied in endless circular patterns a year ago were quickly either rejected or assigned to a team. Teams were formed and disbanded as needed. Everything moved . . . well, faster. Decisions didn't seem more slipshod, just more well paced. There was less hemming and hawing and more deliberate decision making. Time all of a sudden was seen as a competitive advantage. They had a new concept. In statistics, type 1 error was rejecting the correct hypothesis. Type 2 error was accepting the wrong hypothesis. They had added type 3 error—accepting the right hypothesis too late! Everyone thought about unnecessary delay as contributing to type 3 error. They now saw delay as having a cost just as real as making the wrong decision. Therefore, you couldn't delay endlessly to avoid making the wrong decision.

Their company now thought about and neared, they hoped, world-class speed and quality.

Flexibility-Requirement for Mix Changes

As the drive for speed and quality picked up, Mark had heard a common problem. Without the queues of work to buffer scheduling, periodically the incoming *mix* of tasks in a department didn't meet the skill or re-

Over a week or longer, the average load is equal to the total resource availability.

	Insertion: Skill A requirement for setup Maintenance/repair per week	Wave Solder: Skill B requirement for setup/maint/repair per week
Availability of each skill in jobs in jobs/week	100 jobs/week 10 people @ 2 jobs/day = 20 jobs/day = 100 jobs/week	150 jobs/week 10 people @ 3 jobs/day = 30 jobs/day = 150 jobs/week

But daily, sometimes the demand (for repair or maintenance or setup or processing) is not balanced

		Day 1	Day 2	Day 3	Day 4	Day 5
No. of jobs	Skill A Req.	10	10	10	30	30
	Skill A Avail.	20	20	20	20	20
No. of jobs	Skill B Req.	40	40	20	25	25
	Skill B Avail.	30	30	30	30	30

Figure 6.5 Lower Cycle Times Eliminate Workload Buffering

source mix available, and load imbalances occurred with concurrent idle and overloads (Fig. 6.5).

For example, they had ten technicians who were responsible for maintenance, repair and setup of the insertion equipment and another ten who focused on wave solder. The two operations were quite different so specialization had been natural. The staffing levels were nearly exactly matched to the typical average workload over the week. At insertion, there were typically 100 setup, repair, or maintenance tasks per week, or about 2 per person per day.

Wave solder was more complex but had fewer machines. There were about 150 jobs/week or about 3 per person per day.

When there was a backlog of maintanence and repair activities, technicians worked off of a schedule that was endless. They simply did the task at the top of the list, and magically, more appeared over the week. Their queue of work never ran dry.

But when speed became a focus along with value-added utilization, there was enormous pressure applied not to leave equipment down, awaiting repair or idle awaiting setup. The maintenance group worked overtime and weekends to bring the queues down to zero. Now they could work on tasks as they arose. But demand was not level and constant. On Monday and Tuesday, the insertion equipment had some major runs that required no setup changeovers. By Thursday, they required significant maintenance and changeover.

Wave solder ran the opposite way that week. A series of smaller runs, inserted the previous week, caused the load on Monday and Tuesday to exceed staffing levels.

So the technicians responsible for the insertion area were idle half of the time on Monday, Tuesday, and Wednesday, and then backlogged Thursday and Friday. Wave solder ran in reverse. The technicians had a backlog of work on Monday and Tuesday that wasn't eliminated until the end of Friday.

This didn't surprise Mark. As he had learned, it's not the average that will kill you—it's the variance. The average temperature in Death Valley is 70°, 120° during the day and 20° at night.

When there were long backlogs or queues of existing work, daily or even hourly fluctuations in new workloads (fluctuations in demand) could be leveled out. Now there was no such buffer. In fact, this problem led to considerable discussion of whether it was *economical* to run such short leadtimes, with people sitting idle while others were busy. Periodically, some staff members started to suggest backing off of their aggressive attack on leadtimes. Others suggested that perhaps there was a curve of "total cost/reward" of reducing cycle times and now they had gone too far the other way (Fig. 6.6).

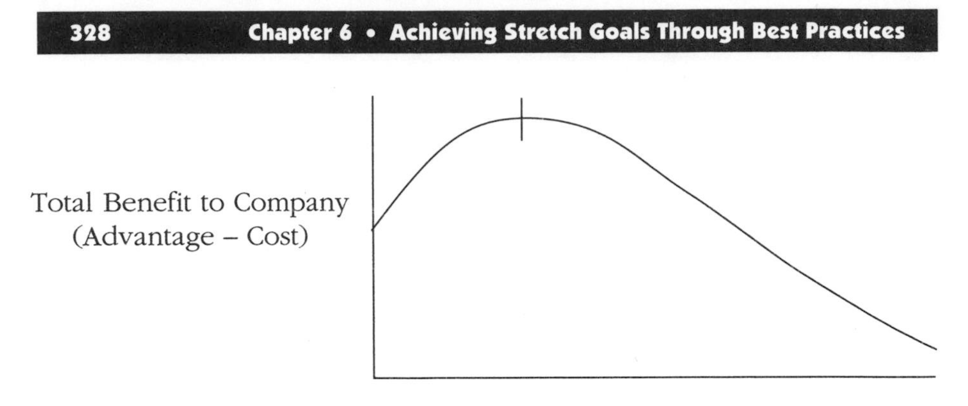

Figure 6.6 Possible Curve of Net Value of Reducing Cycle Time

Mark had found that change always brought both some new problems and resistance or inertia that had to be overcome. In this case, however, he could see what they were talking about clearly. Without buffers or queues of work, any "variation" in demand was fed directly through the system without smoothing or buffering (Fig. 6.7).

When WIP had been eliminated from in front of manual insertion, the station periodically ran idle. Here there is no work arriving for the first three hours, and then two four-hour jobs arrive two hours apart. With a minimum buffer inventory (or Kanban), the idle can be greatly reduced. If we keep one job representing four hours of work in queue, we can buffer the variation and eliminate the idle.

The result of eliminating queues was idle periods followed by overloads. This meant that leadtimes varied as *queues* moved up and down (unless incoming demand was nearly uniform and constant, a situation found only in textbooks). It seemed sensible to increase queues slightly to affect a major increase in utilization.

This, in fact, did make sense in their *old* way of thinking. But Mark was determined to look at new ways of thinking and not let leadtimes start slipping back to old levels (as they had a tendency to do if not watched with vigilance). He asked a small group to look at ways to address this *real* problem and report back their alternatives. He used representatives from maintenance and production, the two most vocal parties, as well as finance and customer service, to help analyze the effect on costs and the customer.

Incoming Load/Hour
(in Hours of Work)

Work Load Arriving in Hours of work	0	0	0	4 (job 1)	0	4 (job 2)	0	0
Hour in Shift:	1	2	3	4	5	6	7	8

situation 1: "no buffering"

idle: 3 hours
busy: 5 hours

time in queue: wait time: job 1: 0, job 2: 2 hours Mean 1 hour

situation 2: buffering (say 4 hours)

idle: 0 hours
busy: 8 hours

time in queue: wait time: job 1: 1 hour

job 2 : 3 hours
mean = 2 hour

Figure 6.7 Without Buffers, Variation Is Felt Directly in Utilization

That was the new way they worked. Management did not unilaterally make all decisions—they set direction, reviewed recommendations for completeness and soundness of thinking, aided teams, and acted more as traffic control and mentor, making sure all the teams and departments were moving forward and conflicts were resolved. One of their new key activities was determining the makeup of teams!

A Report on Flexibility

The team, in fact, took a little longer to report back than Mark had expected. They had done the normal tasks—agreed on the definition of the problem: finding the root cause of the load variation they now observed, finding the alternatives to correct the root cause and agreeing on the correct measures on which to evaluate those alternatives, gotten agreement from the concerned parties on their definition, interviewed people internally, consulted some of their partners (customers and suppliers), researched any literature (aided by a consultant), and then prepared a presentation and recommendation. However, this area was not as well understood as most, and they had had to grapple with new territory for most companies. Their findings were interesting and were as follows:

The problem discussed was, in fact, real. As the usual "buffers of uncertainty or variability" in a company (a.k.a. queues of work) disappeared, the affect was to cause the workload to vary *directly* with the variability of the demand or "input." They had found four approaches to dealing with this load variability that they labelled planned, buffered, adaptive leveling, and surge capacity.

Planned leveling. The approach that they had seen in Japanese companies was to try to *eliminate* the variability or uncertainty in the demand. For example, the Japanese approach to maintenance and repair was to set a planned maintenance shift between production shifts where maintenance schedules and maintenance staffing could be *planned* to be in balance. This maintenance schedule was set to be rigorous enough to avoid unscheduled maintenance needs, and hopefully, breakdowns during the production shifts.

Similarly, in production, the plant often set a schedule for the week that balanced the load on the plant by shift that did not throw the plant into imbalance. In a sense, this approach leveled loads by allowing "overproduction" or "early production." By *not* trying to meet maintenance requirements and demand at the exact moment they arose, they could keep leadtimes relatively low and, as importantly, fairly constant while leveling use of capacity.

This approach made a lot of sense when it was possible to "negotiate" demand for the week (as in setting car shipment levels to distributors or maintenance schedules). It leads to lower capacity utilization (a dedicated maintenance shift meant all production equipment was down, whether it needed maintenance or not).

It did not make sense when there was a *high* degree of uncertainty that could not be controlled, such as when equipment was still too unreliable to plan on scheduled maintenances or when customer demand was "not negotiable," as in a job shop.

Buffered leveling. The approach that made sense when demand was still uncertain or highly variable was to allow a *buffer* of work that smoothed out the load on the resources. This made sense when capacity was expensive and reducing leadtimes ever further did not add a significant competitive advantage. This meant using the *minimum* buffer levels required to balance leadtimes and utilization costs.

Adaptive leveling. Another approach used in Japan when leadtimes were critical and mix was not stable was cross-training/substitution (or *flexibility*) of resources. In this case we allowed the *mix* of demand to vary and responded by changing the allocation of resources to *match* it. In other words, the mountain (of resources) went to Mohammed (the demand). This approach relied on cross-training of individuals so that they could do *multiple* tasks. In that way, we only had to make sure that the *total* demand matched our total resource (Table 6.10).

Referring back to our example of having two types of jobs (one requiring skill class A and one requiring skill class B) if our operators are cross-trained, then we can adapt to changes in *mix* with relative ease. From a planning standpoint, we only needed to staff for *total* hours of workload which, in general, has much lower variability than *individual* skill class requirements.

TABLE 6.10 Flexibility Allows Our Resources to Better Match Demand Mix Variation

		Day 1	Day 2	Day 3	Day 4	Day 5
Number of Jobs—	Skill A + B requirement	50	50	30	55	55
	Skill A + B available	50	50	50	50	50
	Overload	0	0		5	5
	Underload	0	0	20		

The same flexibility is also desirable in equipment—to be able to rapidly set it up for a *variety* of products or tasks to again maximize our ability to rapidly respond to variation in demand. This is at odds with the tendency to fixed, dedicated work cells. As a result, the newest trend is movable equipment that can be shifted to be assigned to actual demand.

Surge capacity. If leadtime performance is critical and capacity is relatively cheap, we can also keep surge or excess capacity to *handle* mix variation. While we may at first think of that as wasteful, in retrospect that may be an antiquated way of thinking of a possibly critical competitive advantage. The world-class company looks for a competitive position that is profitable. If a customer prefers fast response time to lower cost, then we must adapt to the customers' or market requirements. Ironically, we are so used to thinking in terms of *budgets* that we are often victims of narrow thinking.

By allowing surge capacity and adaptive leveling, we could in fact, maintain minimum leadtimes at a slightly higher cost. In this case, shown in Table 6.10, we could keep either a capacity of 55 units or a surge capacity (through overtime or subcontractive, for example) of 5 units.

Indices of Cross-Training

Clearly, adaptive flexibility through cross-training of personnel, multiple purpose (possibly movable) equipment, rapid setup/changeover capability, and flexible facilities is the clear response to a need for short leadtimes under mix or load variability. The next question the team addressed was how much cross-training was needed. In technical capability or quality, they had seen the concept of the variance of a process and measured that variability or process capability to compare it to the range within which the product could be manufactured without problems and get a process capability ratio, C_{pk}.

Now they developed a new concept called *design for flexibility* and a flexibility capability they named C_{f_k}. C_{f_k} was the ratio of the available resources cross-trained for a task divided by the maximum requirement for a resource under normal conditions. The latter could be tracked over time just as they tracked process variability or demand variability. Obviously, they wanted to decrease the maximum or peak requirement by looking at *controllable* causes of it. But now they could see where they were most vulnerable to *inflexibility* in their resources.

In addition, they looked at surge capability as C_{s_k} where C_{s_k} is the ratio of the *total* available resource over all related tasks (where they could consider cross-training personnel or substituting equipment) over the maximum *total* load requirement for that resource.

For example, suppose five units of both resources A & B are cross-trained. Then in the examples, for skill class A

$$C_{f_k} = \frac{25}{30} = \frac{\text{(maximum available resource of A skills)}}{\text{(maximum requirement for A)}} < 1$$

for skill class B

$$C_{f_k} = \frac{35}{40} = \frac{\text{(maximum available resource of B skills)}}{\text{maximum requirement for B}} < 1$$

and the surge capability

$$C_{s_k} = \frac{50}{55} = \frac{\text{maximum available resource A + B}}{\text{maximum requirement}} < 1$$

In all cases, they needed to improve "flexibility" if they wanted to meet the potential variability of the load. Otherwise, we need to reduce the peak loads of requirements for skill classes A and B, and the total by 5 units. The first two could be remedied by cross-training an additional 5 units. However, without an additional surge capability of 5 units (in the worst case), even cross-training wouldn't handle the peak load.

Note that cross-training beyond the 10 units required was laudatory (and useful given turnover, promotions, transfers, and other realities of corporate life) but had no immediate value to the team. Developing these C_{f_k}'s gives the company a rational indication of the greatest need for cross-training or flexibility. If we then prioritized based on "cost" of waiting (which skills' tasks affect the bottleneck equipment, personnel, and tooling), they could set up the highest payback cross-training program.

Measures of "Flexibility Failure"

Another postscriptive way they found to detect a flexibility problem was to look at simultaneous idle and overload (queued) resources within a class of resource. For example (Table 6.11), they periodically saw technicians with a queue of repairs and maintenances waiting while others were idle. Similarly, they saw jobs piled up at several operations while others were idle

TABLE 6.11 Simultaneous Under- and Overutilization (or Load Imbalances) Suggest Opportunities for Improvement Through Cross-Training/Flexibility

	Maintenance Queue of Work for Technician Skill Class 1 (2 Technicians)	Queue of Work for Technician Skill Class 2 (2 Technicians)
9 AM	Repair machine 3	
	Repair machine 7	
	Maintenance machine 4	Maintenance machine 12
	Maintenance machine 2	
	Example: production work center 6 (3 people)	Work center 9 (3 people)
9 AM	4 jobs in queue	2 jobs in queue
	A123	A114
	A126	A117
	A127	
	A129	

with operators waiting for jobs to arrive, not cross-trained to run the other operations. In each case these were reactive measures (instead of the *proactive* flexibility capability measures suggested) but would point out areas where cross-training might alleviate imbalances.

There were several more types of flexibility desired in a world-class company such as: flexibility to "customize" orders; flexibility to respond to changes in demand levels; flexibility to adapt new processes, technology, and equipment, which the group had not yet attempted to measure.

Adding Flexibility as a Measure

When the group had finished their report, Mark and the others he had invited were impressed and educated. They had never really thought about flexibility. In fact, quite the contrary, they thought of specialization and many labor grades or classes as a measure of employee advancement. It was clear that as their "needs" changed (or their customers' needs and competitors' capabilities changed), they continually needed to rethink many of these daily operating assumptions and characteristics—from layout to organization to measures of performance to requirements in the personnel they hired and equipment they bought.

Flexibility certainly required rethinking. They didn't even *measure* it, much less have desired levels for it. Perhaps their human resources database had a file on skill levels for each employee. Again, it continued to amaze Mark how so many things that had totally dominated his "vision" or "management view" now seemed secondary and how many things he had never seen or thought of before now loomed in front of him. He had begun to think of speed and quality instead of budget. Then he'd discovered technical capability over quality and flexibility instead of specialization.

One thing was clear to him. Manufacturing really did have a rich theory or body of knowledge. He had always run the business by budgets and product and process engineering. He had never personally had any training in the "science" or theory of manufacturing. Nor had anyone on his staff. That was a situation he was going to correct. If they made it through this challange, they would educate all managers in the theory of manufacturing, in the best practices they'd studied—if he could find a group or institute that offered such training.

But while it was a laudable thought for the future, time was running out before their six-month presentation to corporate. Mark had reached a certain calm and acceptance of their situation. For the first time, he realized how much he had learned, they all had learned, over the last six months. He was finally prepared to run a competitive manufacturing operation in the rest of the nineties. If they let him.

7

Achieving Stretch Goals: Best Practices in Plant Floor Systems

Empowerment had taken over at PCB Co. Each department, from production to maintenance to finance, had teams working to make their products and services "cheaper, better, and faster." While there were people who still believed that this was another fad and others who mainly felt threatened, the majority were both excited to improve their department and secretly relieved that management was doing something to ensure their job security. Most of the workers knew people that had been laid off and wondered how they could avoid that possibility. Seeing a program to make them more competitive, implemented with urgency, made sense to almost everyone.

However, there had been one unanticipated side effect. The number of complaints about MIS had reached epic proportions. In the past week, Tom, the director of MIS or IT, had had five requests for purchase of stand-alone systems. Each requestor had railed long and hard about how poorly MIS was meeting their needs. He had never heard so many complaints before. In the past, MIS had been a support service that was rarely discussed for more than the obligatory hour at annual division budget meetings. Every year they met, reviewed what MIS proposed, and then cut it back to a small increase over the past year. Few people really listened to the plan. Everyone knew that MIS was the last place they'd spend money, so why even discuss it? The joke went "in the hierarchy of the company there was engineering, then production, then maintenance, then facilities, then nothing, then animals, rocks and plants, then nothing, then amoebas and other single-celled animals without intelligence, then nothing, and then MIS." So what had happened to make MIS such a hot topic?

And hot it was. At the management meeting Tom had finally called to discuss the flood of requests MIS had received (they had never had a management meeting just to discuss MIS!), the room had been a jungle of guerilla warfare. Those currently without systems wanted to go out and buy their own without any MIS oversight. Those supported by MIS wanted immediate attention or their "allocated" budget costs for MIS put back under their own control—meaning they could negotiate with "alternate suppliers." Others wanted to just turn off the systems and just do it manually. "Forget systems," they said. "In a JIT/TQC world, we don't need any systems."

Finance was having an apoplexy over all this and kept reminding people loudly that financial reporting needs were not voluntary. There were governmental and SEC requirements even beyond their internal financial reporting needs. This drove the other departments even wilder with frustration. "You mean we have to pay to give you your data." Tom just stood there bewildered. Every plan he had proposed for years based on user input had been cut back so that all they could do was maintain basic systems and essential services. Why were they suddenly in this fire storm of criticism?

Mark Ritchards was also confused. Each department wanted information that he had thought their current MRP II or ERP system already provided. They had certainly spent enough money and time buying and installing the latest state-of-the-art package. At the time no one had said they didn't need systems. Now here several people were saying they actually wanted to go back to pencil and paper, that Japan didn't use computer systems, and look at what they had accomplished. That did make him wonder why they were buying systems. The meeting had ended under a cloud of emotion. Now, as he sat back in his office, several things were clearer to him. Firstly, empowerment and best-practices programs had really changed the informational needs of the departments and teams. Most of their current production or shop floor systems seemed aimed at reporting financially oriented results at fixed intervals—at the end of a shift, day, week, or month. Second, there was no agreement on how to make the transition from where they were to where they needed to be. What they needed to do was look at their information needs more rationally before each group went charging off individually.

Empowering an MIS Team

Mark's solution was now their modis operandi. He empowered Tom to form and lead a team of stakeholders to recommend what approach the company should take to their plant floor or manufacturing information systems.

As Tom had seen his peers do, he went out to bring in an experienced plant floor systems expert to facilitate their process. Certainly internally he had little expertise in what best practices were in plant floor information systems, and opinion inside PCB Co. seemed divergent. Some insisted no systems were the best systems. Others wanted to proliferate "inexpensive" PC packages. Some simply had complaints.

He found a consultant who had been recommended by a vice president of IT at one of their suppliers. The consultant was from the MES Institute, a group set up to offer training and consulting in plant floor system best practices (much like the Ollie Wight Group did for MRP/ERP).

With a task force focused on fixing the problem, some of the emotion abated—until their first meeting. Then it began all over again with a series of complaints.

The consultant went around the room asking what each member or stakeholder hoped to achieve in these meetings—what they were looking for as an outcome. The comments flew like bats.

- "Our data is inaccurate, late, suspect, so we do not use it even though we spend most of our time collecting and reconciling it."
- "Our leadtimes are too long and we have poor visibility. We do not know what's on the floor or in inventory, so we keep lots of WIP to keep the line running."
- "There is very poor coordination between groups. The right hand does not know what the left hand's doing. We start batches when we have product already available elsewhere."
- "Product costing is difficult or impossible. We do not know what our products really cost to make; we do not capture the right data at the right level of detail. We do not believe in our standard costs."
- "Customer shipments are late and/or require expediters, extra setups, and other costly means of hand-holding."
- "It's hard to maximize process performance. We do not know the exact, as built, data to correlate with our final product test data. And that's getting more critical as the specifications get tighter each year."

- "There is no integrated commmunication between the customer and the plant, so we do not know the latest orders/changes to specifications."
- "We've downsized our staff but haven't taken the work out of the system for recordkeeping and compliance. We're overwhelmed with paperwork and complex systems."
- "Information is considered one of our corporate assets, but too much of it is spread over file cabinets on paper which is not easily accessible or integratable to our corporate supply chain and time to market initiatives."
- "The current system no longer meets our new business needs. It came out of the standard cost/financial world of the '70s and '80s. I get daily/weekly reports on where I was, not where I am, or how I can get to where I want to be."
- "It is difficult to change and costly to maintain—adding products, processes, or custom orders in the system is a pain."
- "It is not based on our architecture of the '90s—we are investing in the past, not the future."
- "I do not have a consolidated view of manufacturing that allows me to compare departmental productivity and capability, see the complete status of the plant, or ensure standardized systems, procedures, and interfaces. In fact, I have nearly as many different plant floor systems as departments."
- "The systems are not all user friendly; any changes require submitting a change request, MIS resources, months of time, and analysis."
- "If I only had the plant floor data in a relational database, tied to easy-to-use graphical tools, I could run this plant so much more effectively."
- "My process is under control but I cannot prove or document it easily. The data I need is spread over six departments, several of which use manual systems."
- "I want to meet OSHA and ISO 9000 standards, but the latest safety and operating documents are not available to our people."
- "I want tried and true solutions designed for manufacturing that are feasible and relatively simple to use—that can be implemented before my problem changes. Current ERP solutions take years to implement from what I hear. I want a solution with a real return that gives me a tool for improving manufacturing."

As the consultant listed their remarks, he didn't try to stem the flow of complaints but simply listened to them while Tom winced. He could see that an outside facilitator certainly hadn't discouraged candid remarks.

When they were done venting, the consultant looked at the list and seemed to be reading it, deep in thought. He turned to the group and ticked through some of the comments. "What we see here is really a set of symptoms. The root cause is something else. There was a time when your current systems were adequate, or should I say, you perceived them as adequate, or I should really say they supported your current manufacturing practices which you perceived to be adequate.

"What's interesting is that these complaints you've listed have always been true. Your data was probably always late and inaccurate. Your systems were always inflexible and therefore hard to change. Your systems never suggested where you could improve operations. But it didn't matter to you before now.

"So what changed? Your manufacturing targets changed. Customers and competitors raised the height of the bar. Quality, speed, cost, and flexibility goals changed by factors of 10 to 10,000. But your plant floor systems didn't change with them. Your infrastructure lagged behind your needs, as it usually does. The targets changed, but not the tools. That's why you're so unhappy right now. Your plant floor systems and tools are no longer congruent with your task. And your task is a hard one. You're being asked, from what Tom said, to meet world-class benchmarks. It makes sense that you'd want world-class tools to support your efforts.

"So I'd like to move you to more of a systems re-engineering process and less of a gripe session. I'd like you to develop a requirements analysis by talking about the 'as is' or the 'as was' way you ran your plant and the problems you're having meeting your goals with the current systems. Maybe manufacturing could begin."

Manufacturing began.

"When our leadtimes were measured in weeks instead of days (or even hours in our new cells), getting information once a day about yesterday's 'successes and failures' was adequate. Jobs didn't move that quickly. There was plenty of work in process, so it was easy to schedule. Quality was measured in percentages; so again, getting rework causes and rates the next day was fine. We'd look every morning in our daily meeting at where the jobs were, how our quality was, which equipment was up or down, which orders were late or hot, and if we were on track for total output and units processed. Then we'd set daily production rates to smooth or rebalance the line and meet activity targets, assign engineers to track down quality problems, assign priorities to maintenance, and decide if we needed overtime or to renegotiate any order delivery dates.

"If equipment broke down, the operator would tell the lead operator or supervisor, who told the production clerk, who called the maintenance department clerk.

"If someone wanted to research a quality problem, they'd gather up all the paperwork—the run cards or batch travelers for the jobs with problems and/or the equipment logs and/or the work instructions—and try to understand what had happened. (Just getting all the paperwork was a real effort, so we did this only for a real disaster.)

"Most of the reports we got covered yesterday's results. So if we wanted to track our activity or any problems that shifted, we'd have the operators fill out a manual report so we could see 'real-time' activity or progress.

"That just doesn't work anymore. We're moving too fast to work with *yesterday's* data. With much lower work-in-process (WIP) levels, we need to do scheduling in *real time,* based on where the jobs are currently, which machines are available and how they're set up, which operators are available, what raw materials are in stock, and what orders are arriving. We can't do that effectively now since we have only some of that information, and then it's available only once a day, as of two hours ago.

"With goals of parts per million defects and 99.9% orders delivered on schedule, we need to *instantly* detect a problem—in quality, machine status, productivity, or raw material late delivery—and notify the responsible person, without the person-to-person and telephone communication lags we have today.

"When we have to diagnose a quality problem, we want to be able to see the history of the jobs, the equipment, the work instructions, and the raw materials immediately, so we don't have to wait to find the paperwork that's now spread out over four or five departments and their filing cabinets.

"When we change a work instruction, I want every operator to be notified on a terminal when that job arrives that there is a new work instruction and have it displayed automatically.

"I want our floor material tracked by back flushing, so that when each job is done, we decrement the floor inventory and order replenishment automatically when inventories fall below specified levels.

"As a manager, I want to see our productivity, quality, and uptime by operator, equipment, supervisor, or technician so that I can see our best and worst performers and catch any trends immediately.

"And I can't afford an endless supply of clerks and secretaries to calculate all this information for me. What frustrates me is that we actually have all this data somewhere. I'm paying to collect it—on run cards, equipment logs and records, operator logs and time sheets, exception reports, and inventory records. It's all there; it's just worthless to me in its current form!

"When we ran the factory by weeks and percentages, I didn't need very accurate and timely data. Now, I have to scramble to meet my goals. I heard a good analogy recently. When you're walking, you don't need to look at where you are every second—things don't change that quickly. But when you drive, you have to see where you are every second. And when you fly, you need systems to help you by keeping track continuously of all the key causal variables (altitude, speed, direction, and wind currents) and alerting you if one is going off track. We may not be flying yet, but certainly we're driving and need more accurate data faster.

"We're not just trying to meet budgets anymore. We're trying to improve everything. That requires much more and different data as well as much more flexibility in linking that data to scheduling, statistical analysis, inquiry, and reporting tools. I don't always know what analysis I'm going to do or what data I'll need. Our system has fixed reports and *no* flexibility. And getting a new report from MIS is a matter of weeks or months if it's *ever* available. As I see it, these systems and reports were really designed for a different era, when status quo, not improvement, was the issue; when everything was measured financially to standards, not to absolutes; and when we accepted weeks of leadtimes and percentage of losses, not hours and parts per million. We've changed—no, that's too weak—we've turned our world upside down, but we're still trying to run this new world with the old systems. It's like trying to run a worldwide order management system still using carrier pigeons for communication. I can't believe there isn't something more relevant to a "faster, better, cheaper" continuous improvement world. And somehow I doubt it's throwing out our computers. I don't see anyone else throwing out computers. I just think we haven't found the right way to use them yet."

One by one, each manager, operator, scheduler, and technician explained their problems with trying to operate in their new world with their old systems. Most of the complaints seemed to center on not having real-time data, instant communication, instant alarming or exception detection and notification, flexible reporting and analysis tools, or systems and tools that actually supported their new tasks and approaches. Their desired solutions still varied over the map, from MIS "fixing it" to trying to do it all manually "like the Japanese," to buying stand-alone systems ("I can get exactly what I need on a PC today!"), to a few who thought their MRP system could do it if they used it properly.

This approach to the problem had been more useful from Tom's vantage point. By looking at their business processes, he could clearly understand the discomfort with the current systems and the urgency to correct the situation. Again, the approach the consultant was taking was a root cause analysis: looking at the gap between their desired business practices and business systems.

The consultant continued. "Now, I'd like to move to the 'to be' side—more of the vision. What supporting systems for your new business practices are needed? Let's just go around the room and list them as they occur to you. Then we'll see if there are any commonalities."

The Requirements

The group started brainstorming on their requirements by making a long list of the specific business practice/supporting system each member wanted (Table 7.1). They did not try to get into the lowest level of detail. Production started with their requirements.

"As we go to shorter runs, more products, more process improvements, and more job-by-job customization, our operators are having trouble knowing exactly what processing each job requires and any special processing instructions for that customer. We'd like the system to either print out the exact instructions, or alternatively, display them for the operator, highlighting any changes or exceptions. All these process improvements are great, but they do mean that you have to keep track of all the changes to the work instructions; and we'd like the system to either print out the tooling

TABLE 7.1 Detailed Requirements

	Requesting Department
1. Show operators the exact instructions on how to process jobs—any nonstandard operating procedures, work instructions, correct equipment, and tooling—and validate the operator, equipment, tool status, and certification for the job.	Production
2. Show the maintenance technician the instructions on how to repair, maintain, calibrate, and set up equipment. Show equipment history—quality and productivity, as well as operator comments. Detect trends in equipment performance.	Maintenance
3. Schedule the shift's activities—based on jobs to be done and resources available—to meet due dates and utilize resources most efficiently.	Production maintenance, production control
4. Provide an engineering database of product and process specifications and a history of changes, all batch records, all process conditions, real time SQC and SPC charts.	Engineering
5. Provide a status, history, and future schedule for jobs, equipment, tooling, operators, facilities, materials, and vendors. Show performance to date on any measure for any of these resources. Find best and worst performers for all.	All
6. Provide a dispatching function for jobs based on *current* status of jobs and resources.	Production
7. Provide scheduling and dispatching tools for maintenance, quality, facilities, and other support functions.	All others
8. Support the quality analyses: quality to event, quality event cycles, correlations of product quality to process conditions.	All
9. Detect and report exceptions, impending scheduled events, and key news, and instantly notify the responsible people.	All
10. Support standard and activity and improvement-based costing/utilization.	All
11. Report on leadtimes/speed for all jobs and all indirect activities.	All
12. Provide flexible and ad hoc reporting and inquiry capability in addition to all standard performance reports.	All
13. Be easier to use than the current MRP II/ERP system and cheaper to support than the current shop floor data collection and reporting system.	All
14. Support management practices and tasks—meetings, communications, exception lists, to-do lists.	All
15. Integrate with other applications (MRP II, automation, lab data, product and process design tools); have potential linkages to suppliers and customers.	All

and equipment to use or validate that what they're using is legal, so no one can use the wrong resources or resources in an 'illegal' state."

That requirement went up on the list as requirement number one. And the others followed rapidly. Maintenance asked for the same capabilities for their technicians—to be able to see the latest instructions on repair, maintenance, setup, and calibration, as well as the equipment history and parts inventories—all on-line or printed out in real time. That would save them time in tracking down recurring problems, finding out whether they had the parts they needed or had to order them, and making sure the latest instructions were used. And they wanted the operators to jot down notes on equipment performance as well as be able to see equipment trends for predictive maintenance. If the system could indicate machines starting to go out of spec, that was even better.

Third was production again, asking for a shift-scheduling tool that would take into account the current work in process, the job due dates, equipment, and operator and raw material availability to develop shift schedules for the next few days that kept the line balanced, utilized, and on schedule—a system that would predict their cost and schedule performance. Production control requested that capability as well.

Maintenance asked for the exact same thing—a scheduling system for their technicians that assigned them to maintenance, repair, setup, and calibration tasks that best utilized the people and met schedules. In fact, in the best of all worlds, the scheduling system would handle both production and maintenance scheduling simultaneously so as to best utilize the plant's resources.

Engineering got a turn. They wanted a complete engineering database on-line—all the product specifications, production processes, and a history of all changes to them; all the batch records for each job with the quality and lab data collected by job; all the process conditions associated with the processing; and so on. This database would be the key to supporting all quality analyses best operating conditions, failure analyses, and SPC and SQC trends as examples. Those were the heart of their improvement program.

Everyone chimed in on requirement number five. They all wanted a comprehensive real-time view of the factory—to be able to see the current status of any job, operator or technician, equipment, tool, facility sensor, or raw material and its history, as well as any schedule for it and its performance to date in productivity, quality, speed, technical capability, cost, and utilization.

Sixth was production again. They wanted a real-time dispatching system that changed the production schedule instantly as equipment broke down, jobs went on hold, processing took longer than expected, a rush job was launched, or other unexpected variances arose. Otherwise, there was little guidance on what to do in real time to correct a schedule that was now obsolete.

All the others thought this would be useful to them as well, and that became requirement number 7.

Engineering came back for number 8. They wanted the quality statistical analysis tools—quality to event cycles, linear and nonlinear correlation, and factor analysis capability that they had seen but didn't previously have the data to support! Now that they had the data, they could really work on improvement.

Points 9 through 15 seemed unanimous. A popular request was being instantly notified of any problems, impending schedule changes, events, or news that you needed to know (9). Everyone was interested in trying to look at their value-added utilization and both standard activity and improvement-based costing (10). Similarly, everyone wanted to understand their leadtimes and speed (11). Flexible reporting and inquiries would supplement preset performance reports and help analyze the data (12).

The next point, a system easier to use than their MRP/ERP system (13), led to a chorus of "yeah, you bet, that's for sure." The memory of trying to install their MRP II system was still fresh in everyone's mind. There was no doubt that this group wanted a system that was a lot easier to install, maintain, modify, and learn than their MRP II system. Getting that implemented had been like wrestling with a bear. Some days even when you were winning, you felt like a loser. If possible, they'd wanted it to be cheaper than their current costs of filling out, filing, and using the mountain of paperwork now in the factory. However, more important was that it provide the return on investment seen as possible.

One supervisor had a request that again everyone echoed. It seemed to her that their systems handled production data but ignored the "management" aspects of their job—report writing, communications, setting up meetings, and following up on problems. Her desk was covered with paper work associated with her job but not assisted by the production management system. She wanted a system with electronic mail, schedules, personal to-do lists and report preparation, and "problem" or "exception" lists. This was a big winner as far as the group was concerned (14).

Finally, it had to integrate with their other systems: MRP II, automation, labs, product, and process design. They wanted to be able to compare facilities worldwide if they expanded overseas. It potentially had to integrate with suppliers and customers, so that they could see the status of materials coming from vendors and let their customers view their order status (15).

Afterwards, they looked at the list and shook their heads. They were sure they could never get what they wanted. It all sounded good, but so did a ten-cent cup of coffee and same-day mail delivery. What was the point of listing all these requirements if they couldn't achieve them for less than $10 million and five years?

They all instinctively knew what their MIS department would say. If they were willing to specify exactly what they wanted, pay for it, and wait for it to be delivered in phases, anything was possible. It was only a question of time, effort, and money.

People were already starting to back away from the list. "I think we're being a bit grandiose here. This is like asking a starving man what he wants to eat, listing a twelve-course meal with finger bowls. Why don't we prioritize this and be a little more realistic?"

This deflated the group, but no one could really argue with this logic. They all knew they couldn't even get a new report today in less than 12 weeks. But yet, to do the job they wanted to meet the stretch goals that they had set, they really needed these capabilities. The consultant again stood up.

"You've listed a great set of requirements. Instead of worrying yet about whether you can get them or not, why don't we first take a cut at summarizing the characteristics of the business processes and supporting systems you're requesting. What you're saying is that first you want a factory and factory systems that are:

1. *Predictive.* Requirements 3, 5, and 7 all relate to a factory that can translate demand into a schedule for all supporting activities—pro-

duction, maintenance, labs. That schedule should also translate into a prediction of utilization, cost, and delivery against that demand. So our first requirement is foresight of prediction: a planning capability.

2. *Compliant* (*and self documenting*). Requirements 1 and 2 relate to a factory that executes all specifications or standard operating practices as stated, whether it's a job being processed; a machine being set up, maintained, repaired, or calibrated; or a lab sample being run on incoming raw material. Each activity is alone when it should be, how it should be, and if needed, documented for compliance. So our second requirement is compliance.

3. *Productive.* Requirements 1, 2, 7, 8, and 14 relate to a factory that "automates" or provides productivity tools for each task to maximize efficiency. Each task is designed and supported for minimum wasted or non-value-added time or effort. We provide all data needed for a task instantly without search. We provide tools that do need calculation analyses: reports, schedules, or performance to budgets. We support actual automation of the manufacturing process itself. So our third requirement is productivity.

4. *Vigilant.* Requirement 9 (and part of requirement 2) relate to a factory and systems that instantly detect all deviations—breakdowns, equipment running slowly, degrading quality trends, jobs not moving, WIP piling up—or upcoming events—scheduled maintenance, operators who need to be recertified, tools that need calibration. We alert the factory to problems or potential problems as if we had an ever-vigilant, ever-present supervisor so that we can remedy them as quickly as possible. So our fourth requirement is vigilance.

5. *Adaptive.* Requirement 6 hints at a factory that is self-correcting—real-time scheduling or dispatching that reprioritizes jobs as equipment breaks down or runs slowly; real-time maintenance and repair prioritization that changes as bottleneck equipment adaptive process control that resets recipes or set points as product or operation characteristics drift. In any case, once we detect a deviation, we need to correct it. So our fifth requirement is to be adaptive.

6. *Measuring value-added* (*or absolute*) *performance of key business drivers.* Requirements 5, 10, and 11 all relate to a factory that monitors drivers. As we all know, final cost or budget performance is really driven by the value-added utilization achieved on all key resources. Customer

service performance is really driven by our value-added leadtime or speed and the distribution of yield and leadtime variance. Our system has to measure what drives our competitiveness—our value-added utilization, speed, quality, flexibility, and so on—not just total up the checks we write to our suppliers. So our sixth requirement is to measure value-added or absolute performance of key business drivers.

7. *Improving.* Requirements 4 and 8 relate to a factory and systems that analyze its best and worst performance of equipment and operators in productivity, quality, speed, uptime, and operating conditions that give rise to best performance. They relate to a factory that measures root causes of all deviations and looks to eliminate them. They are about finding the root causes of random and systematic error so that both can be corrected. So our seventh requirement is to improve the factory.

8. *Integrating.* Requirement 15 relates to a factory that is tightly coupled to the enterprise—to your MRP or ERP system, to the equipment, to the lab systems or facility monitoring system or design systems. Today, we are all focused on enterprise-wide business processes. A factory that is not integrated into the enterprise's processes and systems automatically means an enterprise with less speed and added cost, as customer orders cannot be instantly translated into factory orders or vice versa, factory orders instantly translated into projected availability, new product designs or specifications instantly transmitted to the factory floor for production, or factory data instantly viewed by the R & D team to access manufacturability and help debug problems. So the eighth requirement is integration.

"You've also added one (ease of use) and left one off that most IT groups would have listed (compliance to your corporate IT standards). And I think you've forgotten one—being anticipatory. Let's look at each one briefly.

"What requirements 1 through 8 relate to is return on investment or value or benefits. We should be able to quantify the value or return from each of these capabilities (and we'll try to do that next). The flip side is the cost of the system. Too often we see companies comparing plant floor systems purely on a cost basis. The assumption is that they will all produce the same return. That really is a backwards analysis. We need to begin with the value we can attain—the return. That must bed the driver for a change.

"Once we can see the value, then we can turn to the cost. Cost comes in tangible forms such as initial purchase price, initial training, and implementation costs (which usually run one or two times the hardware and software costs). But there are also intangible costs which relate to the difficulty of getting system and paradigm: the harder the system is to use, the harder it is to get users to switch; the higher the training costs, the higher the number of errors made; the higher the support costs, the longer the implementation. So our ninth requirement is the general area of "low" cost of ownership, but relative to its benefits.

"A tenth requirement is compliance to your corporate IT standards. That is typically an IT responsibility that end users rarely raise. However, it is a major issue of IT cost. The more diverse the operating platforms supported, the higher the IT support costs. Again, though, we have to be sure that we put the focus on the benefits (1–8) and not the platform. If a competitor could achieve 5 to 10% higher equipment utilization by using plant floor tools not available on your IT standard, you'd be at a significant competitive disadvantage. Standards lower cost; they don't increase benefits.

"But there is one category, our eleventh requirement, that wasn't raised yet—anticipatory systems. In most factories, running state-of-the-art manufacturing processes and equipment, it's impossible to eliminate uncertainty. We have yield uncertainty, lead time uncertainty, and equipment uptime uncertainty. As a result, our plans will forever be a moving target unless we anticipate the uncertainty and plan for it. This means setting planning yields and cycle times that anticipate variance and setting WIP levels that buffer us from equipment breakdowns.

"And when we put those together, we have a factory and a supporting or enabling system that is anticipatory, predictive, compliant, productive, vigilant, adaptive, measuring, improving, and integrated, which translates into a flexible and lean factory. It's lean because we focus on eliminating all non-value-added activities. It's flexible because we can react to changes—predicting what is possible, adapting as needed to problems.

"So, for those who thought plant floor systems are merely tracking—you notice that we don't even have that capability listed. That's because tracking is a necessary capability, but without using the data collected for one of these categories, tracking by itself is waste. Tracking is a way to collect data, not a way to run a factory. What we've described is the way we want our factory to run, and therefore what we expect our systems to support in operation (Table 7.2) or our best-practices manufacturing execution system (MES)."

The group broke now, still uneasy. They had come to the meeting expecting to simply choose a system. Instead, they were taking three steps back talking about general requirements. To the more reactive members of the team, this whole first meeting had been a waste of time. They simply wanted permission to go out and buy the systems they still clearly needed. But others were curious where this was all headed. They hadn't heard a word about alternatives.

Tom was feeling reasonably pleased. At least the salvos aimed at IT had abated. He still had a sinking feeling, though, knowing that he was very unlikely to get significant budget approval for new plant floor systems. If anything, this was a leaner world. So at the end of the day he'd once again end up the bad guy. But, at least for one day, the barking dogs were called off.

Their next meeting was a week later. The consultant again listed the requirements they'd generated last time, now organized by his categories of "factory behavior," and then asked, "Who is ready to buy a system?"

Several hands shot up. The cautious members who knew a trap from long experience sat silently.

"What would you buy?"

"A scheduling system," a production control member offered. "IT would clearly make us more predictive and more productive."

"A statistical analysis package," an engineer volunteered. "IT would make us an improving factory and more productive."

"A maintenance package," the maintenance tech said somewhat hesitantly. "I think it would make us compliant and maybe more productive."

The instructor nodded. "Those are all good suggestions and maybe even great ones. But . . ." And he paused as the more cautious smiled at their foresight. "That's going to cost a lot of money. And unless you're very different from my other clients, money is a scarce resource—almost as scarce as time. So unless we have some very rich or very lenient bosses, we'll be allowed to maybe buy one of these or pilot one or perhaps get it budgeted for next year. So what I'd like to do today is look at how to calcu-

TABLE 7.2 Best Practices in Manufacturing Execution Systems

Mode	The Worldwide Virtual Factory/MES System: Strategic Capabilities
Planning	1. Predictive System Optimization/Prediction/What-if • Schedule performance • Budget performance • Resource utilization • Leadtimes • Equipment maintenance • Inventory reorders
Execution	2. Compliant System Self-documenting/Data Collection • Direct activities—manual/automated • Indirect activities—manual/automated 3. Productive System Automation/Productivity Tools • Direct activities • Indirect activities
Closed-Loop Execution	4. Vigilant System Alarms and Alerts • All trends • Exceptions • Time-based events 5. Adaptive System • Reschedule • Adaptive process control • Repair/calibrate/maintain
Management	6. Measurement/Integrated Data-based • Absolute performance • Technical capability/process capability • Best/worst performance
Improvement	7. Improving System Root Cause—Best/Worst Performance
Integration	8. Integrating System Enterprise Processes • Supply chain • R&D • Financials • Product transfer
Cost of Ownership	9. Intuitive, Low-cost System • Look and feel—intuitive • Purchase cost • Training cost • Maintenance cost • Integration cost

TABLE 7.2 **Best Practices in Manufacturing Execution Systems (*Cont.*)**

Mode	The Worldwide Virtual Factory/MES System: Strategic Capabilities
IT Standards	10. Compliant to IT Standards • Platform • Database • GUI • Message bus (CORBA)
Planning	11. Anticipatory System—Variance Anticipation • Equipment downtime • Yield variance • Leadtime variance

late the ROI from these categories so that you can clearly present the real ROI or competitive opportunity they represent.

"The way to justify any manufacturing execution system is to relate the return to the strategic initiatives or problems your company is facing. Most companies are trying to reduce their costs and cycle times and improve their quality, on-time delivery, and flexibility. But each business has a few key business drivers that differentiate the manufacturing leaders and produce the real competitive advantage.

"If an industry is capital intensive, the key business driver is usually value-added equipment capacity utilization for direct and/or indirect labor—what percentage of the time your labor resource produces customer orders.

"If an industry is raw material intensive, the key business drivers are usually conversion rate (or yield) and leadtime—how much value-added utilization we get of the raw materials in quantity (yield) and time (leadtime).

"If an industry has few barriers to entry, the key business driver is usually cost (as above), speed, and customer service. What percentage of the leadtime in producing a customer order is value-added. If an industry is technology driven, with constant changes in products and processes, the key manufacturing driver is usually time to reach sustaining yields or quality on new products, processes, or equipment.

"In each case, best performance on the key business driver usually translates into competitive advantage and sales revenue and/or margin advantage.

"So how do our factory and MES capabilities translate, specifically, into ROI?

"Let's look at the ROI from each of our best practices in manufacturing execution systems, starting from number one: predictive scheduling.

"The capability for planning or prediction is extremely important for high-mix, capacity-limited operations. If we run a dedicated line or cell, the line is balanced for a certain production rate. We simply run it up to that capacity. There is little planning required beyond ensuring adequate raw materials. However, as we run multiproduct, multiprocess plants, correct scheduling has significant ROI in two areas—equipment utilization and cycle time/inventory reduction—beyond the strategic goal of improving customer service.

"I know that your company has already studied both equipment utilization and scheduling improvement programs. If we look at the root causes of lost utilization at the bottleneck (see Figure 4.6), some of the key categories include: idle, no product (the machine is up and available but has no work to process); idle, no operator (the machine is either ready to be loaded or unloaded, but no operator is there to load or unload it); setup time (the time spent in setup); and in process, but running below the 'rated' speed (the job is processed slower than the standard rate due to equipment problems or use of a less efficient piece of equipment for processing).

"In each case, a scheduling or predictive capability can improve or actually eliminate each of these categories. For idle, no product, a scheduling capability should:

1. Load or release enough work into the factory to keep the bottleneck continually utilized

2. Prioritize jobs to keep work flowing steadily to the bottleneck

3. Set buffer or in-process inventory levels in front of the bottlenecks to buffer breakdowns or slowdowns at the feeding operations to the bottlenecks from disrupting bottleneck utilization.

"If these are done properly, 'idle, no product' lost utilization is totally eliminated.

"For idle, no operator, a scheduling system predicts or projects equipment completion times so that an operator is ready to load or unload the equipment with the next job. Too often, equipment may sit idle as operators are not aware that equipment has nearly completed a job until the machine actually stops, and a flashing light or signal is activated. Then the operator must find the next job to load. This delay robs capacity utilization and speed. Again, this category can virtually be eliminated with predictive scheduling.

"Setups cannot be eliminated without significant equipment re-engineering. However, the number and total setup time lost can be decreased by careful scheduling to minimize the number and size of changeover.

"Finally, scheduling can optimize the use of the most efficient equipment for each job. Often there are alternatives for processing. Machines are rated at different speeds or tolerances. Without scheduling, an operator may not find the best allocation scheme.

"So, when there is a bottleneck, we can estimate the return from a predictive capability as:

$$\begin{bmatrix} \text{The number of hours at the} \\ \text{bottleneck(s) lost to: idle—} \\ \text{no product, idle—no operator} \end{bmatrix} \times \begin{bmatrix} \text{The number of} \\ \text{units/hr produced} \\ \text{at that bottleneck} \end{bmatrix} \times \begin{bmatrix} \text{The} \\ \text{margin} \\ \text{per unit} \end{bmatrix}$$

"It is likely that we'll have additional return from decreased setup time and lost efficiency due to poor job assignment. However, these savings must be estimated case by case; they cannot be eliminated totally.

The second major area of ROI from predictive scheduling is cycle time/inventory reduction. Again, in a single product or process cell or line, there is minimal WIP. However, as we move to multiproduct, multiprocess plants, there is a natural tendency to use WIP to buffer operations or areas from their feeding operations. This is particularly true when there is little visibility across areas of when a downstream operation is running out of work or has a growing stockpile due to a slowdown.

"In a simple 'flow' shop, where there are few alternative process flows, a Kanban pull system minimizes inventory while keeping operations fully utilized. In complex operations, Kanban will not optimize setups or equipment utilization as I know you discussed in your scheduling improvement project (Chapter 3). For those cases, a predictive system can optimize utilization and customer service while minimizing inventory.

"Again, it is hard to definitively determine the ROI without specific data, but in general, scheduling will drive WIP levels toward their minimums. If current levels exceed 2.5 times the 'theoretical minimum' (the time to process a lot through an empty factory), we will probably see WIP drop to this level or below. Many companies we visit have WIP levels of 5 to 10 times the theoretical minimum or a value-added leadtime of 10 to 20%.

"So we could make a rough estimate of the potential ROI from inventory reductions as:

$$\begin{array}{c}\textit{Potential Savings}\\ \textit{from Inventory}\\ \textit{Reduction}\end{array} = \begin{bmatrix}\text{Current number of days}\\ \text{of WIP inventory} - 2.5\\ \times\ \text{theoretical leadtime}\end{bmatrix} \times \begin{bmatrix}\text{Average daily}\\ \text{production}\\ \text{rate}\end{bmatrix} \times \begin{bmatrix}\text{Average}\\ \text{holding cost/}\\ \text{unit/year}\end{bmatrix}$$

Example:

If the current leadtime = the 5 weeks (35 days) and the theoretical leadtime = 73 hours (about 3 days), and the average daily production rate = 6,000 boards/day, and the average holding cost/unit/year = $2.50, then the potential savings = [35 days actual – 7.5 days theoretical] × [6,000 boards/day] × [$2.50 holding cost/board/year] = $412,500/year savings.

"Obviously, in predictive scheduling, we are focused on meeting customer service. What we've seen is that, in addition, we can also improve utilization and inventory levels.

However, achieving these benefits will require more than a scheduling system or engine. It requires an overall scheduling or predictive system comprised of the scheduler and a "data engine," a "reporting engine," and an "exception engine." Many companies have bought scheduling packages only to be disappointed at the results attained. Where did they go wrong?

"A scheduler is like a car. It doesn't go very far without fuel. The fuel for a scheduler is timely, accurate data on job location and status, equipment state, yield and rework rates, equipment productivity and available hours, labor availability, setup times, and so on. When a scheduler is not integrated with a tracking data engine, the schedules generated are based on inaccurate or old data. The factory floor soon learns to disregard the schedules generated and create their own.

"Similarly, a scheduling system has to be self-improving. Without a reporting engine, we cannot see the actual performance to schedule, equipment utilization, and inventory levels. These, reported by root cause category, lead to further improvement programs.

"Finally, all schedulers depend on assumptions about yields, rework rates, equipment availability, and job status. When these assumptions are no longer true, we may miss the schedule. For example, if yields are running lower than planned or if the bottleneck production rate is lower than planned, our schedule is now invalid. We could wait until the customer schedule is missed, or alternatively, our 'exception' engine could instantly detect a problem and notify us. This allows us to take action at the earliest corrective point.

"So, clearly, a predictive system depends on more than a scheduler. It depends on an integrated view of the factory.

"Now let's move to the return from compliance, area 2. This is one of the easiest ways to justify a manufacturing execution system (MES).

"If we look at our direct *and* indirect manufacturing activities—processing batches at each operation, setting up equipment, repairing equipment, and maintaining equipment, we'll find that there is a human error rate implicit in all our manual processes. Typically, people will have an error rate of 1 in every 200 tasks. This rate increases as the variety, complexity, and stress increases. So if each job is unique and difficult and must be completed under time pressure in a stressful environment (noisy, hot, noxious), the error rate will be considerably higher. This means that typically, .995 of the time, each operation will be done correctly, and .005 will have an error.

"We can also greatly decrease the chance of error by reducing the variety or complexity (as in a dedicated low line or cell) and carefully designing the operation manufacturing process and work area. For example, we can color code parts and their insertion area on a board to ensure, unless you're colorblind, a perfect match.

"But if we stay at a typical operation, a mistake is made once in every 200 attempts. When we look at a product going through multiple operations, say 20, then the probability of producing good product through to completion is $(.995)^{20}$—the probability of producing good product at the first operation multiplied by the probability of producing good product at the second operation and so on for 20 operations. $(.995)^{20}$ is .90—meaning we'll have a combined rework and yield loss rate of 10% (some mistakes can be reworked—such as at insertion, some are 'fatal,' such as at wave solder). If our error rate were 1 in 2,000, our combined yield and rework loss rate would be approximately 1% (Table 7.3). In fact, we probably do not have an equal error rate on every operation. Some operations are undoubtedly more error prone than others. But we can estimate our own average error rate by reverse calculation from our observed yield and rework rate. Our average error rate per operation would be one minus (the observed successful completion rate for jobs to the root of the number of operations). So if our observed successful job completion rate is 95% (yield and rework rate of 5% for completed jobs) for a 20-step process, our success rate per operation is $\sqrt[20]{95}$ or .9975. This is a simplistic analysis which ignores that operations may not be independent but serves to illustrate a general point—that all manual tasks have an inherent error rate, no matter how small.

"This is true for processing jobs at each operation, recording the data on a batch sheet (as our medical products boards require) or indirect tasks, such as setup, calibration lab tests, incoming inspection, and repair. That is why your batch records (in the medical products area) require so much time in batch record release. If there are 100 entries on the batch record, and you expect a 1 in 200 error rate, we'd need to 'rework' nearly half of the batch records—39% to be exact. And that is remarkably similar to the observed batch record rework rate.

"So how does an MES produce ROI? By ensuring compliance (category 2) it eliminates almost all sources of human or 'random' error from not

TABLE 7.3 Probability of Job Completion Without Error as a Function of the Number of Steps (Entries) and Error Rate

Number of operations/ probability of error	1	10	20	40	100	200	400	1000
.995 (1 in 200)	.995	.95	.90	.82	.61	.37	.13	.01
.9995 (1 in 2000)	.9995	.99	.99	.98	.95	.90	.82	.61

following specified operating procedures. We call this capability electronic Poka Yoke, or foolproofing. The MES checks that the operator is doing the right job with the right materials with a legal tool and machine in a legal state with the correct specification and recipe at the right time and so on. We can check for all of the 28 conditions your quality team discussed as the resources of random error (see Table 2.6).

"Now we can actually calculate the ROI. It is the cost of rework and scrap for boards caused by human or random error. As a simple example, if your current scrap rate is running at 1.5%, your average cost per board is $20, and your output volume is two million, then we'd expect a return of $600,000/year.

"In fact, in another more complex industry, semiconductors, in every paper-based semiconductor plant in which we installed an MES, they immediately saw a 3 to 7% increase in their yield and therefore output. In every pharmaceutical company in which we installed an MES (or electronic batch record system), we saw the batch record rework rate drop from over 45% to nearly 0%. Both of these industries have a large number of operations (semiconductor) or data entries (pharmaceutical), but the principle is identical for every plant run using paper run cards to track production.

"In indirect activities the results of technician random error are usually more subtle. We see their error rate in the meantime between failure at equipment and in unplanned downtime caused by maintenance done incorrectly or schedules ignored. We may also see indirect error rates in product quality—off-spec, yield loss, and rework rates—when setup and calibrations are done incorrectly. We see the error rate in late orders and missed shipments when the wrong jobs are moved by scheduling.

"As opposed to in production, where random errors are eventually seen in scrap, rework, or off-spec product, indirect random errors are often invisible to us. They are often accepted as part of the manufacturing variability of equipment, tooling, facility conditions, or parts. Yet, in reality, indirect random errors are as preventable as direct (production) random errors and may be even more costly when equipment utilization is involved.

"We can see the impact of indirect random error in two studies customers of ours performed. In the first, a major electronics manufacturer looked at when equipment breakdown most commonly occurred. It was right after a scheduled maintenance! In other words, many maintenance activities were being done incorrectly and actually causing breakdowns. Another customer used an MES to assist repair personnel by providing repair 'protocols'—step-by-step repair tasks and checklists. They found that the mean time between failures improved by 40%!

"What both studies illustrate is that random or human error is as real an issue in indirect activities as direct activities and equally as preventable. And again, we see the need for an MES to support all production activities, not just direct manufacturing. Compliance is a key capability for any factory and MES, for all tasks performed.

"The ROI from ensuring compliance in direct manufacturing can be calculated as:

$$\begin{bmatrix}\text{The yield loss by product}\\ \text{due to operator error per year}\end{bmatrix}\times\begin{bmatrix}\text{The cost per unit}\\ \text{produced of that product}\end{bmatrix}+$$

$$\begin{bmatrix}\text{The rework quantity by product}\\ \text{per year due to operator error}\end{bmatrix}\times\begin{bmatrix}\text{The cost/unit to}\\ \text{rework that product}\end{bmatrix}+$$

$$\begin{bmatrix}\text{The off-spec quantity by}\\ \text{product per year due}\\ \text{to operator error}\end{bmatrix}\times\begin{bmatrix}\text{The lost margin/unit}\\ \text{from producing off spec}\\ \text{of that product}\end{bmatrix}$$

"Note that this calculation grossly underestimates the real cost of direct manufacturing noncompliance, as it ignores the effect of yield loss and rework and off spec on all the indirect activities such as scheduling, planning, customer service, supervision, and engineering. Each time we have losses, we need to replan and reschedule production—to restart material as needed, to reallocate production to customers, and to project out shortages or overages. We need to report on the root causes of losses and often investigate with both production and engineering resources. We are not even considering all these associated indirect costs of noncompliance in our calculation.

"The ROI from ensuring compliance in indirect activities is often harder to calculate because few companies have acccurate data on random errors in those activities. The easiest focus may be on equipment repair and maintenance. We can again study the occurrence of breakdowns to see if there is evidence of a pattern similar to what our customer observed: that it is most common immediately after maintenance or repair. We can also estimate the rate of random error in general by looking at how often we find "lemon" repairs or maintenance—how often machines immediately breakdown after a maintenance or repair. Usually, these will connote random error. The ROI from compliance in equipment maintenance and repair will be:

$$\begin{bmatrix}\text{The number of random errors}\\\text{(estimated as the number of ``lemon'' repairs)}\end{bmatrix}\times\begin{bmatrix}\text{The average}\\\text{cost of a repair}\end{bmatrix}+$$

$$\begin{bmatrix}\text{The number of random}\\\text{errors at a bottleneck}\end{bmatrix}\times\begin{bmatrix}\text{The mean time to}\\\text{repair at that bottleneck}\end{bmatrix}\times$$

$$\begin{bmatrix}\text{The number of units/hour}\\\text{produced at that bottleneck}\end{bmatrix}\times\begin{bmatrix}\text{The margin}\\\text{per unit}\end{bmatrix}$$

"These terms represent the wasted maintenance or repair cost (how many activities are performed that are 'lost' due to random or human error times the average maintenance or repair cost) and the lost production due to the added unscheduled downtime at the bottlenecks, the number of preventable breakdowns due to random or human error times their average duration times the number of units that could be produced during that time × their margin per unit).

"The next best-practice category of factory management is productivity. A factory system should support direct labor productivity by minimizing the non-value-added activities performed. In general, however, these are eliminated by re-engineering the workplace and actual work task. In addition, a factory system must be able to support full automation (the replacement of a manual task with an automated one). Capabilities should include download of a recipe (of the process instructions) to a piece of equipment, interaction with a material transport system (to direct loading or unloading of raw materials, tooling, or the job itself), upload of quality or equipment data from the equipment, and alarm handling.

"However, in most factories today, far greater ROI can be obtained by improving productivity of the indirect activities in the factory. Most companies have already re-engineered their direct production activities. Far fewer have significantly examined their indirect factory. This typically is the second most easily quantified return from use of an MES—improved indirect labor productivity.

"Obviously the raw materials of direct manufacturing are the ingredients or parts we use in manufacturing. The more difficult we make it to obtain those raw materials and jobs to work on, the lower our direct productivity of operators. That is why we typically ensure a steady flow of raw materials and jobs to work centers—to avoid delays due to no work or materials.

"For indirect labor, our raw materials are typically data. Their productivity depends upon the ease of access to information: on the history, current status, or schedule for the five elements of manufacturing in any form—from individual detail to aggregated rollups. A scheduler needs to know the current location of all lots, which are behind schedule, the status of equipment and schedule for upcoming maintenance, the availability of raw materials, and so on to produce a shift or day schedule. A technician repairing a broken machine needs to see its complete history, the availability of any parts needed for repair, the quality of the product it has been producing, and so on to diagnose and repair it. An engineer hunting down the cause of a quality problem with a particular lot needs to see its full lot history and any changes or events at the likely operations where a problem may have occurred—equipment repairs or maintenance, raw material changes, and so on (see Figure 6–5).

"In each case, this person's productivity will be directly impacted by how long it takes that person to find this data. When we work with a 'paper-based' manufacturing system (lot travelers, equipment logs, specifications), we have minimal access to the data needed for indirect tasks. The schedulers, engineers, technicians, and supervisors literally have to hunt down the needed paperwork, manually assess status and location, or count or total up activity across individual records.

"When the information is on islands of disparate or stand-alone computer systems—inventory on MRP or ERP equipment records on an equipment maintenance system, quality on an SQC system, and specs on a document management system—the results are only slightly better as he or she must sign onto system after system, find the necessary data, and pull it into a common database; print it out, recode, and retype it into a common database; or simply print it out.

"In general, in any paper-based manufacturing environment, we've found that indirect labor will spend 15 to 30% of their time getting their 'raw material' (the data needed to perform their daily tasks). With integrated MES, the data is instantly available on one system, typically linked to a statistical package, report writer, and query package. The person can see all the data needed in multiple windows concurrently displayed.

The user can easily extract data (by filling out tables) from the full database.

"When we also tie the data to 'productivity tools' for scheduling, planning, reporting, and alarms, we get a second productivity improvement. The actual amount of time your indirect personnel spend getting their data can easily be estimated through a brief time study. They can simply record each day and the time they spent getting different types of data (either at the end of the day or as they perform their tasks). A full industrial engineering study is not usually required but certainly will give even more accurate data.

"The ROI from providing data access is then:

$$\text{By indirect labor type}: \begin{bmatrix} \text{The percentage of time currently} \\ \text{spent getting data that would} \\ \text{be available in an MES} \end{bmatrix} \times \begin{bmatrix} \text{The total budget} \\ \text{for that indirect} \\ \text{labor cost} \end{bmatrix}$$

"In our experience, these are the three easiest returns from using plant floor systems to quantify. Many other ROI categories depend on whether you will make 'better' decisions with an MES (categories 5 [adaptation], 6, and 7 [measurement and improvement]) or find problems sooner and correct them—category 4 (vigilance). While this supposition is probably true, it is much harder to prove. That is why we initially focus on quantifiable improvement for estimating minimum ROI.

"So let's go back to buying systems. Who wanted that scheduling package?"

A hand went up.

"But I don't want it without the data," the production controller added. "It has to be integrated to a factory tracking and reporting system which also handles exception reporting."

"And who wanted that statistical analysis package?"

"I'd still like it," the engineer replied, "but it also has to be fully integrated to the factory data. And I'd like to also ensure full compliance on all direct and indirect activities. A lot of my job will be easier if I can focus on real process improvement instead of correcting processing errors or maintenance errors."

"And now, hopefully, we see how to justify these packages. But let me add another justification we haven't covered: reducing your current IT expenditures. Unfortunately, most of us already have legacy systems. We may have an ERP system which tracks WIP location. We have a stand-alone SPC system. There may be a time and attendance system. We may use a stand-alone maintenance system. In each case, we'll be paying to support, maintain, and enhance these systems. And there's an endless list of enhancements no one can get to, frustrating users—am I right?"

Tom had to nod, along with all the other team members.

"Ironically, you're often paying more to support, maintain, and enhance these disparate, stand-alone, isolated systems and databases than you'd pay to replace them with a single integrated system—with a unified view of your factory, a unified (if distributed) database, and a common set of tools integrated to the database.

"In addition, as our customers become global, they want a single plant system to implement worldwide to ensure: common nomenclature for products, operations, and tasks; consistent measurements of yield, cost reliability, and customer service to allow comparisons across plants; common best practices leading to best performance with a hodgepodge of systems in each plant. Our most advanced customers are standardizing on a set of best-in-breed applications they are rolling out worldwide.

"So often it's cheaper to replace a collection of loosely related systems on the plant floor. An integrated system also reduces operational costs by ensuring that data is only entered once, stored once, set up once, and archived once. With all your plant floor systems currently in use, how many times does an operator have to enter data on the same job (once into the ERP system, once into the SPC system, and once on the operation log sheet)? And each one has to be set up. And what do you do when the systems don't reconcile, when the work reported by the ERP or costing system doesn't match the operations logs? You work extra hours to see where definitions weren't consistent; data was entered in one system but not the other; or data was entered after shift end in one but before shift end in the other—waste, waste, waste.

"So I've been justifying these plant floor or manufacturing execution systems based on ROI. But, in fact, you're paying for them now, only usually without most of the benefits. Almost every one of our clients collects every piece of data imaginable on log sheets, PCs, ERP shop floor control, time and attendance systems, and maintenance records; they simply can't use it.

"They pay for collecting it. They pay for storing it. They just don't see any real return. These systems, on the whole, are meant to feed corporate systems—for costing, finance, and planning. They take data from the factory floor and use it for corporate purposes. Unfortunately, they seldom improve manufacturing. This is true for every ERP vendor's plant floor management system. It ends up being a data engine to support the ERP modules rather than improve the factory.

"So now let's look at what your alternatives are to support or enable your best practices. We have the list of best practices. Let's list the alternatives that I've heard from you. Some of you believe that your new ERP system will support these. Others want some stand-alone PC tools. I've heard a few of you suggest that no system at all is required—you can manage the factory manually, with visual tools like signal lights and Kanbans. Someone suggested we look at work cell controllers or supervisory process control systems. Well, that's a good list to start with. Why don't I let you critique each one as a group.

"My suggestion is that you review each tool by first looking at what it tracks or sees. This will indicate the real focus or visibility the tool will bring to you. Then you can add what it doesn't track, what remains invisible to it. Second, I'd look at what measures of performance it reports versus those you now want. Third, I'd look at what assumptions it makes about your manufacturing operations. Fourth, I'd look at what it improves through its use. By then you'll really understand the essence of that tool—what its real purpose and value are. Then I'd grade it versus your best practices capabilities."

The group discussion flowed easily. Team processes were a way of life now at PCB Co. Someone was elected a facilitator, another the scribe. Interestingly, a common realization had arisen during the discussion of the best practices they wanted to support on the plant floor. Before they even began grading the alternatives, they discussed what they had learned so far in the process and what had really stood out to them or changed their thinking about plant floor systems and management.

"I never thought about systems before in terms of a vision of how I wanted a plant to run. I've always viewed them simply as a way to get some data or reports. That checklist of best practices was very useful in seeing what we really want."

"I never thought about quantifying the ROI carefully. I knew I needed a system but I couldn't tell you how much it might return."

"I never thought that there was a real need for any systems or data collection on the plant floor. It never occurred to me how much we might be able to learn from the data we collect if it really were accessible. I spend my time looking at problems, not best performers."

"I was focused on the technology, not the return. I just wanted to be sure it met our technology goals. I assumed all these shop floor tools were about the same. So I figured you bought the cheapest one that met your technology goals. Now I can see how to evaluate alternatives from the company's viewpoint."

"I just wanted a scheduling system. It never struck me how useless it would be with the data or tracking systems to feed it."

"I never realized how much duplication of data there is, how many times we collect the same data and how little use we get from it on paper."

These last two themes seemed common to everyone in the room.

The thing that had struck them all was that all the tools they desired were *useless* without the data being available in a timely and accurate form. And if they could manipulate that data flexibly with easy-to-use packages, they'd be ten times better off than they were now with very limited data that was late and often inaccurately reported by an inflexible system. It made them realize just how poor or limited their current systems on the factory floor were. What they wanted was a "real-time window" onto the factory floor (to all the jobs, equipment, people, work instructions, facilities, and raw materials) that they could see over time—to view current status or history or future schedule and flexibly report or inquire on. Without that database or data repository, all the tools they'd requested were less valuable or even useless.

It finally struck them that to make manufacturing an improving "team sport," to really *integrate* manufacturing, you needed to integrate the data and systems around a common view of the entire factory. To schedule and dispatch, you needed to see the status and schedule of the jobs, the equipment, the personnel, and the raw materials. To understand and improve quality, you needed to understand the "state of the factory" when a job was processed. To understand speed, you had to see the job at *all* times. To understand utilization, you had to see *all* the resources, all the time.

The whole team realized that they were trying to manage a factory in real time when they were "blind" to much of it most of the time. They were trying to manage a race from periodic "snapshots," every eight hours or so, that took an hour or two to "develop" with their current batch computer system!

So whatever tool they chose had to, at a minimum, see their entire factory, all the time. Otherwise, they'd have to make operational decisions without data. They'd assume instead of know. With this in mind, they decided to do a quick grading of all the alternatives before they plunged into in-depth evaluations to see if any really had an integrated, continual real-time view of the factory.

Right off the bat, however, each of their alternatives flunked this very first critical requirement (Table 7.4). Their ERP only tracked the inventory and work in process in any detail. The manual systems weren't integrated, and worse yet, no one had real access to the data. It simply went into large filing cabinets and soon disappeared. Its accuracy was very questionable, so no one really used it. The same problem existed with the stand-alone individual systems. In both cases, the data existed in stand-alone silos with no integrated view of the factory, no integrated data collection. The work cell controllers or supervisory process control systems only saw automated equipment and their recipe. The rest of the factory didn't exist as far as it was concerned.

Now with that overview, they were ready to critique each system as their instructor had suggested. They were curious now to understand what these tools really did since none seemed to address their full needs on the factory floor.

They started with their ERP (a.k.a. MRP II) system.

When they had bought their ERP or MRP II system, they had assumed that it would solve all their manufacturing needs. After all, it was billed as a "manufacturing" system and had been sold to them as a complete, integrated solution. In fact, that was one of the reasons they had bought it—they had been assured that it would be the only system needed. But as they started to discuss what it really did or saw in the factory, as the instructor had suggested, another picture arose (Table 7.5).

The only part of it that related to the actual manufacturing process were the shop floor control module and the equipment maintenance mod-

TABLE 7.4 Grading the Alternatives

Tracking Visibility of:	WIP Inventory	Equipment Status/History	Personnel Status/History	Facilities	Recipes/ Specifications	Integrated
ERP	X					No
Manual systems	X	X (individually)	X	X	X	No
Stand-alone systems	X	X (individually)	X	X	X	No
WCC/SPC		X (automated)			X (for automated equipment)	No

Comments

1. MRP II	■ Tracks only WIP—doesn't have a real-time history, status, or schedule for equipment, personnel, facilities, work instructions ■ Doesn't handle execution control ■ Is a planning system, not an execution system ■ Keeps status quo, doesn't improve
2. Manual systems	■ No access to data to use in improvement tools ■ No information, communication across physical distances ■ No management efficiency tools
3. Stand-alone systems	■ No integrated view of the factory for use in reporting, scheduling, quality improvement tools, and execution control ■ Duplicate data setup, entry, reporting
4. WCC/DCS	■ Focus on automated equipment ■ Doesn't see the whole factory—only the cell ■ Really another form of stand-alone system—the *island* or work center by itself

TABLE 7.5 Grading ERP/MRP II and ERP/MRP II Tracking

What it sees of the factory	• WIP—after each operation is done; quality/material and labor hours used • Equipment—standard earned hours; maintenance work (parts/labor hours) • Operator—time and attendance; standard earned hours • Tools (fixtures)—assets
What is invisible to it	• WIP—lot prior to processing (to check compliance); lots on hold (status) (don't see it continually); quality measurements • Equipment—don't see it continually (7 × 24); in all states in real time; through all activities; with history processing, idle, broken, in repair, setup, etc. with goal data • Specifications/recipes—no • Operator—doesn't see it continually • Facility conditions—no
What it reports/knows	• WIP—quantity completed, quantity at each operation • Equipment—standard earned hours (based on number of units processed); actual maintenance costs/work order • Cost variances based on standard vs. actual activity
What it doesn't report/ know	• WIP—SQC by operation; quality data/lot • Equipment—actual time in state; complete equipment history; quality data/process data • Recipe—actual recipe/lot; instructions • Facilities—facility data • Operator—history by activity
What it assumes and doesn't improve	• Standard yield/cycle time/rework rate/units per hour equipment rates (quality/speed/utilization)
What it improves	• Level of raw materials required to run the production plan • Planning adequate standard capacity for planned load • Detecting budget variances

ules. The first saw work orders *after* they were completed at each operation and collected the quantity completed (or moved out), as well as the direct labor and material cost used in the work order. The second saw work orders for maintenance and collected the parts used and labor hours required for the maintenance or repair. So the ERP system saw, tracked, or gave visibility on work orders after each operation was completed and the cost of equipment maintenance determined.

What was invisible? They didn't see the lot or work order prior to processing to verify it was at the right operation, or was using the right tooling, equipment, specification, etc. They had no status or history of equipment, only the maintenance cost, so they could only collect standard earned hours or utilization, not the actual value-added utilization. They had no status or history of operators. Again, they only had earned hours or standard utilization. The asset system had a listing of equipment and tooling used to calculate depreciation and asset value. It was great for budgeting depreciation but not used in manufacturing. There was no document, specification, or recipe management system. There was no facility management system or data.

So what the ERP system saw was WIP, inventory quantities, cost variance, equipment earned hours, and maintenance costs (with variances). It didn't see value-added utilization or time. It didn't have process and product quality data. It didn't have detailed lot, equipment, tool, operator, or facility histories.

As they discussed ERP or MRP II, it soon became clear that it was an enterprise's view of manufacturing, a macroview that had little visibility into the details of the plant. It was not a window into the factory. It saw only a small part of the factory and only for a small part of the time—jobs after completion, maintenance after completion. It was really *finance's* view into the factory (for measuring performance to budget and standard costing) and *planning's* view into the factory (for planning job releases and tracking job location and quantities against schedule).

In fact, as they moved on to what it assumed and what it improved, it soon became clear that in manufacturing, it was really more of a planning and finance system than an execution system. It planned production to meet demand with minimum store room inventory. It used your current or started leadtimes and yields for planning purposes. For example, if your boards took 6 weeks to manufacture, it planned for 6 weeks of leadtime and started jobs 6 weeks before they were due (if there was adequate capacity). If yields were at .92 and an order required 100 boards, it started 109 boards to meet the order requirement (100/.92). In other words, ERP assumed all that was required for world-class manufacturing was availability of raw materials and capacity.

It was becoming clear to the group that ERP was not an improvement or execution system. It was a model of manufacturing of how much input is required for how much output and the cost of transformation.

Now Tom and the others were beginning to understand why the JIT/TQC revolution had sprung up outside of or in spite of ERP. As they looked at how it met their best-practice goals, it became clear that their ERP system would address only a few of them, if any (Table 7.6).

It was not going to be truly predictive without a finite capacity scheduling system (it assumed infinite capacity and only indicated capacity overloads), actual equipment utilization figures, or actual equipment status.

TABLE 7.6 MRP Versus Best Practices

Predictive	• Requires linkage to Finite Capacity Scheduler (FCS)—not in MRP • No data on equipment status/schedule to drive FCS—only WIP and raw mat data (Chapter 3)
Compliant	
• Direct	• Can't ensure correct equipment/state, tool/state, operator prior to processing
• Indirect	• Doesn't handle indirect events (Chapter 2)
Productive	
• Direct	• Doesn't support automation—download of recipe, upload of data
• Indirect	• Doesn't support all data required (partial) by plant schedulers, engineers, maintenance
Vigilant	
• Trends	• Doesn't see SQC, changes in real time in equipment status, productivity, etc.
• Time based	• Doesn't project job completion, missed maintenance and take actions as a result
Adaptive	
• Scheduling	• No A.P.C.
• Adaptive process control	• No automatic rescheduling based on triggered exceptions
Measure	
• Value added	• Doesn't measure actual value added utilization or time in state
• Best/worst	• Doesn't report on quality, productivity, uptime by individual equipment, operator, etc. (Chapter 1)
Improving	• Not linked with extracts of product quality data to process condition data to a statistical package (Chapter) • Doesn't see best/worst equipment
Integrating	
SCM	• Yes
Automation	• No
Anticipating	• Doesn't plan required buffer inventory • Doesn't analyze planning yields to use (Chapter 3)

It wasn't going to be compliant for lot processing if it didn't have the specifications, the equipment, and tool status (to check legality); if it didn't have the ability to store and download recipes (for automation); and if it didn't "see" the operation *before* processing (it had only a move-out or work credit transaction).

It wasn't going to be compliant for indirect activities such as maintenance, repair, setup, or calibrations, since there was no ability to define indirect activity specification and required resources and then verify that they were being used and in the correct state.

It wasn't going to be as productive for direct processing since it didn't support automation, operator instructions, or required data analysis (SQC, SPC, quality data collection, equipment tracking). It wasn't going to be as productive for indirect activities since it didn't have all the data needed by engineers, schedulers, QC personnel, managers, and supervisors tied to tools for scheduling, quality analysis, SPC, and so on.

It wasn't designed to be a comprehensively vigilant system—triggering alarms on quality, rework, equipment status, or problems, for example. It mainly found cost problems once a shift, day, or week in performance to standards and/or budgets.

It wasn't adaptive in recentering a process or rescheduling automatically as needed.

It didn't support value-added measurements or reporting on productivity, quality, or speed by individual equipment, operator, or tool. It wouldn't support improvement programs without the product and process quality data or the ability to find individual best and worst equipment, operator, and tool performance.

It didn't anticipate yield or leadtime or equipment uptime variance; it assumed fixed and constant yields, leadtimes, and equipment productivity, and simply reported variances.

The group was a little stunned over how poorly their ERP system had fared in their evaluation. If you had asked most of them a priori how well it supported manufacturing, each would have said that it didn't really meet their own needs, except for performance to budget (where they relied on it) but assumed that it met everyone else's! It was only now as a group that they saw how inadequate it was for their vision of their factory. The advantage of using their best-practice requirements and the ROI analysis was that they could rationally determine if their objections were objective or emotional.

Tom was concerned now. The ERP system had been extraordinarily expensive and the pet project of their corporate CFO. PCB Co. had not really been given a choice. Diversified had simply mandated its use and told them it *would* fit and it *would* handle their manufacturing needs. He wasn't sure, ROI or no ROI, fit or no fit, that he could go back to corporate and explain that the ERP system did *not* meet their needs in manufacturing. In fact, he was sure that he couldn't. He made a note to ask the consultant about this problem at a break—a long break.

None of the other alternatives took as long to grade.

The manual system or "no system" advocates were next. Unfortunately, they had already backed off after they had developed the list of requirements and thought about how they currently did factory floor management. They already had a manual system—that was the way they currently operated! They had a paper job ticket/batch sheet, paper equipment records, and paper SQC charts. Its inadequacies were already clear—there was almost no access to the data. Data on paper was useless. They needed to get it into a database that everyone could access; into tools for quality analysis, reporting, and scheduling; and into monitoring for instant detection and notification of floor problems.

"No systems" sounded good but didn't provide the tools or data needed for quality improvement, problem solving, or scheduling simulations. It now seemed like a step backwards. What they needed was to use technology *properly* instead of avoiding it.

But they still graded it carefully in preparation for all the eventual questioning they'd face. Many people still had the "no system is the best system" attitude.

In terms of what it tracked theoretically, it could track anything. Unfortunately, in practice, since the data was rarely used, the accuracy and timeliness of the data collected was very suspect. In fact, many in the room believed that the paperwork was filled out at shift end "to the best of each person's recollection." In theory, nothing was invisible but in practice, *everything* was invisible, since no one had access to the data entombed on paper. It was only under great duress or in an emergency that anyone wanted to attack the mounds of paperwork to try to analyze or summarize it. And even if they attempted to use it, no one really believed that the data was accurate. So, for all practical purposes, the factory was completely invisible. "Data, data everywhere but not a datum fit to use" would be a paraphrase of the situation.

Again theoretically the manual systems could report on anything, but in practice, it was far too time consuming and painful to do so. The result was that almost nothing was reported on except total unit counts of work. No one knew which were the best machines or their real value-added utilization, even though there were extensive machine logs. No one knew where the jobs were at any given moment or the WIP levels. So they employed expediters to answer those questions. All that data and no information was their summary.

They weren't sure what a manual system assumed. One way suggested it assumed that data had no value or use and assured that no one would be able to use it. It certainly wasn't seen as improving anything. It was simply a recording mechanism.

If possible, it fared even more poorly on the best-practice requirements. As purely a recording tool, it really couldn't ensure use of any best practice. It certainly wasn't predictive. Far from it, the manual factory appeared to be a black hole. Jobs disappeared into it until magically they emerged. Unless you ran short cycle time cells, you had no real visibility. They weren't truly compliant because the paper system didn't know the status of the resources you were going to use. It couldn't prevent or warn you if you were going to use a machine that was down or parts that were on quality hold. You could set up manual systems to try to avoid these problems, but they were frequently error prone—as a maintenance tech forgot to change the sign on the equipment to indicate that it was down.

They weren't productive. In fact, writing down the data was a waste of effort if it wasn't going to be used. They certainly didn't help the data needs of the indirect functions.

They weren't vigilant in general because calculations, trend charts, and statistics were too laborious to do by hand. They also weren't adaptive. They were passive systems.

They had the data best-practice measurement but not the access to it, so, realistically, they got a zero here as well. The same situation held for supporting improvement programs. The data analyses were impossible without getting the data into a computer. Certainly they were not anticipating anything except more data to be collected.

The people who had been so strong on "no systems" had moved from a total "downsizing" of systems to a "right sizing" mentality. They now called for the need for systems *where appropriate.* So in a work cell, they saw the need to track equipment uptime and quality problems but not WIP location (because the cycle time was so short).

A more confusing issue was the stand-alone systems. Everyone had a positive feeling toward the new typically PC-based tools. They looked great and they were reasonably priced (at least for a small number of users). They ended up being discussed at great length. Finally, the instructor called a halt to the meeting (which was now dragging on past 6:30). He suggested that they break now and actually go look at these systems tomorrow at a local trade show to understand how'd they work at PCB Co. in the factory.

Tom thought about approaching him now over the corporate ERP issue. However, it was late, and they were all tired from working through the analyses.

The point solutions were discussed at great length on the trip the next morning. This seemed to be a big trend now—separate systems on PCs or work stations for every application. There were systems for maintenance, WIP tracking, SQC, scheduling, electronic batch records, specifications/work instructions, automation interfaces, facilities monitoring, time and attendance, and so on. What was wrong with using these? They all were on the same hardware platform. They could track all five elements, just separately. The brochures made it sound like they were integratable.

After a morning of demo's at the local trade show, they saw the pros and cons much more clearly. Each system by itself looked excellent. The problem was they were stand-alone systems that didn't share data definition/set-up, data entry, data storage, and reporting/analysis. This meant that they supported each department and function as if it were the *only* one that existed. You had to set up the same route in the WIP tracking system and the scheduling system. You had to set up the same equipment in the automation system and the maintenance system. You had to set up the same products in all the systems. You had to look on the scheduling system to see the next job to do, track it on the shop floor system, look at the work instructions on the spec system, update machine status on the maintenance system, enter quality data on the quality system, and so on. Each system had a *different* user interface, command structure, terminology, and set of reports. You could, similarly, see WIP on one system, quality on another, maintenance schedules on the third, equipment status at automated machines on the next, and so on. It was as if manufacturing were a set of fiefdoms totally independent of each other—a break up of the United PCB Co. into a set of Baltic departments. It was like trying to play basketball where each player has their own system showing only themselves. With their goal of an integrated manufacturing world—schedules that took into account materials, personnel, and equipment, that related quality to all five elements; that supported execution control based on checking and displaying information on all five elements; that provided integrated reporting (point solutions were another hindrance)—actually a step backward rather than the solution. It further isolated each department.

They didn't spend as much time on the analysis against their best practices because they had ruled out any system that didn't have an *integrated* view, management, and control of the factory. That integrated approach was intrinsic to every best practice. It was okay if the data or applications were distributed in architecture, but the application had to be integrated in user functionality—in its complete view of a factory. The stand-alone systems clearly violated this principle. Everyone was moving to integrated office suites of applications and to integrated enterprise planning systems. The factory was no different in its need for an integrated approach. The stand-alone systems were seen more as a stopgap measure. They were used by departments or functions who couldn't get integrated tools and preferred some support or tool to none. But they just continued and exacerbated the problem of islands or silos of information that couldn't communicate or work as one business process.

That left work cell controllers or supervisory process control systems to examine as a solution. These systems controlled an automated area. Again, after some discussion, it seemed that these were just another set of stand-alone systems. Instead of breaking the factory up into departments, it broke it up into independent work centers. You couldn't see across work centers to schedule the whole factory or coordinate front- and back-end operations or view how product quality at one work center test related to process data at another.

In addition, their main focus was on equipment. These systems rarely tracked the raw materials, tooling, personnel, and manual work instructions. Even the equipment history recorded often left out maintenance history detail. These seemed more like specialized *execution* and *monitoring* systems—to ensure that automated equipment ran correctly and any anomalies were detected instantly. They weren't factory scheduling or costing or management systems. In fact, they really seemed one level below the factory execution system (control systems). Their full analysis also ruled them out as a total solution.

By now, their view of a complete manufacturing solution involved three functions (Fig. 7.1). There were planning systems, such as MRP II or ERP, which told the factory what to make and when to make it, and focused on standard costs and budgets. There were automated systems that controlled/monitored automated equipment or cells. What they had not yet found was an integrated factory floor management system that worked with those two and controlled, monitored, and synchronized the *entire* factory—that saw *all* the resources and events/tasks/conditions, recorded them, and reacted appropriately; that scheduled the jobs and tasks taking the entire factory into account; that recorded product quality and correlated

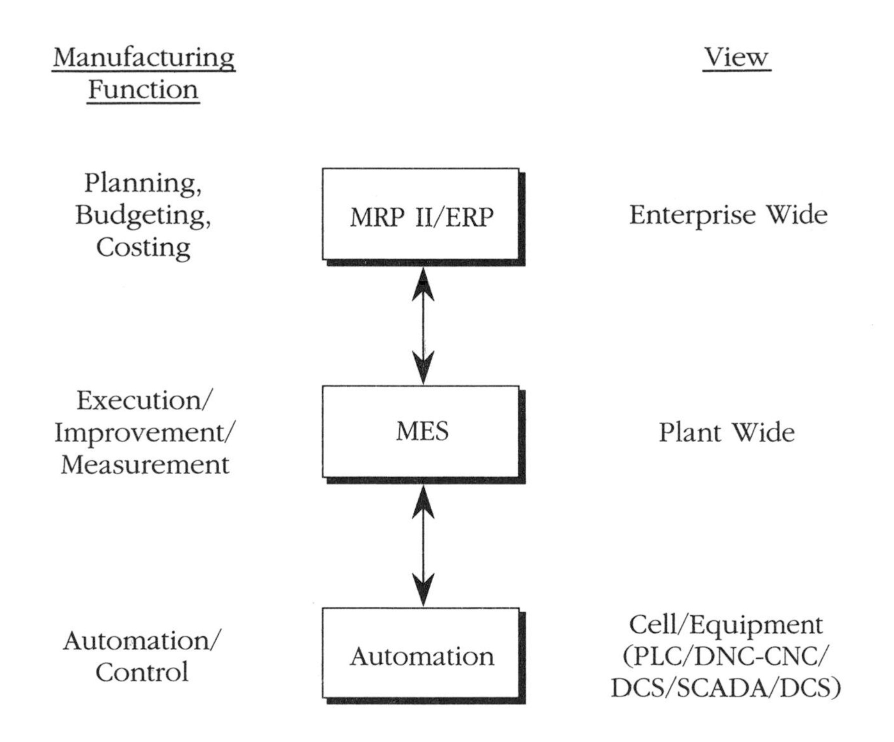

Figure 7.1 A Total Solution for Manufacturing Encompasses Three Functions

it to the underlying causes or factors (the process conditions) that affected quality; that reported on value-added speed and utilization, flexibility, and supported management tasks of communications, reporting, meetings, and analysis.

There had to be such a system simply because there was such a need for one! All they had seen so far were pieces—individual stand-alone systems they might have been interested in *prior* to their drive to achieve stretch goals. Their hurdle levels had been raised by a factor of 10 to 10,000. They were expected to improve quality problems from percentages to parts per million, leadtimes from weeks to days or even hours, and customer service from 75 to 80% to 99% or better. These had to be a new tool, something designed specifically to help them on the plant floor execute and improve to these world-class levels.

At this point, the instructor re-entered the discussions. He was extremely pleased with the group's progress. Their analyses of their current potential plant floor management options had been accurate and sophisticated. In addition, rationality had ruled over emotion. No initial advocate for any option had become so defensive or stubborn that the process ground to a halt. Each had argued the pros of their option assiduously but fairly. The result: the group had come to a useful shared view of their requirements and how each option stacked up against them.

However, now it was time to discuss manufacturing execution systems, an option with which the team members were not familiar. These were the fourth generation of what were originally called shop floor control systems in MRP or ERP systems. PCB Co. had not heard of them but they were used extensively in the semiconductor industry; being widely adopted in the health care industry, expecially pharmaceuticals; and implemented by leaders in over 20 other industries from aerospace and defense to metals, food and beverage, personal care products, textiles, fibers, specialty chemicals, plastics, films; and so on. He began to explain to the team the history of MES.

Take 1: Shop Floor Control

"In the '60s and '70s plant floor management was viewed as a part of MRP systems. The shop floor control (SFC) module viewed the shop floor as a black box through which jobs navigated automatically (and mysteriously) once parts or raw materials were supplied. In practice, many people thought of launching jobs into 'the great void' or 'the black hole' (as the plant floor was referred to) and then, realistically, having to expedite them when they fell behind. SFC tracked each job through this 'netherworld' collecting its direct costs—labor and material—as it finished each operation. The reason (it seems obvious now) that the plant floor was a black box to users was that we didn't see most of it on our system! The equipment, operators, tools, facilities, and specifications were invisible! We used SFC to track our standard costs/WIP valuation, plan job releases, and locate jobs that were falling behind or static.

"In the MRP shop floor module, all that existed were the work orders or jobs (Fig. 7.2). There were no other objects or entities defined or tracked. Therefore, the factory's behavior was mysterious, since we couldn't see the forces (resources) that impacted its performance!

"However, that wasn't the purpose of the MRP SFC module. It was used in cost accounting (to track standard inventory valuations and manufacturing cost variances) and in planning (to see how many units were already released to the factory floor). It was never meant to support manufacturing. It was fixed in the MRP II view of a company—planning and financial records of all the company's transactions.

"This is why the JIT/TQC revolution sprang up *outside* MRP II systems. Unfortunately, most purchasers thought they were also buying a manufacturing execution system. However, by now we understand that any execution system must see the whole factory to be effective in execution, improvement, measurement, and our other best practices.

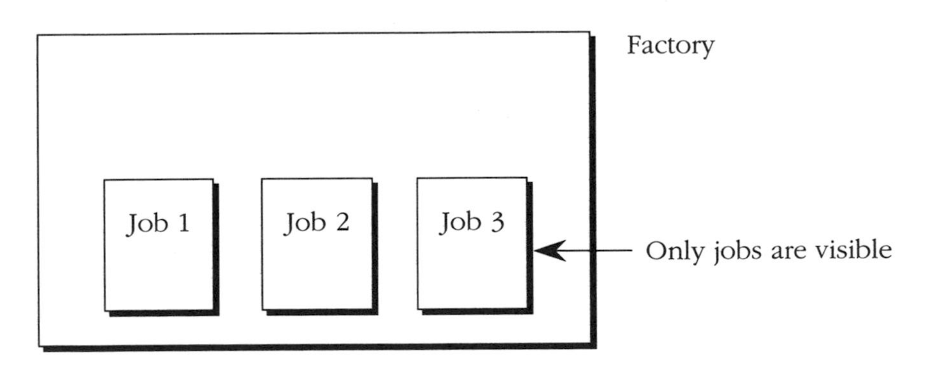

Figure 7.2 The "Invisible" Factory

"To see this point made through a simple analogy, suppose your secretary told you that she had purchased an open airline ticket for you to Chicago because of a great discount fare available. When you ask if she's booked you a hotel, she says no, there are plenty there. The same goes for a car. You ask if she's set up any appointments in Chicago and arranged agendas for those meetings. She again says to just call for appointments when you get there. Obviously, you'd say it may be a cheap airline ticket, but it's going to be a miserable business trip. You'll have to wait at the airport and hope they have a car. Similarly, if there's a convention in town, you'll have to stay outside of town. And, probably, most people will already be busy and not able to see you."

While no one would plan business trips that way, that's exactly what MRP II systems did for manufacturing jobs. They sent them off with their airplane ticket—a box of kitted parts or raw materials with enough "aggregate" capacity. And it should be no wonder that jobs waited in front of bottlenecks, and a generation of expediters became a standard part of our overhead cost structure and operating procedures. We did by hand what our manufacturing system ignored—we used paper records for operator logs, machine logs, maintenance records, WIP status, quality records, and process/product documentation. The floor was "awash" in a sea of paper and jobs. However, data on paper rarely became information; it instead became a data tomb. Management was by reaction—when a job *didn't* come out on time, when a machine had been down for hours, and when operators were reviewed for performance problems—and not by anticipation or plan. Systematic error analysis and "success" data bases were nonexistent. No one had the clerical help to record performance by individual machine, operator, vendor, or process spec, or find the best performers on speed, quality, utilization, or customer service. In fact, many of those weren't even measured as actual values but accumulated as standard earned values based on the number of units produced. Other values were only measured upon job completion, days or weeks too late to detect and correct a problem.

Data, data all around us and not a drop of information.

Take 2: The Paperless Run Card and SQC

With the emphasis on manufacturing quality that began in the '80s, companies went in one of two directions to upgrade their execution systems. Either they added a stand-alone quality system used in addition to the SFC module (such as an SQC package), or they added a "paperless traveler or electronic batch record" system (EBRS) to replace SFC and feed the MRP II system. The EBRS recorded and displayed all the data on the batch traveler and also displayed the work instructions to the operator. This basically automated the batch paperwork on the factory floor so that all the work instructions and data were now in a central database. These systems started in complex manufacturing industries (such as semiconductor and disk manufacturing) and soon spread to regulated industries that needed paper records of the "as built" or "proof of conformance" documentation (such as aerospace and defense and pharmaceutical/medical products companies). They were usually referred to (incorrectly) as paperless manufacturing systems.

These systems simply replaced the existing cumbersome paper system with an electronic one that provided "execution control" and an electronic trail. Operators were taken through each step of the manufacturing process with all their data instantly validated and available on the computer. This eliminated many of the random errors by operators and met certain regulations regarding availability of all safety-related documentation to the operators required by OSHA.

People thought that if they simply had access to all the data they wrote on their batch travelers or run cards, they'd be able to improve manufacturing. Similarly, they often added stand-alone SQC or SPC systems that produced X-bar, R, or p charts for any of the quality measurements they took and plotted manually. Again, they thought that if they could automate the quality calculations, they'd be done more accurately and timely than in their manual approach.

While these systems seemed exactly what people wanted, they had a major failing: They didn't have the data that explained the underlying cause of problems (and so could be used for improvement) or the data needed for scheduling or utilization/costing and speed analyses. The problem was that they *still* didn't see three-fifths of the factory. They were blind to the rest of your resources—personnel, equipment and tooling, or facilities. They had only added work instructions to our view, and in general, only for product manufacturing, not for maintenance, calibration, repair, etc. When an SQC violation was detected or perfection attained, there was no information available on the state of the five elements that *caused* it, the data needed to improve the manufacturing process.

In a sense, the factory was only slightly better off than before. It had the data that showed the symptom or problem on a computer, but not the data needed to find the underlying cause of the problem. It had computerized the recording of manufacturing results, but not the data that might explain why over half of the factory (its cost, quality, and speed) was still invisible.

Therefore, computerizing the run card was less valuable than previously thought, much as the SFC module had been less valuable than expected. The reason: The underlying *use* of the data, the underlying manufacturing paradigms or practices, hadn't changed! We were just automating the current practices and so only got the potential advantage of *cost* savings through automation of data collection. In a sense, the electronic run card was the *status quo* system—the automation of the "as is" instead of the replacement with the "to be" built around best practices.

Similarly, this phase did nothing to improve our scheduling and dispatching capability. Without the real-time status and planned availability of the equipment, operators, and tooling, finite capacity scheduling was impossible. Companies that bought stand-alone scheduling systems were rarely successful with their new tool. Either they had to enter an enormous amount of data prior to each scheduling run or the schedules had to assume average capacities and availability. In either case, they soon fell into disuse.

Similarly, we couldn't analyze our costs. With three-fifths of our resources still invisible to us, we couldn't see *actual* equipment utilization or energy costs or indirect personnel utilization.

In fact, we couldn't see our indirect tasks at all! Therefore, we couldn't see our indirect speed and how to improve it.

In practice, plant floor management certainly appreciated the new ability to now see their production—activity and quality—in real time and be alerted to any problems instantly. However, the improvement or analysis phase remained as frustrating as ever. We knew of some problems sooner but still didn't have the tools to analyze what caused them, and therefore, determine how to eliminate them. We remained in a "status quo" state, with more data available faster but not as much information or ROI as we may have thought we were going to get.

We had eliminated many *direct* random errors but had not touched the larger problems (90% of the total waste) of systematic errors and inefficiencies. Our stand-alone SQC systems detected variances instantly, but still didn't speed their analysis.

It was only a matter of time before the definition of the "paperless factory" was extended.

Take 3: Integrated Plant Floor Management/ Manufacturing Execution Systems

To meet the real needs of plant floor personnel, the first *integrated* plant floor management system, or manufacturing execution system (MES), was invented by Consilium in 1984. This system finally tracked *all* the resources on the plant floor: the work in process and materials/parts, personnel (both direct touch operators and indirects) equipment and tooling, facilities (energy, gases, water, facility conditions), and work instructions/specifications/recipes. Finally, the entire factory was visible. In fact, these systems could be thought of as a "window" onto the factory floor (Fig. 7.3).

They recorded all events or tasks as they occurred—whether processing of jobs or indirect supporting tasks such as preventive maintenance, repairs, and cleaning—and updated the relevant resources' status. In that way, we had a current status of the entire factory that could be used for scheduling and monitoring as well as a history that could be used for costing, improvement analyses, and reporting on all measures (speed, utilization, etc).

These systems extended our concepts of tracking. In an SFC or paperless run card system, we defined the "objects" or "entities" we wanted to track (called work orders/lots/batches/jobs) and tracked them through the individual "events" or "tasks" that were used to manufacture them (called operations). After each operation, we updated the work order status (location, cost, quantity) and recorded the operation history.

In these new systems, we also define additional objects or entities—all the *other* resources in our factory we want to track. We define equipment, tools, personnel, facilities, etc. These resources also are tracked through the events or tasks that used them or occur to them. For example, we will track equipment through its use in processing and tasks or events that occur to it such as preventive maintenance, cleaning, breakdown and repair, calibration, and so on. Just as in our WIP tracking approach, after each task or event, we update the status of the resource and record its history.

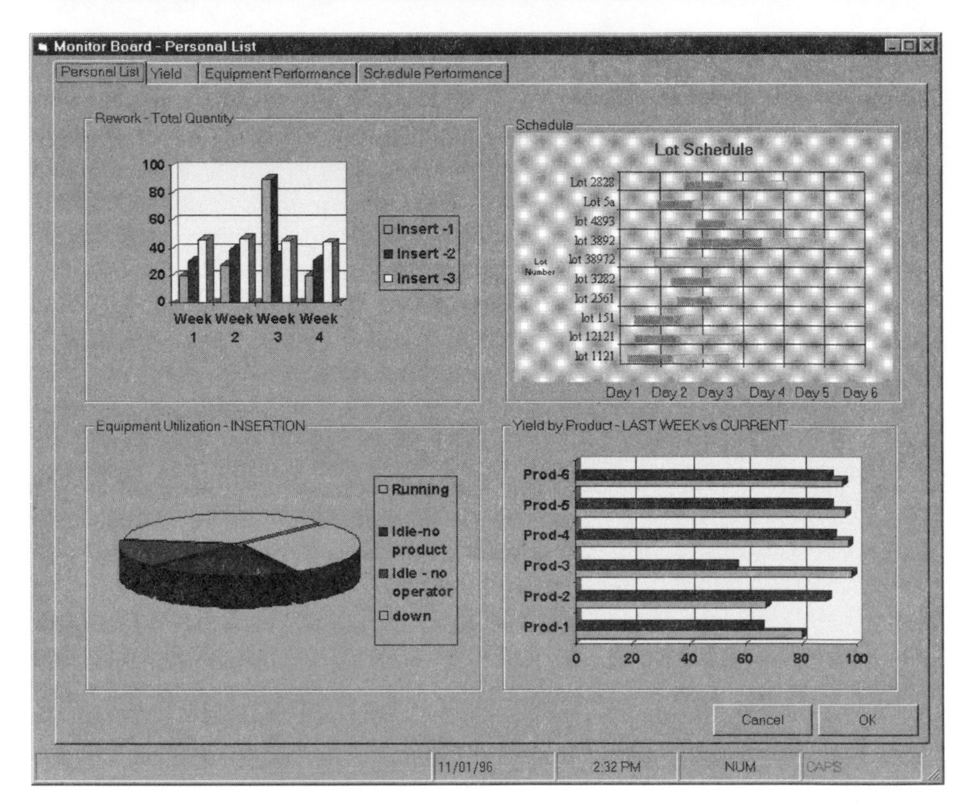

Figure 7.3 Plant Monitor Board

In addition, just as we can set schedules for each work order at each operation, we can concurrently define preset schedules for these tasks based on time or usage or "triggers" that trigger their occurrence. For example, we can define a preventive maintenance schedule for an inserter by a fixed time interval/elapsed time (say every four days), or by a usage counter/elapsed activity (say every 10,000 insertions), or by a trigger (any SQC violation).

Now we finally have the entire history, status, and future schedule for the entire plant—our first *integrated* view of the factory over time. Now, we can add the execution reporting/detection, improvement, and productivity tools for each plant floor team member. We have the data repository to support scheduling, dispatching, SQC, quality improvement analyses (cycles, quality event charts), costing/utilization, speed analyses, and so on. For the first time we can have a true *paperless* factory. All the data has a structure for collection, replacing the endless set of operator, equipment, facility, and tool logs. But more importantly, we can move away from status quo systems to continuous improvement systems and world-class performance.

The typical functions in a plant manufacturing execution system are given in Table 7.7.

The Advantage of Integrated Over Stand-alone—Integration of the Team

I want to cover the review of stand-alone packages again, since many of my clients get hung up over buying individual solutions for each function or department. Undoubtedly when you go back to your own departments, you'll again get pressure to buy what you yourself need, independent of the rest of the team. You won't want to have to wait for other departments to make a decision.

Some of your colleagues will say, or strongly suggest, "The logic of wanting to see all five elements makes sense, only why not do it with independent stand-alone or point solutions that you can add over time on the same computer? That way we can start with a WIP tracking system, then add an SQC system, perhaps scheduling, then preventive maintenance, and so on. We can start small, not wait for consensus on an integrated solution, and keep adding until eventually we have an integrated view of the factory. As long as they all run on the same computer, they'll be "compatible," and we can grow at *our* rate of need. So why can't we simply use a set of stand-alone systems to achieve the same results? All the same data will be there to support the factory either way."

In order to understand why integrated solutions will ultimately dominate over stand-alone or point solutions, it's necessary to look at the usage of a system over time. Initially, we need to set up the system—then define our processes and procedures to it. Next we need to train the users in its usage. Then the users actually use it to collect data, support analyses, communicate with one another, prepare reports, and coordinate activities. Finally, we need to improve our processes and practices as we learn the underlying causes of waste. Integrated systems allow a *single, consistent, integrated* execution of each of these steps.

TABLE 7.7 Manufacturing Execution System Functions

Tracking and Management of:
- Raw materials/ingredients (inventory)
- Equipment
- Tooling
- Facilities (conditions)
- Personnel/labor

Planning and Scheduling
- Real-time dispatching (jobs)
- Shift scheduling
- Activity, material, and capacity planning/projected customer service (feasible daily plans)

Costing
- Standard or activity based
- Actual

Management Reporting/Inquiries
- Standard
- Ad hoc
- Exception (alarms and alerts)

Quality Data
- SQC/SPC
- Quality data collection
- Correlation/statistical analysis
- Adaptive process control (advanced)

Industrial Specific Modules
- A&D: mil spec (labor tracking)
- Health care: batch record
- Semiconductor: reticle tracking

Specifications Management
- Documents/specifications/work instructions
- Recipes (for automation)

Interfaces to:
- ERP
- Automation (DCS/Equipment)
- LIMS (laboratory information management systems)
- Material transport systems
- Man–machine interface (MMI)

Let's look at using stand-alone systems for WIP tracking, quality, and scheduling versus an integrated system. In the first step, we need to set up each of our stand-alone systems. Each one requires us to re-enter or redefine a route (of the operations to be tracked and scheduled), the products to be made, and even our list of valid users. Typically, each system also has its own training, manuals, user interface/look and feel, and installation and support procedures. In other words, we have a setup and maintenance cost that is often double or triple that of one package where we defined any object or its properties once (the products, routes, operations), and the system has a single look and feel and integrated documentation.

Worse yet, we always risk that these packages could get out of sync—that without one definition of a route, product, operation, equipment, etc. when conditions change, we may update some of our stand-alone systems but not others. Then our views of the factory are not consistent.

It gets worse in actual operation. At the time of operation, the operator on the plant floor must now use three separate screens or windows or, in the worst case, terminals to see the dispatch list (on the scheduling system), move the work order in and out of the operation (WIP tracking), and then enter quality data and see the SQC chart (quality system). Each may have separate look and feel, commands and edits, or even hardware requirements (PC versus workstation versus terminal). We may have to enter the same data three times (the work order, number, the quantity, the operator, the equipment ID, etc.).

The data now ends up being stored in three separate databases! When we want to analyze it, we have to find out how to extract it from each into a common reporting tool (if we want a single report on WIP status, activity, quality, and schedule performance). Otherwise, we end up with three separate reports—on schedule performance, activity, and quality.

When the definitions of objects and states diverge, then we end up with "irreconcilable differences."

"My system says our equipment downtime is 8%."

"Well, mine has it at over 12%."

"But you defined downtime to include setup and our maintenance system sees that as in production."

"Yes, but you include operator acceptance of the repair in production time, and we classify it as down."

The next step then is to build a front end to the setup and definition for all three systems so that they can be defined simultaneously. Similarly, we can build a front end for data collection that takes an "integrated" operator transaction and feeds each stand-alone system the data. Finally, we can use extract and reporting tools to build a common view of the data in the three databases.

These *system integration* projects usually cost two to five times what the original packages cost! In addition, they have to be maintained endlessly as the packages themselves come out with new releases at different times.

The point: We want *one* view of our factory, whether we are defining it, entering data, viewing data, communicating with each other, reporting on it or understanding it, or scheduling it. Stand-alone packages can never provide this integrated capability and so end up costing considerably more in operation! We can clearly see this in three systems already well established—MRP II, CAE, and PC tools.

MRP II or ERP systems replaced a melange of stand-alone corporate systems for financing, planning, purchasing, inventory, shipping, and so on. Why did MRP II/ERP systems arise from this "clutter of capability"? For the exact same reason that we wanted to have one complete and consistent financial and logical view of our company. An order should be entered, planned, tracked, shipped, and paid for without appearing multiple times or being lost periodically. Raw materials needed should be planned, ordered, received, and paid for without gaps and overlaps. We need this consistency to ensure our *financial* integrity. Similarly, we need this same consistency on the plant floor to ensure our *manufacturing* integrity. That ensures that the products and processes we defined are manufactured/executed as defined, scheduled appropriately, costed accurately, quality tracked, and improvements managed correctly.

Over and over again, we see that integrated manufacturing processes and practices require integrated tools. If manufacturing takes place at a workstation using all five resources, then we need to see all five simultaneously on *one* system. If our costs are the sum of the usage of five resources, we need to see all five resources and track and measure them consistently.

In CAE, the current movement is toward integrated frameworks, for exactly the reasons we've discussed. Users don't want to have to repeat set-ups in multiple tools, replicate data in multiple tools, or find that tools are inconsistent and can't be used together for concurrent engineering.

In PC tools, as discussed, the overwhelming trend in businesses is toward integrated office suites!

In each case, integrated eventually surpasses stand-alone as the strongest vendor keeps extending their reach to solve the "real" problem in its entirety.

Take 4: Best Practices/Integrated Frameworks—Maximizing Net Benefits Per Unit Time With Flexibility

We now have over ten years of experience with MES in over twenty countries, in over twenty industries with over 50,000 users. Our findings are that most companies do not know how to use their MES to gain maximal return on their investment. In fact, far too many users simply duplicate their existing paperwork, procedures, reports, and analyses without understanding the best practices that can be enabled by MES. This means that they do not obtain the maximal benefit or ROI possible from MES. They do not obtain the full competitive advantage versus another company not using MES. They are at a competitive disadvantage to another company using MES to its full advantage.

That is why we started the MES Institute´ and wrote the first book on the best practices enabled by MES: *Achieving Stretch Goals—Best Practices in Manufacturing* (Prentice Hall, 1996).

The fourth generation of MES (and MES vendors) support these best practices in their packages, documentation, and training courses. The correct way to evaluate an MES and MES vendor is on whether it supports the best practices that would provide you a competitive advantage. As we discussed earlier, these may focus on increasing equipment utilization in capital intensive industries, increasing indirect and direct labor productivity in labor-intensive industries, increasing speed in all industries, and/or supporting the fastest ramp-up of new products, processes, and factories. Unfortunately, most MES and MES vendors have merely copied the ideas of the original MES and not understood or introduced the use or support of best practices. They have remained at the third phase where they support tracking but not improvement.

I am often asked whether MES will be subsumed into ERP or process control systems. The answer is only if those suppliers develop a core competency in the theory behind MES—the understanding of the best practices, the uses of MES that really create the return on their investment. Otherwise, their MES offerings will be nicely integrated with their real core offering (their ERP or DCS system) but not offer significant ROI. This will appeal greatly to the IT organization who otherwise must support separate ERP, MES, and DCS offerings. But manufacturing, who must offer the lowest cost, highest quality product on time with the lowest leadtime will still not have the best tools to do so. The company as a whole will be at a disadvantage. Companies who make this mistake usually focus on IT concerns versus ROI/competitive advantage issues. They may view all MES products as identical though, ironically, they have vociferously insisted on a specific ERP or DCS vendor.

To date, no ERP or DCS vendor has created that core competency. The ERP vendors are still offering enterprise centric MES—an updated shop floor control system that still is meant to feed corporate planning (now called supply chain) and financial applications. Now you have a grading system to test any new ERP/MES or DCS/MES offering and ascertain if it really will provide a competitive advantage.

Luckily, however, there is another alternative solution called frameworks. Frameworks revolved around agreements about how systems will interact—sharing data and communication or alarms or interrupts through object messaging. More sophisticated frameworks go beyond messaging to common agreement on the objects that need to be defined in the integrated solution and their major properties or elements. For example, both MES and ERP systems required a bill of material or formula. If all ERP and MES vendors agree on a common definition of a bill of material or formula and its characteristics (with individual vendor extensions allowed), then integration becomes much easier.

Several organizations have begun to define these frameworks for manufacturing systems, including SEMATECH with their CIM Framework Specification model and Open Applications Group (OAG) with their Open

Applications Integration model. CORBA (or common object request broker architecture) promises to support a uniform messaging protocol between applications.

If these frameworks are successful, they will allow users to combine best-of-breed ERP, MES, and DCS applications to maximize their ROI.

What the Japanese Did

Let me cover one last area. Often clients point to the Japanese as evidence that manual or no systems on the factory floor are required. It is important to understand their approach to systems in the context of their manufacturing and product strategies.

Strategically, the Japanese originally concentrated on producing a small number of high-volume, high-quality products. In this strategy, dedicated lines, or focused factories, were the logical approach. Such factories were similar to our original assembly lines. They really needed little in the way of software support. Scheduling meant setting a line rate. By producing a small number of products, quality could be improved with manual tools. The real focus was on continuous improvement (eliminating all forms of waste) and then on automation (to streamline the direct manufacturing process and replicate it automatically). Indirect functions were supported manually. With a small number of products produced in focused lines, this worked successfully.

Today, however, the Japanese are moving into product proliferation and customization. There is a strategic move to compete *differently*, based on rapid introduction of many new and customizable products. This change in strategy has also changed their focus on factory floor management systems.

Toyota, the inventor of the famed Kanban or JIT system, is now looking at more computerization of the factory floor because more frequent product changeovers and introductions complicate scheduling and quality improvement. Toyota is designing an MES that integrates with automation and links to the planning system.

In a recent study of Japanese businesses' key priorities to accompli[illegible] their manufacturing strategies, the number one priority was the integra[illegible] of information on the factory floor to allow rapid customization and product introduction!

The Japanese have always excelled at hard automation. With a fixed production line, they have automated material handling and data collection to allow lights out/peopleless operations. With the move to multiproduct manufacturing, they will also excel at flexible manufacturing, allowing production of many products on the same line.

The Japanese were slow to use computer systems due to the difficulties of using a Kanji keyboard to enter data and a social prejudice that typing was women's work. Now the introduction of PCs and workstations with graphical user interfaces will only further speed the acceptance of MES.

Another factor accelerating adoption of these systems is globalization. Japanese companies run plants around the world—usually located in Europe, the Americas, the Far East, and Japan. To manage these, Japanese companies want systems that support all facilities and allow a global view of inventories, capacities, and costs. This also supports technology transfers between plants.

The concept of the manual or "no system" CIM system is rapidly disappearing in Japan, replaced by global MES.

So, in summary, there is a tool that was designed specifically to support your vision of a factory and its best practices. That tool is a manufacturing execution system, or MES, as it was named by Advanced Manufacturing Research (AMR). Gartner Group calls it MOM—manufacturing operations management.

We've discussed six ways that people go wrong in their approach to MES:

1. They duplicate their existing paperwork-based system instead of reengineering to best practices (eg, they collect all the data on how the job is done after it's processed [MRP II approach] versus edit and verify that the right five elements are being used prior to processing to avoid random errors); they get shiftly or daily reports instead of exception alarms if quality or productivity is falling during the shift.

2. They approach MES piecemeal—by department or function, not allowing a single integrated view needed for best practices in quality, scheduling, measurement, etc.

3. They don't build their MES implementation program around best practices adoption rather than new pure tracking and display capability attained.

4. They focus their purchase decision on IT technology platform standards and costs instead of best-practice functionality and ROI.

5. They don't train the users on the best practices to the adopted.

6. They ignore the current costs of supporting, maintaining, enhancing, and integrating the current legacy systems when justifying a new integrated MES. These expenditures are often made on systems that are technologically obsolete.

"So now that's it. You're ready to go out and study MES vendors' offerings and see if you find one that rings your ROI bell."

The group, given a starting list of requirements, felt very comfortable with this next task. They called their local hardware platform vendors for recommendations. They called their own vendors and customers who had a few additional choices. They even signed up for a service with both Gartner Group and AMR advising them on MES options. After they had narrowed the choice to three vendors, based on industry fit and experience, they asked for demos. What PCB Co. asked was for demonstrations and/or explanations of how their system supported their required best practices. Two of the three were not sure why PCB Co. was asking for this approach and simply showed basic tracking functions. The third had clearly designed their product with best practices in mind and demonstrated it accordingly. The team unanimously went with the choice that had shown them the most potential for ROI.

Tom never asked the consultant for advice on fighting the corporate "ERP uber alles" decision that it would also run their manufacturing operation. He didn't have to. Their team had carefully evaluated using the ERP shop floor/plant management offering versus their selection and shown millions of dollars of savings with their choice. The team was aligned on the need for MES that returned value to manufacturing, not to corporate IT, finance, or planning alone. When the team presented the ROI calculations for the use of an MES instead of the ERP shop floor system, return on investment and reason were victorious.

And so PCB Co. had another tool in its fight to survive and thrive. But time was winding down. The six months were nearly over, and while progress was extraordinary, they had not yet met their stretch goals. Realistically, it would take them a year to double output with the same capital equipment and indirect labor force.

Mark Ritchards was staggered by the progress they had made—with few naysayers left. More than that, he was proud of the capabilities they had demonstrated: to learn, to change, to work as productive teams, to interact with vendors and customers, to become a fast, speedy, learning organization. The organization had truly woken up and now performed managerial processes at a different level of expertise. But was it too late? His meeting with Diversified was next. Would they accept the current improvement level coupled with their rate of change? Would they applaud their progress or grade it a failure?

Mark obviously wouldn't know until he presented to them. This time the CFO was not willing to coach him. Clearly, Diversified was still under great pressure to show a higher rate of return on their assets and raise their stock price. Sometimes Mark wondered if they had deliberately set a stretch goal that was unachievable; if they basically did not see manufacturing as a necessary core competency but rather as a too capital-intensive operation that hurt their return on assets. If so, they had succeeded—so far. Without more time, he couldn't achieve their targets.

But, strangely, he wasn't nervous going into the Diversified review meeting. He believed, deeply, that the team and their manufacturing capability had grown immensely—to world-class levels. That success had given him new confidence. He now knew what to do to run any facility. This experience had given him a Ph.D. in the art and theory of plant management. He just wished he'd had a real course that had covered what they'd learned—years earlier. Or perhaps, without the current pressure to change, he'd have ignored it. The instructors they had employed had been phenomenal.

No, he wasn't nervous. He felt prepared to run a plant anywhere. It was a professional confidence he had never thought about in the days of single company careers.

He opened the door to the conference room and prepared to present his case.

Epilogue

Epilogue: 24 Months Later

Mark couldn't believe how successful the initial public offering had been, the wild celebrations within the now independent company, the grins on the faces of his management team, the tears of joy on his mother's face, and the glee in his family's expressions as the PCB Co. symbol (PCBC) flashed across the stock ticker board.

It seemed like years ago since his meeting at Diversified to review their progress against their stretch goals. It had been a long, heated meeting as Mark had come with charts showing their progress on hundreds of measurements they now monitored in their improvement programs. He recounted all their efforts and gave them roadmaps that showed when they would hit their milestones, attaining their stretch goals. With the addition of an MES, they actually saw areas in indirect labor exceed their stretch goals and support more automation than planned.

But the board at Diversified had not wanted to wait. As passionately as Mark argued, the board remained resolute. Diversified could not wait for promises. Their progress had been impressive, and it certainly would make it more likely that a buyer would keep them open—but not Diversified. They were negotiating with a contract assembly house for all their board manufacturing. Final agreement just awaited the results of their meeting.

Mark was calm. He and his team had also prepared for this possibility. These days they were rarely surprised. They had learned the best practices of factory management and applied them to all their management processes. He had anticipated this as a potential outcome. In fact, some of his friends at corporate had tipped him off that Diversified was in serious negotiations to outsource its manufacturing of PCBs. That was merely vigilance on his part. And he'd planned.

"If you are going to divest yourself of this operation, our team would like to buy it—in a leveraged buyout. We've arranged financing but want you to carry a note for five years at prime plus one interest for 65% of the purchase price. That will give you a second source for all your boards if your overseas contractor has any problems. In fact, we can move on this in 30 days and have the asset off your books. That will save your carrying it while you search for a buyer." Mark's intelligence also suggested that they had not yet started shopping the facility to prospective buyers. They wanted to keep maximal pressure on their potential vendors and see if Mark and his team could meet their goals for survival.

What Mark had not told Diversified was that he had already spoken with their outside customers about the potential for additional business if he were able to cut his prices, reduce leadtimes, and further improve customer service. Each, already impressed with the measurable improvements in quality, delivery, and attitude, had expressed real interest. He believed that he could replace and surpass Diversified's business in the time it would take to transfer it to a contract house.

Diversified was noncommittal until Mark said that his management team was resigning. Diversified had not anticipated this. The president became more receptive and suggested that as long as their offer was at fair market value, there probably was a win–win situation. After all, he thought, who would want to buy a failing operation that was losing the majority of its business to an overseas contractor.

Mark's team, to a man, participated in the LBO along with almost every employee. Mark wanted everyone to have a stake in their new enterprise—even if it was merely symbolic. Bob's brother-in-law came through, as advertised, with the LBO funding.

If they thought they had worked hard before, the next twelve months were a true hardship. Mark's whole team each adopted one or more customer(s). Each customer was treated like an honored partner. With their new MES, they actually let their customers view, on-line, the status of their orders and generated a certificate of assurance to eliminate incoming inspection and testing.

Within a year, business was expanding beyond what their new levels of value-added utilization could support. They discussed whether to borrow more money to expand, to limit expansion to what cash flow could fund, to ask customers to provide expansion capital in exchange for equity, or to consider an initial public offering.

And so, 24 months after one of the worst days of his professional career, Mark Ritchards and his team had one of their best. It seemed like a dream, but one he could get used to. And he did.

Index